Genetics of Industrial Microorganisms

Committees

Organizing Committee

D. Perlman, Chairman
A. L. Demain, Vice-Chairman
O. K. Sebek, Vice-Chairman
W. M. Stark, Secretary
J. Berger
S. Bernstein
R. E. Cape
C. A. Claridge
R. P. Elander
B. Lago
A. I. Laskin
J. Lein
C. Vézina
D. I. C. Wang
R. W. Sarber

Program

A. L. Demain, Chairman
J. Berger
R. E. Cape
C. A. Claridge
H. O. Halvorson
D. A. Hopwood
J. Lein
J. Nüesch
J. Shapiro
C. Vézina

Finance

R. E. Cape, Co-Chairman
R. P. Elander, Co-Chairman
W. Charney
L. Day
P. Lemke
R. W. Thoma

Local Arrangements

G. P. Peruzzotti, Co-Chairman
D. Perlman, Co-Chairman
S. Bernstein
J. Davies
R. S. Hanson
M. E. Howe
E. W. Marty
T. C. Nelson
H. J. Peppler
G. Reed
M. Sussman

Publications

O. K. Sebek, Chairman
A. I. Laskin

Genetics of Industrial Microorganisms

Proceedings of the Third International Symposium on Genetics of Industrial Microorganisms

held at

The University of Wisconsin
Madison, Wisconsin
4–9 June 1978

Edited by

O. K. Sebek
The Upjohn Research Laboratories
Kalamazoo, Michigan

and

A. I. Laskin
Exxon Research and Engineering Company
Linden, New Jersey

The Symposium was sponsored by the
American Society for Microbiology;
Society for Industrial Microbiology;
Division of Microbial and Biochemical Technology,
American Chemical Society;
International Association of Microbiological Societies;
and United States Department of Commerce

AMERICAN SOCIETY FOR MICROBIOLOGY

Washington, D.C.
1979

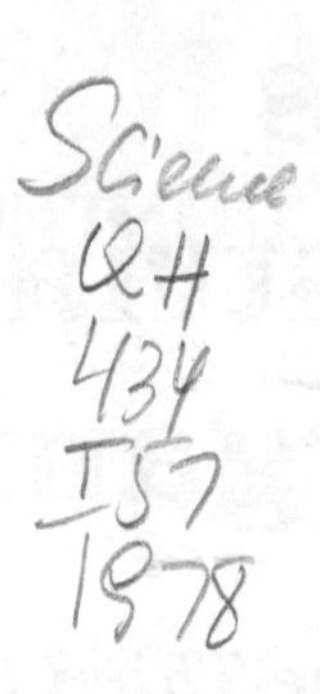

1913 I St., N.W.
Washington, DC 20006

Library of Congress Cataloging in Publication Data

International Symposium on Genetics of Industrial Microorganisms, 3d, University of Wisconsin–Madison, 1978.

Genetics of industrial microorganisms.

Includes index.
1. Microbial genetics–Congresses. 2. Industrial microbiology–Congresses. I. Sebek, O. K. II. Laskin, Allen I., 1928– III. American Society for Microbiology. IV. Title.
QH434.I57 1978 660'.62'015751 78-31511

ISBN 0-914826-19-0

Printed in the United States of America

CONTENTS

Evolution of New Metabolic Capabilities

Preface

This book contains extended abstracts of the material given in the lecture and panel sessions of the Third International Symposium on Genetics of Industrial Microorganisms (GIM 78) held at the University of Wisconsin-Madison, 4–9 June 1978.

The symposium was sponsored by the American Society for Microbiology, the Society for Industrial Microbiology, the Division of Microbial and Biochemical Technology of the American Chemical Society, the International Association of Microbiological Societies, and the U.S. Department of Commerce. It received generous financial support from 38 U.S., European, and Japanese companies and organizations. The chairman of the Organizing Committee was D. Perlman, and W. M. Stark served as the secretary. The excellence of the program was due to the efforts of the Program Committee, chaired by A. L. Demain.

The content of the book reflects the advances that have been made in microbial genetics since the first symposium of this series in Prague, Czechoslovakia, in 1970 and, more particularly, since the second one held in Sheffield, Great Britain, in 1974. This is especially true of some of the new tools of microbial genetics (recombination by restriction enzymes, protoplast fusion) which had not been put to use at the time of the Sheffield symposium and which offer exciting possibilities of constructing new organisms with superior and predetermined properties. The importance of the material discussed may be measured by the realization that there are at present about 200 U.S. and approximately 500 corporations worldwide which use fermentation technology for making various products useful to man (antibiotics, organic and amino acids, solvents, vitamins, enzymes, alkaloids, steroid hormones, fine chemicals) or for utilizing unusual materials (hydrocarbons) as fermentation substrates.

The material in this book is a blend of theoretical considerations and practical applications of microbial genetics. It covers a broad range of topics and includes papers on mutation, gene cloning, genetic aspects of primary metabolism and of antibiotic synthesis, and genetic approaches to new products and processes. It discusses genetics of actinomycetes and fungi, regulation of fermentation processes, genetics applied to hydrophobic compounds, resistance development to antibiotics, genetic engineering and government regulations, and evolution of new metabolic capabilities of microorganisms. Impressive results that are here reported may serve again to demonstrate the importance of applying new advances in molecular genetics and metabolic regulation to microbial synthesis of products which are of considerable benefit to mankind.

It is with anticipation of new progress in these areas that we look forward to the fourth symposium (GIM 82) of this series, which will take place in Japan in 1982.

O. K. Sebek
A. I. Laskin

Acknowledgment

The Symposium organizers gratefully acknowledge the generous financial support received from the following companies and organizations:

Abbott Laboratories
American Chemical Society, Division of Microbial and Biochemical Technology
Ayerst Research Laboratories
Beckman Instruments, Inc.
Bristol Myers Company, Industrial Division
Cetus Corporation
CIBA-GEIGY Corporation
E.I. du Pont de Nemours & Company
Eli Lilly and Company
Exxon Research and Engineering Company
Farmitalia
General Electric Company
Gist-Brocades, n.v.
Hoffmann-LaRoche Inc.
Lederle Laboratories (American Cyanamid Co.)
Meiji Seika Kaisha Ltd.
Merck & Co., Inc.
Miles Laboratories
New Brunswick Scientific Co.
Novo Industri A/S
Panlabs, Inc.
Pfizer, Inc.
Rhone-Poulenc S.A.
Roussel-UCLAF S.A.
Schering Corporation
G. D. Searle & Co.
Shionogi & Co., Ltd.
Smith Kline Corporation
Society for Industrial Microbiology
E. R. Squibb & Sons, Inc.
Stauffer Chemical Company
Standard Brands, Inc.
Sterling-Winthrop Research Institute
The Upjohn Company
Union Carbide Corporation
Warner Lambert/Parke-Davis
Wyeth Laboratories, Inc.
U.S. Department of Commerce

Remarks of the Chancellor

Delegates to the Third International Symposium on Genetics of Industrial Microorganisms, Honored Guests:

It is my pleasure as Chancellor of the University of Wisconsin-Madison to bring you greetings from our faculty and students. We are pleased and honored that you have come to our University to hold your symposium. You come at a time when there is considerable activity here in your field of science, and both faculty and students will benefit greatly from your visit.

Industrial microbiology, or, more precisely, applied microbiology, on our campus currently involves some 21 departments and schools as well as 5 research-oriented institutes and centers. This large effort has grown in the past 60 years starting with the studies by E. B. Fred (who later became President of the University) and W. H. Peterson on the acetone–*n*-butanol fermentation process in World War I. From these beginnings—with high priority given to solving "practical problems" of primary interest to industry—longer-range studies of a fundamental as well as an applied nature were established, and many have continued up to the present time. Although some 40 years have passed since E. F. McCoy showed that her isolate *Clostridium madisonii* had superior potential for production of *n*-butanol over other solvent-producing strains, it was a milestone in our use of genetics in selection of "more efficient microorganisms" for industrial processes.

The cooperative program on microbial production of penicillin started in 1943 on this campus under the leadership of W. H. Peterson and M. J. Johnson of the Department of Biochemistry. The affiliated team of M. P. Backus and J. F. Stauffer and their students rather quickly developed the so-called Wisconsin group of penicillin-producing strains of *Penicillium chrysogenum* and gave a practical direction to the fermentation industry to produce thousands of tons of this therapeutic agent. I am glad to note that you will have an opportunity to mark this 35th anniversary by K. B. Raper's lecture on Tuesday afternoon,[1] and that Professors Backus, Johnson, and Stauffer will be present for this recognition.

The more recent interest in the genetics of antibiotic-producing organisms has focused on the role of plasmids (studied by J. E. Davies and B. Weisblum) and other efforts in finding antibiotic-producing organisms and methods of inducing antibiotic production (including the work of D. Perlman). Since the solution of many of the problems in preparing antibiotics and other therapeutics on an industrial scale requires the joint participation of microbiologists, chemists, and pharmacologists, mention should be made of the value of team research as practiced on our campus. The research group led by microbiologist/biochemist C. J. Sih has devised several practical syntheses, using both chemical and enzymatic steps. Other problems of current interest include the methane-producing fermentations (studied by R. S. Hanson), the hydrolysis of cellulose to glucose, and the degradation of lignin (studied by J. G. Zeikus and T. K. Kirk of the Forest Products Laboratory), and fundamental studies on nitrogen fixation (studied by W. J. Brill) and food toxins (studied by Professors Bergdoll, Chu, and Sugiyama of the Food Research Institute).

With more than 60 years of activity and commitments to the study of industrial

[1] Editor's note: Professor Raper's lecture was not included in this volume at his request.

fermentations and related problems on this campus, it is indeed fitting that your symposium is being held here. We will be extremely interested in the scientific reports made in the presentations at your sessions. Although the fermentation industry is not at a crossroads as far as its future is concerned, I am advised that the recent developments in your specialized science make possible a "new era" for the fermentation industry. As a chemist, I envy you in being present for the important developments that are expected in your field.

Once again, I welcome you to our campus. We hope that during your visit you will have an opportunity to explore and discuss with our faculty and students many of your common interests, and we look forward to having you visit with us again whenever possible. Thank you for coming.

Irving Shain
Chancellor
University of Wisconsin-Madison

Opening Address
The Many Faces of Recombination

DAVID A. HOPWOOD

John Innes Institute, Norwich NR4 7UH, England

A REVOLUTION IN THE GENETICS OF INDUSTRIAL MICROORGANISMS

The word revolution has tended to be debased by overuse, but it is no exaggeration to say that a revolution has occurred in our subject in the 4 years since the Second International Symposium on the Genetics of Industrial Microorganisms (GIM74). There are two aspects to this revolution. On the one hand, both academic microbial geneticists and industrial people have experienced a dramatic change in attitude. A significant number of the leaders in microbial genetics are now interested, *directly and personally,* in possible applications of their discoveries; at the same time, the imagination of industrial managers has been caught by the dramatic advances currently taking place in microbial genetics. These developments have started to close the gap between the sophistication of the most advanced research and what is currently feasible with industrial microbes, thereby beginning to fulfill the hope of the farsighted and courageous organizers of GIM70 in Prague when they initiated this series of symposia. The other aspect of the revolution is the scientific advances themselves. They have given us the potential both to put sophisticated genetics into organisms that make useful products and to put useful products into organisms that have sophisticated genetics.

The series of these advances that I wish to highlight here can be united under the heading of recombination, with that term used in its broadest sense to include any process which helps to generate new combinations of genes that were originally present in different individuals. In this broad sense, recombination embraces two areas: (i) events that bring genes into the same intracellular environment, that is, mating phenomena and their artificial equivalents, and (ii) processes which join parts of these DNA segments together, including the classical mechanism of crossing-over between homologous DNA sequences and also the various devices, both natural and artificial, which lead to "illegitimate" recombination. There have been exciting advances in both these areas since GIM74, which have given us a most impressive array of new recombinational tools for the construction and analysis of microbial genotypes.

In the sheer versatility of these tools, procaryotes—eubacteria and actinomycetes—are, for the moment, ahead of eucaryotic microbes. For this reason, I shall illustrate the revolution in the tools of genetic manipulation by citing exclusively procaryotic examples. However, the possibilities for manipulating eucaryotes will undoubtedly increase, especially since steps have been taken to transfer some of the new recombinant DNA technology to fungi such as yeast (1a, 20). Moreover, fungal genes have been transferred to and expressed in bacteria (33, 39, 42). Thus, the boundary between procaryotes and eucaryotes is being broken down.

"NEW" RECOMBINATIONAL TOOLS

Table 1 includes five apparently diverse phenomena which all help in the construction of recombinant genotypes. The dates of their discovery, or the start of their intensive study, are given in the table. Amazingly, knowledge of all these systems was developed after GIM74, or not long before it, so that they had little or no impact on that meeting. Let us briefly review this series of phenomena before returning to consider each in more detail.

Item 1 refers to the P1 R plasmids. P stands for *Pseudomonas*, but it could equally well stand for "promiscuous" because these sex factors will transfer themselves by mating between virtually any gram-negative strains and so can be used to put sex into many kinds of bacteria—often those responsible for useful industrial processes—that do not otherwise have the benefit of a generalized recombination system. These plasmids turned up in 1970 in *Pseudomonas* strains isolated from hospital situations. Their uniquely wide host range was soon realized (14) and they were found to mobilize chromosomal genes (37), but the general utility of these sex factors for chromosomal mobilization and mapping in gram-negative bacteria has been demonstrated only within the past 2 or 3 years.

The transposable genetic elements (item 2, Table 1) are quite diverse. Their study has only recently become one of the major growth areas in bacterial genetics, even though some of the first observations were made some time ago. For example, bacteriophage Mu was discovered as early as 1963 (40), and insertion sequences were described in several laboratories in the late 1960s (reviewed in 27, 38), but the real impetus to the use of all these special DNA elements as agents for the in vivo rearrangement of bacterial genes developed only after the definition of transposons in 1974 (19).

The most recent addition to the list comes from the demonstration in 1977 (10) that bacterial restriction enzymes actually function as recombination enzymes in vivo. This discovery (item 3, Table 1) developed out of the in vitro use of restriction enzymes to manipulate DNA sequences (item 4, Table 1). The first report of such artificial construction of hybrid molecules was in 1973 (13), and the first cloning of foreign DNA was described in the following year (9). Since then, this field has undergone a dramatic development. To a considerable degree, this has been responsible for the changed attitude toward genetics in industrial companies because these techniques offer the hope of making genetic manipulations which are completely outside the scope of traditional mutational schemes (21).

It is a mystery why protoplast fusion in bacteria (item 5, Table 1) was described only in 1976 (17, 35), years after the artificial fusion of cells or protoplasts of animals and plants had become almost routine (see reviews in 15, 32) and some time after fungal protoplast fusion was achieved (16). After this late start, the technique has shown promise as an extremely valuable addition to the range of recombinational tools available for manipulating the genetic constitution of bacteria.

TABLE 1. *"New" recombination tools*

Item	Year
1. Conjugation by promiscuous sex factors	1971
2. In vivo rearrangements by transposable genetic elements	1974
3. Recombination in vivo by restriction enzymes	1977
4. In vitro recombinant DNA techniques	1973
5. Protoplast fusion in bacteria	1976
6. Combinations of 1–5	

This, then, is the collection of new recombinational tools which have been forged or sharpened since GIM74. Let us now consider each in the context of industrial microorganisms.

Conjugation by promiscuous sex factors. Table 2 lists some gram-negative bacteria in which chromosome mapping has been made possible by the use of various P1 R plasmids. In some strains, a "wild-type" plasmid not only transfers itself between bacteria with high frequency but also promotes the transfer of chromosomal markers often enough to generate recombinants, and so to allow linkage mapping, with comparative ease. In other cases, the frequency of chromosomal recombination is very low, hence the value of a variant sex factor (R68.45) described in *Pseudomonas aeruginosa* in 1976 (18). The variant has since been shown to differ from the parent R68 plasmid by the insertion of an extra 1.4 megadaltons of DNA (25). This segment presumably was derived from the *P. aeruginosa* chromosome, yet it confers on the plasmid the ability to mobilize chromosomal genes of a variety of other gram-negative bacteria, apparently in a more or less random way.

The sex factors mentioned in Table 2 (with the exception of R772 in *Proteus mirabilis:* reference 11) lead to the transfer of rather large segments of chromosome from donor to recipient. This makes them very useful tools for long-range mapping by selecting for one donor marker and measuring the frequency of co-inheritance by recombinants of other genes (4). In this way, with five or six selections, the entire linkage map can be covered, and circular linkage maps have rapidly been generated in *Rhizobium leguminosarum* (3), in two strains of *R. meliloti* (29, 30), and *Acinetobacter calcoaceticus* (41).

Recombination mediated by such plasmids is not confined to the descendants of single strains. For example, it has been possible, by means of R68.45, to promote chromosomal recombination rather freely among three species of *Rhizobium,* each of which nodulates a different species of legume—peas, clover, and French beans (26). The significance of this finding is the possibility of tapping a larger gene pool in attempting to breed bacteria with superior combinations of agronomically useful characters such as better survival in the soil, increased competitiveness with other rhizobia, or more efficient symbiotic nitrogen fixation. Such quantitative characters are certain to be affected by large numbers of genes, even if a few major genes directly control the specific functions of nitrogen fixation or host-plant recognition (2). In such a situation, a program of crossing between widely divergent strains is likely to be effective, just as in the improvement of quantitative characters in other useful bacteria.

In vivo rearrangements by transposable genetic elements. Table 3 lists a hierarchy

TABLE 2. *Recombination mediated by P1 plasmids*

Species	Wild-type plasmids	Recombination frequency mediated by		References
		Wild-type plasmids	R68.45	
Pseudomonas aeruginosa	R68, R91	10^{-5}–10^{-6}	10^{-4}	18, 37
Rhizobium meliloti	RP4	10^{-5}	—	30
R. meliloti		—	10^{-5}	29
R. leguminosarum	RP4, R702, R1033	10^{-9}	10^{-6}–10^{-7}	3
Acinetobacter calcoaceticus	RP4	10^{-6}	—	41
Rhodopseudomonas sphaeroides		—	10^{-5}	36
Proteus mirabilis	R772	10^{-5}	—	11

TABLE 3. *Some examples of translocating DNA elements*[a]

Class	Examples[b]	Length (base pairs)	Termini[c]
Insertion sequence	IS*1*	768	18/23bp IR
	IS*3*	1,400	
Transposon	Tn*3* (Ap)	4,600	140bp IR
	Tn*5* (Km)	5,300	1,450bp IR
	Tn*7* (SmTp)	13,500	150bp IR
	Tn*9* (Cm)	2,500	768bp DR (IS*1*)
	Tn*10* (Tc)	9,300	1,400bp IR (IS*3*)
Phage	Mu-1	38,000	

[a] From several sources, mainly references 27 and 38.
[b] Ap, Ampicillin; Km, kanamycin; Sm, streptomycin; Tp, trimethoprim; Cm, chloramphenicol; Tc, tetracycline.
[c] bp, Base pairs; IR, inverted repeat; DR, direct repeat.

of DNA elements which share the property of being transposable from one site to another within bacterial cells, between and within chromosomes, plasmids, or bacteriophage genomes, by illegitimate recombination—that is, recombination that does not require exact base sequence homology between the interacting DNA elements.

The insertion sequences (12, 27, 38) appear to be the simplest of these elements. Perhaps more interesting, for their practical applications, are the transposons, because of the selective markers they carry. Some consist simply of a gene or genes coding for antibiotic resistance flanked by a pair of insertion sequences. These may be in the same orientation (a so-called *direct repeat,* like the two IS*1* elements in Tn*9*) or in opposite orientation (an *inverted repeat,* as in Tn*10* with its two IS*3* elements). In other examples the repeats have not (yet) been characterized as insertion sequences. Evidently, these repeated sequences at the ends of transposons are critical for the transposition phenomenon (all transposons have them), and it is probably significant that the insertion sequences themselves have repeated sequences on a smaller scale at their termini. The top member of the hierarchy is bacteriophage Mu, a still more complex entity with a set of genes coding for phage components so that this element has an extracellular existence in addition to moving about from site to site within bacterial cells. Much is now known about its molecular biology (7).

These transposable elements have a host of applications in bringing about illegitimate recombinations and thus the engineering, in vivo, of useful new bacterial strains. Such applications are discussed elsewhere in this symposium (see also 7). At this point, I shall simply mention the use of transposons as a special kind of mutagenic agent, again taking the genetic analysis of *Rhizobium* as an example. To gain an understanding of the complex phenotype represented by a successful symbiosis between the bacterium and its host legume which will serve as a basis for the establishment of nitrogen fixation in other plants, we need to study the properties of mutations affecting the symbiosis. Genetic characterization of such mutations, which are induced by conventional mutagens, requires the testing on plants of considerable numbers of recombinant progeny from crosses between mutant and normal strains. Such tests are laborious and expensive. If, however, the mutations arise by the insertion of a transposon within or close to the target gene, this becomes tagged with antibiotic resistance (28) which can be followed as a normal marker by simple petri dish tests rather than by tests on plants (J. E. Beringer, V. A. Buchanan-Wollaston, and A. W. B. Johnston, unpublished data). Such a use of transposons promises to revolutionize the genetic analysis of complex and not easily selected characters in any of the wide range of organisms into which it is becoming possible to introduce transposons, or in which indigenous transposons might be found.

Recombination in vivo by restriction enzymes. A priori, it was likely that the well-known ability of many type II restriction endonucleases (34) to cut their recognition sites at staggered positions on the two DNA strands and therefore to generate cohesive ends, as first shown for *Eco*RI (31), had a biological significance. This conclusion tended to follow from the fact that there can be nothing about the chemistry of DNA or of endonucleases which makes the generation of cohesive ends inevitable, since certain sites (e.g., CCCGGG) are cut alternatively by enzymes which generate flush ends (e.g., *Sma*I) or by others which generate cohesive ends (e.g., *Xma*I). Thus, staggered cutting is likely to be adaptive, probably because it leads to the possibility of rearranging groups of genes by illegitimate recombination within bacterial cells, as has now been demonstrated (10). It is too early to quote applications of this finding, but applications will doubtless follow as more is discovered about ways to regulate the phenomenon, perhaps by the use of temperature-sensitive mutations in the modification enzymes which normally protect the restriction sites from cleavage (24).

In vitro recombinant DNA techniques. Almost all of the effort in DNA cloning has been in *Escherichia coli,* whereas essentially all of the bacteria used commercially belong to other groups: for example, the bacilli for various enzymes and some antibiotics, the pseudomonads and related strains (including *Rhizobium* and methylotrophic bacteria), and of course the actinomycetes and their relatives for most of the bacterial antibiotics, amino acids (*Corynebacterium, Brevibacterium,* etc.), and certain enzymes. When single exotic gene products, such as insulin, are to be made by fermentation, *E. coli* may be a suitable host for foreign DNA, but for many useful jobs of strain construction, where many properties of existing industrial strains are crucial, we need cloning vectors and host systems for industrial bacteria. These are being developed for bacilli, pseudomonads (8), and streptomycetes (6).

In *Streptomyces coelicolor* we now have two potential cloning vectors, the plasmids SCP2* (5) and SLP1 (M. J. Bibb, unpublished data). SCP2* is a plasmid of some 18 megadaltons with single *Eco*RI and *Hin*dIII sites, five for *Bam*HI, and four for *Sal*PI; SLP1 is only about 8 megadaltons, with single *Eco*RI and *Sal*PI sites and three each for *Hin*dIII and *Sal*GI. The key to the use of these plasmids as potential vectors for recombinant DNA is the detection of transformation, even if it occurs at low frequency, by virtue of a property which we have likened to lethal zygosis in *E. coli* (5). A single spore carrying SCP2* or SLP1 in a dense background of an SCP2$^-$ or SLP1$^-$ strain in an agar plate is revealed as a "pock," from the center of which a clone of the plasmid-carrying strain can be recovered (Fig. 1). This visual detection method is very powerful, being capable of recognizing a single plasmid-carrying spore in a population of at least 10^9, but in fact, under optimal conditions, transformation of *Streptomyces* protoplasts by plasmid DNA, in the presence of polyethylene glycol, followed by protoplast regeneration, can give rise to transformants at the remarkably high frequency of 20% of the total regenerated population (6).

Recombinant DNA techniques in streptomycetes will undoubtedly have significant applications, at least in situations where useful genotypes can be constructed by the transfer of small numbers of genes between genetically isolated strains. Obvious examples are the introduction of genes coding for enzymes that facilitate the breakdown of complex carbohydrates into developed strains, of genes controlling the addition or removal of particular chemical functions which could generate new

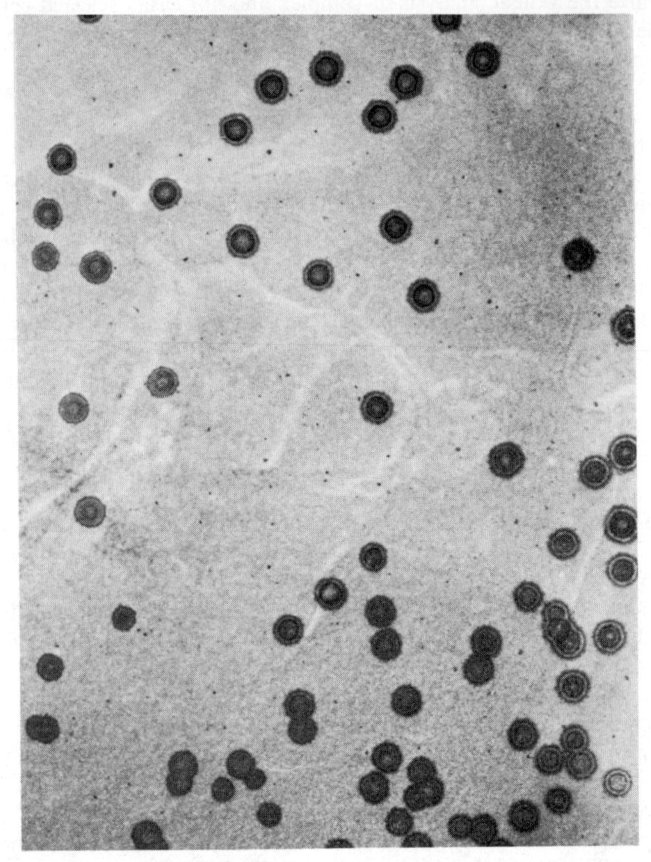

FIG. 1. *"Pocks" produced by single SCP2*-containing spores germinated in a lawn of SCP2⁻ spores. ×3.5 (Photograph kindly supplied by M. J. Bibb.)*

antibiotic structures from parts of existing biosynthetic pathways currently found in different strains, or of genes for the synthesis of potential antibiotic side chains.

Protoplast fusion in bacteria. Recombination through protoplast fusion in bacteria has so far been reported only for the gram-positive genera *Bacillus* (17, 35) and *Streptomyces* (1, 23). It is hoped that progress will be made in developing fusion for gram-negative species, although the presence of an outer membrane in these organisms may present problems. One application of the technique is in the establishment of a recombination system in new strains with no efficient natural means of conjugation. In this capacity, protoplast fusion may come to play, for gram-positive species, a comparable role to conjugation mediated by P1 R factors in gram-negative strains.

The most exciting characteristic of protoplast fusion in *Streptomyces* is the extremely high frequency of recombinants produced, under nonselective conditions and in the absence of known sex factors. These frequencies, as measured in six-factor crosses in *S. coelicolor,* are routinely greater than 10% (22) and commonly as high as 20% (D. A. Hopwood and H. M. Wright, unpublished data), when fusion conditions are optimal.

Since the efficient use of protoplast fusion as a generalized empirical recombination technique in strain improvement depends on obtaining very high recombination frequencies, we tested the idea that recombinant frequencies might be increased by UV irradiation. A UV dose sufficient to kill, say, 99% of protoplasts should eliminate those protoplasts not involved in a fusion event. Selective advantage is given to those protoplasts that fuse and give rise to viable progeny by recombining out the lethal hits in the parental genomes, which will tend to be at different loci in different members of the population. This technique works. If the recombination frequency in an experiment happens to be rather low (for example, 1%), it can be increased ten times by irradiating the two parental protoplast suspensions immediately before fusion; if the recombination frequency is already high (for example, 20%), it can be doubled to about 40% (D. A. Hopwood and H. M. Wright, unpublished data).

Another interesting feature of protoplast fusion is that more than two strains can be combined in one fusion. Up to four strains have been fused together, to yield recombinants inheriting genes derived from all four parents (22).

In its application to strain development, protoplast fusion may be regarded as complementary to gene cloning: instead of being applicable to qualitative changes involving few genes, one of its chief uses is likely to be in breeding for a complex quantitative character such as yield of an antibiotic. For example, in the mutation-screening pedigree in Fig. 2, there is little chance that the same series of yield-enhancing mutations will have been selected in the two main branches, because yield is influenced by many genes, not simply those coding for the structural and regulatory genes of the biosynthetic pathway itself but also genes affecting the supply of primary metabolites or cofactors, the operation of competing pathways, the control of permeability barriers, etc. Thus, recombination between the topmost strains in the two branches will generate a very large number of genotypes with respect to yield mutations, many of which will not have arisen before, and some of which may show synergistic interactions: that is, particular new combinations will have higher yields than those predictable from the effects of the individual mutations.

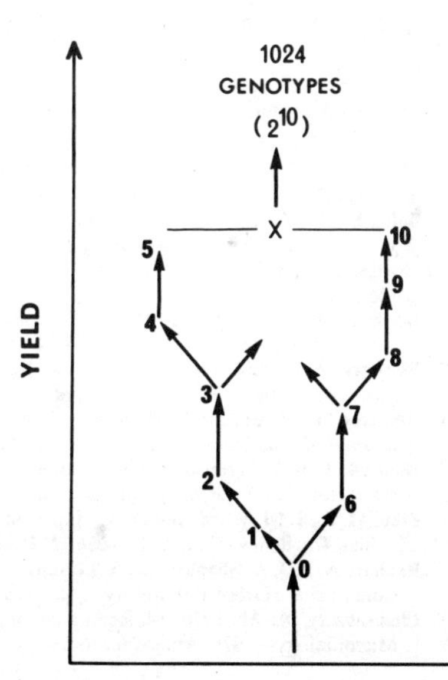

FIG. 2. *Part of an idealized strain-improvement selection pedigree. It is assumed that each yield-enhancing step, after divergence from strain 0, represents mutation in a different gene, so that 2^{10} genotypes are produced by crossing strains 5 and 10.*

It follows that the ability to introduce rounds of recombination easily into a strain improvement program is useful. Protoplast fusion has an advantage over conjugation. Since recombinants are so numerous, it is feasible to screen a random sample of the total progeny of a protoplast fusion for improved strains. In contrast, when natural conjugation is used to generate the recombinants, counter-selectable markers must be introduced into each parent before each round of recombination in order to select samples of rare recombinant progeny for screening (21), a much more laborious procedure.

CONCLUSION

The past 4 years have seen a dramatic development of new methods for bringing about recombination in industrial bacteria. We are only on the threshold of applying these methods, but it is already clear that they are not mutually exclusive: useful results will come from combinations of different procedures. For example, we could imagine cloning a restriction-modification system into two species that did not possess it, before making an interspecific protoplast fusion, relying on the in vivo operation of the restriction enzyme to achieve desirable new combinations of nonhomologous DNA segments from the two species. In this way, with sufficiently powerful selection, we could hope to produce a complex new genotype which could not be constructed rationally because its components could not be identified in advance. Or we could clone a useful gene into a transposon in vitro, transform the plasmid vector into a recipient, and allow the transposon to insert itself in vivo into the chromosome of the recipient to make the foreign gene a permanent and stable part of its genome. This has already been done in *E. coli* by J. J. Manis, B. C. Kline, and D. J. Kopecko (personal communication).

The possibilities are endless, and we really are living at an exciting time for the genetics of industrial microorganisms. However, we must guard against the danger of disillusionment when it becomes apparent that revolutionary new processes are still going to take some time to reach commercial fruition. The most futuristic applications are still in the future; meanwhile, let us walk before we try to run.

ACKNOWLEDGMENTS

For valuable discussions during the preparation of this paper I thank John Beringer, Mervyn Bibb, Keith Chater, Andy Johnston, Jan Westpheling, and Helen Wright.

REFERENCES

1. **Baltz, R. H.** 1978. Genetic recombination in *Streptomyces fradiae* by protoplast fusion and cell regeneration. J. Gen. Microbiol. **107**:93–102.

1a. **Beggs, J. D.** 1978. Transformation of yeast by a replicating hybrid plasmid. Nature (London) **275:** 104–109.

2. **Beringer, J. E., N. Brewin, A. W. B. Johnston, H. M. Schulman, and D. A. Hopwood.** 1979. The rhizobium-legume symbiosis. Proc. R. Soc. London Ser. B (in press).

3. **Beringer, J. E., S. A. Hoggan, and A. W. B. Johnston.** 1978. Linkage mapping in *Rhizobium leguminosarum* by means of R plasmid-mediated recombination. J. Gen. Microbiol. **104**:201–207.

4. **Beringer, J. E., and D. A. Hopwood.** 1976. Chromosomal recombination and mapping in *Rhizobium leguminosarum.* Nature (London) **264**:291–293.

5. **Bibb, M. J., R. F. Freeman, and D. A. Hopwood.** 1977. Physical and genetical characterisation of a second sex factor, SCP2, for *Streptomyces coelicolor* A3(2). Mol. Gen. Genet. **154**:155–166.

6. **Bibb, M. J., J. M. Ward, and D. A. Hopwood.** 1978. Transformation of plasmid DNA into *Streptomyces* at high frequency. Nature (London) **274**:398-400.

7. **Bukhari, A. I., J. A. Shapiro, and S. L. Adhya (ed).** 1977. DNA insertion elements, plasmids and episomes. Cold Spring Harbor Laboratory, Cold Spring Harbor, N.Y.

8. **Chakrabarty, A. M.** 1976. Molecular cloning in *Pseudomonas,* p. 579–582. *In* D. Schlessinger (ed.), Microbiology—1976. American Society for Microbiology, Washington, D.C.

9. **Chang, A. C. Y., and S. N. Cohen.** 1974. Genome construction between bacterial species *in vitro:* replication and expression of *Staphylococcus* plasmid genes in *Escherichia coli.* Proc. Natl. Acad. Sci. U.S.A. **71:**1030–1034.
10. **Chang, S., and S. N. Cohen.** 1977. *In vivo* site-specific genetic recombination promoted by the *Eco*RI restriction endonuclease. Proc. Natl. Acad. Sci. U.S.A. **74:**4811–4815.
11. **Coetzee, J. N.** 1978. Mobilization of the *Proteus mirabilis* chromosome by R plasmid R772. J. Gen. Microbiol. **108:**103–109.
12. **Cohen, S. N.** 1976. Transposable genetic elements and plasmid evolution. Nature (London) **263:**731–738.
13. **Cohen, S. N., A. C. Y. Chang, H. W. Boyer, and R. B. Helling.** 1973. Construction of biologically functional bacterial plasmids *in vitro.* Proc. Natl. Acad. Sci. U.S.A. **70:**3240–3244.
14. **Datta, N., R. W. Hedges, E. J. Shaw, R. P. Sykes, and M. H. Richmond.** 1971. Properties of an R factor from *Pseudomonas aeruginosa.* J. Bacteriol. **108:**1244–1249.
15. **Evans, P. K., and E. C. Cocking.** 1977. Isolated plant protoplasts, p. 103–135. *In* H. E. Street (ed.), Plant tissue and cell culture, 2nd ed. Blackwell Scientific Publications, Oxford.
16. **Ferenczy, L., F. Kevei, and J. Zsolt.** 1974. Fusion of fungal protoplasts. Nature (London) **248:**793–794.
17. **Fodor, K., and L. Alföldi.** 1976. Fusion of protoplasts of *Bacillus megatherium.* Proc. Natl. Acad. Sci. U.S.A. **73:**2147–2150.
18. **Haas, D., and B. W. Holloway.** 1976. R factor variants with enhanced sex factor activity in *Pseudomonas aeruginosa.* Mol. Gen. Genet. **144:**243–251.
19. **Hedges, R. W., and E. A. Jacob.** 1974. Transposition of ampicillin resistance from RP4 to other replicons. Mol. Gen. Genet. **132:**31–40.
20. **Hinnen, A., J. B. Hicks, and G. R. Fink.** 1978. Transformation of yeast. Proc. Natl. Acad. Sci. U.S.A. **75:** 1929–1933.
21. **Hopwood, D. A.** 1977. Genetic recombination and strain improvement. Dev. Ind. Microbiol. **18:**9–21.
22. **Hopwood, D. A., and H. M. Wright.** 1978. Bacterial protoplast fusion: recombination in fused protoplasts of *Streptomyces coelicolor.* Mol. Gen. Genet. **162:**307–317.
23. **Hopwood, D. A., H. M. Wright, M. J. Bibb, and S. N. Cohen.** 1977. Genetic recombination through protoplast fusion in Streptomyces. Nature (London) **268:**171–174.
24. **Humphreys, G. O., G. A. Willshaw, H. R. Smith, and E. S. Anderson.** 1976. Mutagenesis of plasmid DNA with hydroxylamine: isolation of mutants of multi-copy plasmids. Mol. Gen. Genet. **145:**101–108.
25. **Jacob, A. E., J. M. Cresswell, and R. W. Hedges.** 1977. Molecular characterization of the P group plasmid R68 and variants with enhanced chromosome mobilizing ability. Fed. Eur. Microbiol. Soc. Lett. **1:** 71–74.
26. **Johnston, A. W. B., and J. E. Beringer.** 1977. Chromosomal recombination between *Rhizobium* species. Nature (London) **267:**611–613.
27. **Kleckner, N.** 1977. Translocatable elements in procaryotes. Cell **11:**11–23.
28. **Kleckner, N., J. Roth, and D. Botstein.** 1977. Genetic engineering *in vivo* using translocatable drug-resistance elements. J. Mol. Biol. **116:**125–159.
29. **Kondorosi, A., G. B. Kiss, T. Forrai, E. Vincze, and Z. Bonfalvi.** 1977. Circular linkage map of *Rhizobium meliloti* chromosome. Nature (London) **268:**525–527.
30. **Meade, H. M., and E. R. Signer.** 1977. Genetic mapping of *Rhizobium meliloti.* Proc. Natl. Acad. Sci. U.S.A. **74:**2076–2078.
31. **Mertz, J. E., and R. W. Davis,** 1972. Cleavage of DNA by R_I restriction endonuclease generates cohesive ends. Proc. Natl. Acad. Sci. U.S.A. **69:**3370–3374.
32. **Pontecorvo, G.** 1975. Alternatives to sex: genetics by means of somatic cells, p. 1–14. *In* R. Markham, D. R. Davies, D. A. Hopwood, and R. W. Horne (ed.), Modification of the information content of plant cells. North-Holland, Amsterdam.
33. **Ratzkin, B., and J. Carbon.** 1977. Functional expression of cloned yeast DNA in *Escherichia coli.* Proc. Natl. Acad. Sci. U.S.A. **74:**487–491.
34. **Roberts, R. J.** 1976. Restriction endonucleases. Crit. Rev. Biochem. **4:**123–164.
35. **Schaeffer, P., B. Cami, and R. D. Hotchkiss.** 1976. Fusion of bacterial protoplasts. Proc. Natl. Acad. Sci. U.S.A. **73:**2151–2155.
36. **Sistrom, W. R.** 1977. Transfer of chromosomal genes mediated by plasmid R68.45 in *Rhodopseudomonas sphaeroides.* J. Bacteriol. **131:**526–532.
37. **Stanisich, V. A., and B. W. Holloway.** 1971. Chromosome transfer in *Pseudomonas aeruginosa* mediated by R factors. Genet. Res. **17:**169–172.
38. **Starlinger, P., and H. Saedler.** 1976. IS-elements in microorganisms. Curr. Top. Microbiol. Immunol. **75:** 111–152.
39. **Struhl, K., J. R. Cameron, and R. W. Davis.** 1976. Functional genetic expression of eukaryotic DNA in *Escherichia coli.* Proc. Natl. Acad. Sci. U.S.A. **73:**1471–1475.
40. **Taylor, A. L.** 1963. Bacteriophage-induced mutation in *Escherichia coli.* Proc. Natl. Acad. Sci. U.S.A. **50:** 1043–1051.
41. **Towner, K. J., and A. Vivian.** 1976. RP4 fertility variants in *Acinetobacter calcoaceticus.* Genet. Res. **28:** 301–306.
42. **Vapnek, D., J. A. Hautela, J. W. Jacobson, N. H. Giles, and S. R. Kishner.** 1977. Expression in *Escherichia coli* K-12 of the structural gene for catabolic dehydroquinase of *Neurospora crassa.* Proc. Natl. Acad. Sci. U.S.A. **74:**3508–3512.

Mutation and Microbial Breeding

G. SERMONTI

Institute of Histology and Embryology, The University, Perugia, Italy

PURE AND "APPLIED" MUTAGENESIS

The most widespread, if not the exclusive, tool for increasing genetic variability in microorganisms is still treatment with mutagenic agents. Although extraordinary progress has been made in the past decade both in the knowledge of the chemical basis of mutation and in increasing the yields of secondary metabolites by means of mutagenesis, the two fields developed independently. The most substantial contribution made by the geneticist to microbial breeding was that of providing new mutagens and directions for their use (17). We can apply to the area of mutagenesis the statement made by Langrish (20) about the relationship between science and technology: "Once a new area has been established, the aim of science is to understand, the aim of technology is to make it work, and industry has been very successful at making things work without too much reliance on understanding."

There are two levels at which the microbial geneticist can contribute to the rationalization of mutational breeding. One relates to the structural nature of variations; the other, to the phenotypic modifications causing or occurring in the improved strains. As the latter belongs more to the area of the biochemist (13; R. P. Elander, Genetics and Molecular Biology of Industrial Microorganisms, Orlando, Fla., 1976; R. P. Elander, this volume), I will restrict my discussion to the first.

MUTATION RATE PROBLEM

The molecular nature of the point mutation affecting a gene is, in general, of no practical relevance. The common mutations usually breed true, irrespective of whether they are missense or nonsense, transitions or transversions. An important point for the economy of mutant screening is the *rate* of mutation attained and this depends on the mutagen used, the dose, the physiological conditions, etc. By changing the mutagen or the environment, the dose can be adjusted to obtain the maximum mutation frequency. It has been observed that a moderate mutagen dose (that is, less than maximum rate) is advisable (3, 4) for obtaining superior strains. Calam (10) cautioned that excessive alterations (mutations) in the chromosome are to be avoided since they may result in metabolic "saturation" (3; Elander, Genetics and Molecular Biology of Industrial Microorganisms, Orlando, Fla., 1976) whereby beneficial mutations would tend to become exhausted and the deleterious ones would prevail. By lowering the mutation rate, one can hope to reduce the formation of multiple mutations and to increase the frequency of the single hits in the desired gene(s). The notion of a changing response to mutagens during strain selection is usually neglected (but see reference 7). To cope with it, a determination of the rate of mutagenesis during progressive strain selection is needed. The rate should not be determined by phenotypic changes such as yield variants or morphological variants because many genes may be involved. Instead, reversion from auxotrophy or mutation to antibiotic or analog resistance should be employed.

MUTATION REPAIR

Mutagens cause mutations by promoting errors in replication or repair of DNA (36). When *E. coli* is treated with UV radiation, a group of emergency functions are induced (derepressed) which promote the survival of the cell by excising the lethal photoproducts and by patching the "excision gap" through new DNA synthesis. This hypothetical repair is called "SOS repair" (25, 26). This simple formulation does not involve mutagenesis. Mutations result secondarily from the fact that the induced DNA repair system is error prone (6). In some mutant strains in which the DNA repair system is impaired (*recA*$^-$, *lexA*$^-$), the killing effect of UV becomes dramatic, but the UV mutability is abolished. This means that the enzymatic repair occurring in these strains is less effective but more accurate (25). SOS constitutive (*spr*) or conditional constitutive mutants [*tif, dnaB,* and *lig*(Ts)] have been reported (26).

Other mutant strains (*uvr*$^-$) show not only a drastically reduced survival after UV treatment but also a high frequency of induced mutations among the survivors. They rely on another repair system, the so-called postreplication repair system, which is apparently far more error prone than excision repair (8). UV-resistant UV-nonmutable strains are also known which have been assumed to have acquired improved ability to perform an error-proof type of postreplication repair (35).

Thus, in its mechanism induction of mutation is more similar to the induction of an enzyme system or of a temperate phage than to the bombardment of a target, as was formerly thought. The same system is involved in induction of mutation by ionizing radiations, alkylating agents, and other agents breaking or cross-linking DNA strands.

This new perspective of mutagenesis may have practical implications for the microbial breeder. The mutagenic efficiency can be governed by the selection of appropriate media for the expression of the SOS system, or by adopting proper pre- or post-treatment conditions. A low level of UV (100% survival) prior to UV killing treatment increases the repair and mutagenic effect (34); incorporation of bromodeoxyuridine makes DNA much more sensitive to UV radiation, particularly to the 313-nm wavelength (5). Chemicals that inhibit excision (e.g., caffeine, acriflavine, 8-methoxypsoralene) increase sensitivity to UV (7). On the other hand, post-treatment with chloramphenicol prevents the operation of the SOS system by eliminating the mutagenic effect of UV (12). It will therefore be of primary importance to consider the intrinsic (genetic) condition of the repair systems in any strain that undergoes mutagenesis. A dose-effect curve for survival and mutagenesis should be known for each industrial strain which is to be subjected to the mutation-selection procedure. Strains refractory to UV can be treated with ethyl methane sulfonate, which does not rely on any error-prone repair system (7).

Nitrosoguanidine was also indicated to exert an indirect mutagenic effect (15). This mutagen seems to produce mutations by altering a DNA polymerase III molecule causing in turn an error-prone duplication in a small section of the chromosome, before it is replaced by an unaffected DNA polymerase (19, 30). The implications of this process in synchronized mutation and in comutation experiments is described by E. Cerdá-Olmedo and R. Ruiz-Vázquez (this volume). Synchronized mutagenesis was exploited in *Streptomyces olivaceus* (21) and *S. limpmani* (15), and comutagenesis was developed in *S. coelicolor* A3(2) (27, 29) and *S. rimosus* (2).

TRANSPOSABLE ELEMENTS

In recent years, new genetic processes leading to inheritable alterations, not referred to as "mutations," have been discovered in bacteria and in eucaryotes. A group of them involve DNA transposable elements entering, leaving, or moving along the chromosome and switching genes or groups of genes on and off. Such transposable elements are known in *Enterobacteriaceae* and are identified as insertion sequences (IS) (33). IS elements are discrete DNA sequences of a defined length, i.e., 800 to 1,400 nucleotide pairs, and may occur in several copies in bacterial, plasmid, and virus chromosomes. IS elements cannot replicate autonomously, but can be transposed from one chromosome site to another independently of the known *rec* functions. Various IS elements have been characterized by a variety of methods. They have also been visualized and purified. Insertion of an IS element may abolish the function of the gene into which it is integrated and impair the function of genes located distally with respect to the promoter (polar mutations). Its excision may produce reversion to the wild type. IS elements may also implement positive control and cause chromosome rearrangements (9). For instance, IS1 causes a high frequency of chromosomal deletion from the site of its insertion (28). Duplications in association with an IS element have been observed in mutants harboring IS2 (1). Structures behaving formally like IS elements, usually containing IS and also bearing additional genes unrelated to the insertion function, are known as trasposons or "jumping genes" (9). They are more than 2,000 nucleotide pairs long and often include genes for antibiotic resistance. By progressive tandem insertions, the IS1 elements also provide a possible mechanism for the amplification of R-determinant segments (23).

The occurrence of IS elements in *Streptomyces* is suggested by some experiments carried out with *S. coelicolor* A3(2) in search of chloramphenicol-sensitive variants (31, 32). Reversible loss of chloramphenicol resistance is frequently (ca.10%) accompanied by arginine requirement and/or other phenotypic variations. Arg^- mutants are very likely deletions, apparently of various lengths (M. R. Micheli, personal communication). The "jumping" of the chloramphenicol resistance element was recently observed in *S. coelicolor* (L. Lanfaloni, personal communication).

Elements similar to "transposons" are also known in eucaryotes, and similar phenomena may be expected to exist in fungi (22).

Some of the novel phenotypes arising at high frequency in *Streptomyces* and failing to map on the chromosome have been attributed to plasmid loss (references in 18, 23). As noted by Hopwood and Merrick, their occurrence may be interpreted in terms of some kind of transposition of genetic information (18). Whatever the interpretation, it appears evident that genetic information for the production of secondary metabolites is, to some extent, carried on elements not stably bound to the chromosome and requires special treatment in both mutation and recombination programs.

In conclusion, recent advances in basic mutation research have profoundly changed our view on mutagenesis. Mutation appears not to be simply a mistake, but a true cellular function, promoted by specific agents and involving inducible enzyme systems. The process may be regulated from the outside or the inside of the cell. The cell also has at its disposal special DNA sequences (IS) which cause variations by switching genes on and off at high frequency. They might prove to be controllable by technical devices.

REFERENCES

1. **Ahmed, A., and E. Johansen.** 1975. Reversion of the gal 3 mutation of *Escherichia coli:* partial deletion of the insertion sequence. Mol. Gen. Genet. **142:**263–275.
2. **Alačević, M.** 1976. Recent advances in *Streptomyces rimosus* genetics, p. 513–519. *In* K. D. Macdonald (ed.), Second international symposium on the genetics of industrial microorganisms. Academic Press, London.
3. **Alikhanian, S. I.** 1970. Applied aspects of microbial genetics, Curr. Top. Microbiol. Immunol. **53:**91–148.
4. **Backus, M. P., and J. P. Stauffer.** 1955. The production and selection of a family of strains in *Penicillium chrysogenum.* Mycologia **47:**429–462.
5. **Boyce, R., and R. Setlow.** 1973. The action spectra for ultraviolet light inactivation of systems containing 5-bromouracil-substituted deoxyribonucleic acid. Biochim. Biophys. Acta **68:**446–454.
6. **Bridges, B. A.** 1969. Mechanisms of radiation mutagenesis in cellular and subcellular systems. Annu. Rev. Nucl. Sci. **19:**139–178.
7. **Bridges, B. A.** 1976. Mutation induction, p. 7–14. *In* K. D. Macdonald (ed.), Second international symposium on the genetics of industrial microorganisms. Academic Press, London.
8. **Bridges, B. A., and R. J. Munson.** 1966. Excision-repair of DNA damage in an auxotrophic strain of *E. coli.* Biochem. Biophys. Res. Commun. **22:**268–273.
9. **Bukhari, A. I., J. Shapiro, and S. Adhya (ed.).** 1977. DNA insertion elements, plasmids and episomes. Cold Spring Harbor Laboratory, Cold Spring Harbor, N.Y.
10. **Calam, C. T.** 1970. Improvement of micro-organisms by mutation, hybridization and selection, p. 435–459. *In* J. R. Norris and D. W. Ribbons (ed.), Methods in microbiology, vol. 3A. Academic Press, London.
11. **Clark, C. H., and D. M. Schankel.** 1975. Antimutagenesis in microbial systems. Bacteriol. Rev. **39:**33–53.
12. **Defais, M., P. Caillet-Fauquet, M. S. Fox, and M. Radman.** 1976. Induction kinetics of mutagenic DNA repair activity in *E. coli* following ultraviolet irradiation. Mol. Gen. Genet. **148:**125–130.
13. **Demain, A. L.** 1973. Mutation and the production of secondary metabolites. Adv. Appl. Microbiol. **16:** 177–202.
14. **Elander, R. P.** 1976. Mutation to increased product formation in antibiotic-producing microorganisms, p. 517–521. *In* D. Schlessinger (ed.), Microbiology—1976. American Society for Microbiology, Washington, D.C.
15. **Godfrey, O. W.** 1974. Directed mutation in *Streptomyces lipmanii.* Can. J. Microbiol. **20:**1479–1485.
16. **Guerola, N., J. L. Ingraham, and E. Cerdá-Olmedo.** 1971. Induction of closely-linked multiple mutations by nitrosoguanidine. Nature (London) New Biol. **230:**122–125.
17. **Hopwood, D. A.** 1970. The isolation of mutants, p. 363–433. *In* J. R. Norris and D. W. Ribbons (ed.), Methods in microbiology, vol. 3A. Academic Press, London.
18. **Hopwood, D. A., and M. J. Merrick.** 1977. Genetics of antibiotic production. Bacteriol. Rev. **41:**595–635.
19. **Jiménez-Sanchez, A., and E. Cerdá-Olmedo.** 1975. Mutation and DNA replication in *Escherichia coli* treated with low concentrations of N-metyl-N′-nitro-N-nitrosoguanidine. Mutat. Res. **28:**337–345.
20. **Langrish, J.** 1974. The changing relationship between science and technology. Nature (London) **250:** 614–616.
21. **Matselyukh, B. P.** 1976. Structure and function of the *Actinomyces olivaceus* genome, p. 553–563. *In* K. D. Macdonald (ed.), Second international symposium on the genetics of industrial microorganisms. Academic Press, London.
22. **Nevers, P., and M. Saedler.** 1977. Transposable genetic elements as agents of gene instability and chromosomal rearrangements. Nature (London) **268:**109–115.
23. **Okanishi, M., and M. Umezawa.** 1976. Plasmids involved in antibiotic production in Streptomyces, p. 19–38. *In* E. Freeksen, I. Tarnok, and J. H. Thumin (ed.), Genetics of Actinomycetales. Fischer Verlag, Stuttgart.
24. **Ptashne, K., and S. N. Cohen.** 1975. Occurrence of insertion sequence (IS) regions on plasmid deoxyribonucleic acid as direct and inverted nucleotide sequence duplication. J. Bacteriol **122:**776–781.
25. **Radman, M.** 1974. Phenomenology of an inducible mutagenic DNA repair pathway in E. coli SOS repair hypothesis, p. 129–142. *In* L. Prakash, F. Sherman, M. Miller, C. Lawrence, and H. W. Tabor (ed.), Molecular and environmental aspects of mutagenesis. Charles C Thomas, Publisher, Springfield, Ill.
26. **Radman, M.** 1975. SOS repair hypothesis: phenomenology of an inducible DNA repair which is accompanied by mutagenesis, p. 355–367. *In* P. Hanawalt and R. B. Setlow (ed.), Molecular mechanisms for repair of DNA, part A. Plenum Press, New York.
27. **Randazzo, R., G. Sciandrello, A. Carere, M. Bignami, A. Velcich, and G. Sermonti.** 1976. Localized mutagenesis in *Streptomyces coelicolor* A3(2). Mutat. Res. **36:**291–301.
28. **Reif, H. J., and H. Saedler.** 1975. IS1 is involved in deletion formation in the gal region of *E. coli* K 12. Mol. Gen. Genet. **137:**17–28.
29. **Russi, S., A. Carere, A. Siracusano, and A. Ballio.** 1975. An operon for histidine biosynthesis in *S. coelicolor:* biochemical evidence. Mol. Gen. Genet. **123:**225–232.
30. **Sermonti, G.** 1978. Extrachromosomal and directed mutations in *Streptomyces coelicolor,* p. 5–12. *In* E. Freeksen, I. Tarnok, and J. H. Thumin (ed.), Genetics of the Actinomycetales. Fischer Verlag, Stuttgart.

31. **Sermonti, G., A. Petris, M. Micheli, and L. Lanfaloni.** 1977. A factor involved in chloramphenicol resistance in *Streptomyces coelicolor* A 3(2): its transfer in the absence of the fertility factor. J. Gen. Microbiol. **100:**347–353.
32. **Sermonti, G., A. Petris, M. Micheli, and L. Lanfaloni.** 1978. Chloramphenicol resistance in *Streptomyces coelicolor* A 3(2): possible involvement of a transposable element. Mol. Gen. Genet. **164:**99–103.
33. **Starlinger, P., and M. Saedler.** 1976. IS-elements in microorganisms. Curr. Top. Microbiol. Immunol. **75:**111–152.
34. **Weigle, J. J.** 1953. Induction of mutation in a bacterial virus. Proc. Natl. Acad. Sci. U.S.A. **39:**628–636.
35. **Witkin, E. M.** 1967. The radiation sensitivity of Escherichia coli B: a hypothesis relating filament formation and prophage induction. Proc. Natl. Acad. Sci. U.S.A. **57:**1275–1279.
36. **Witkin, E. M.** 1976. Ultraviolet mutagenesis and inducible DNA repair in *Escherichia coli.* Bacteriol. Rev. **40:**869–907.

Nitrosoguanidine Mutagenesis

E. CERDÁ-OLMEDO AND R. RUIZ-VÁZQUEZ

Departamento de Genética, Facultad de Ciencias, Universidad de Sevilla, Seville, Spain

N-methyl-*N*′-nitro-*N*-nitrosoguanidine (nitrosoguanidine) owes its widespread application in practical mutagenesis to its ability to induce high mutation frequencies at relatively high survival rates by simple protocols (see 36 for review and references). However, the peculiarities of its action may be put to more elegant and efficient use.

SEQUENTIAL MUTAGENESIS IN SYNCHRONIZED CULTURES

The maximum frequency of nitrosoguanidine-induced mutations of a gene in a synchronized culture of *Escherichia coli* occurs at the time the gene is being replicated (10). This finding has been confirmed in many organisms (3, 5, 6, 12, 16, 24, 25, 27, 28, 33, 50, 56). It thus becomes possible to focus mutagenesis on some parts of the genome and to leave others comparatively unaffected. The succession of mutagenesis maxima for different genes constitutes a replication timetable, or "replication map," of the organism. This replication map can be correlated with the genetic map for studies on the mode and regulation of chromosome replication (15, 20, 44, 51–54).

COMUTATION

Concentrating the high mutagenicity of nitrosoguanidine on small replicating segments of the genome often results in a simultaneous induction of mutations in closely linked genes (18). In *E. coli,* this "comutation" is detectable for genes separated by less than 1.5% of the total chromosome length. Some 100 base-pair changes are spread over a segment about 50 thousand base pairs long. As a consequence, clones selected for mutations in a gene have increased probabilities of carrying mutations in nearby genes, up to 20 times higher than unselected clones. Losses of activity of a gene product to levels that would result in auxotrophy in appropriate genes may thus be found at a frequency of about 1%. More subtle changes are much more frequent.

Detailed studies of comutation have been carried out in *Streptomyces coelicolor* (39). The sensitive segments are longer than in *E. coli,* and all mutations are located in them. Neighboring genes exhibit very high comutation rates: 4% of the clones selected for mutations in *hisA* had lost the function of the *hisG* gene product. Very high comutation rates have also been reported for genes within a plasmid (29).

Comutation may also affect distant genes simultaneously replicated by different replication forks (15, 46). Comutation between closely linked genes has not yet been found in yeast, but there is a remarkable specific comutation between distant genes, indicating rigid scheduling of the advancing replication forks (13).

Comutation has been used to diminish the number of clones to be screened in searches for unselectable mutants, to alter the regulatory elements closely linked to structural genes, to explore the contents of specific chromosome regions carrying at least one gene where mutants can be selected, to distinguish the autonomous and integrated states of episomes, to detect changes in the mode of replication (unidirectional or bidirectional, in organisms where both are possible), and to compare the genomes of different species.

SATURATION AND REPAIR

Another intriguing feature of nitrosoguanidine mutagenesis is that the proportion of mutants among viable cells rises quickly to a maximum and levels off, even when lethality continues to increase (9, 23). This effect may be correlated with the cessation of DNA synthesis in the cells which have been mutated (18, 23).

Nitrosoguanidine action triggers repair activity (7), particularly near replication forks (45), even though methylations are uniformly distributed over the DNA.

Repair plays an important role in survival after nitrosoguanidine treatment, as witnessed by the rapid killing of *uvr* and *rec* mutants (21) and by the comparative resistance of *Micrococcus radiodurans* (49). Error-prone repair is responsible for some of the mutations induced in stationary, starved cells (21), but, in general, the known repair defects make little or no difference in nitrosoguanidine mutagenesis.

Alkylating agents, including nitrosoguanidine, induce an error-free repair mechanism (22, 43), possibly involving DNA polymerase II (35), which counteracts both lethal and mutagenic effects of alkylation. The lack of specificity of this mechanism for many alkylating agents suggests that it is not responsible for the peculiarities of nitrosoguanidine mutagenesis.

The DNA replication is essential for the occurrence of mutations, as shown by the *Haemophilus* transformation assay. DNA of the treated cells contained few mutations, unless it was replicated in the same cells before extraction and assay (4, 26).

SIMILARITY TO OTHER MUTAGENS

Nitrosoguanidine and other nitroso compounds (1, 2, 19, 34, 48) are not the only mutagens acting preferentially on replicating segments. This also seems true of such unrelated agents as base analogs (41), hydroxylamine and formaldehyde (42), UV light (30, 31, 47), near UV light (55), acridines (37), and ethyl methane sulfonate (EMS) (14). The advantage of nitrosoguanidine for sequential mutagenesis lies in the high mutation and survival rates and in the prominent maxima obtained with well-synchronized cultures, while the effects of some of the other mutagens are, at least, unclear.

A NITROSOGUANIDINE-IMMUTABLE MUTANT

The long suspected participation of cellular functions in nitrosoguanidine mutagenesis has been confirmed by the isolation of an *E. coli* mutant which exhibited very low mutation rates when treated with nitrosoguanidine. SE101, an arabinose-sensitive strain (40), was treated with nitrosoguanidine, allowed further growth, and plated on nutrient medium. Seven thousand clones were then examined for resistance to arabinose on arabinose-glycerol plates containing 5 μg of nitrosoguanidine/ml. Eight clones producing few mutants were isolated. All were easily killed by nitrosoguanidine, but one (strain SE43) was resistant to nitrosoguanidine mutagenesis.

Strain SE43 grows poorly, is extremely sensitive to UV and nitrosoguanidine, and undergoes few mutations with either of them. Spontaneous mutations and EMS mutagenesis and lethality are normal. In conjugation experiments strain SE43 was found to be a double mutant. One of the mutations, tentatively named *inm-1*, is characterized by its *i*nsensitivity to *n*itrosoguanidine *m*utagenesis; strain SE53, containing *inm-1*, but not the other mutation, is healthy and is normal for UV lethality and mutagenesis and for nitrosoguanidine lethality. Therefore, the other mutation,

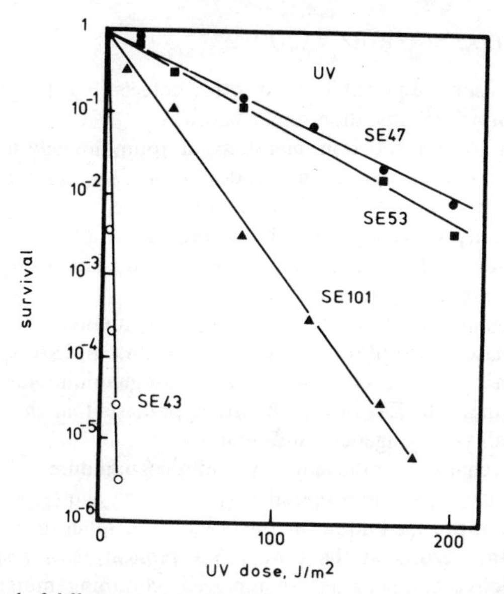

FIG. 1. *Survival of different strains after UV irradiation. SE43 is the original nitrosoguanidine-immutable, UV-sensitive strain, derived from SE101, a wild-type Hfr. SE53 is a nitrosoguanidine-immutable, UV-resistant strain derived from the cross SE43 × SE47.*

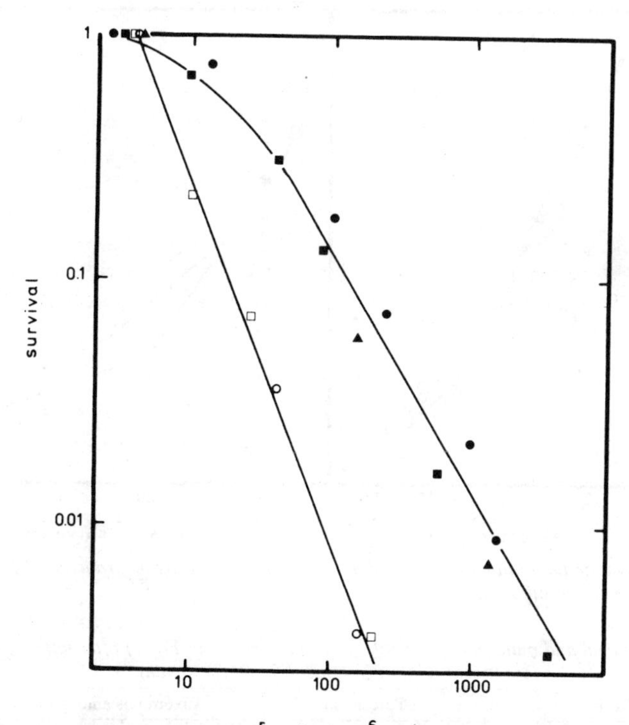

FIG. 2. *Arabinose-resistant mutants induced by UV irradiation in strains SE101 (▲), SE43 (○), SE47 (●), and SE53 (■). UV radiation and ultrasonic oscillation (□) yield the same mutant frequencies at the same survival levels in strain SE43; thus, UV radiation is not mutagenic for this strain.*

unlinked to *inm-1*, must be responsible for the other defects (Fig. 1–3, Table 1). SE53 exhibits a normal, inducible alkylation repair pathway.

From conjugation and transduction, *inm* maps at approximately min 79.5 of the new, 100-min map. The immediate vicinity is devoid of replication or repair-related genes. Probably *inm* represents a heretofore unknown gene.

Mutation *inm-1* sharply distinguishes the mutagenic action of nitrosoguanidine (severely reduced) from its lethality (unaffected) and demonstrates the participation of cellular function(s) in nitrosoguanidine mutagenesis.

Although nitrosoguanidine and EMS induce similar methylations in DNA (32) and share similar base substitution specificities (11, 38), EMS does not seem to produce comutation (17). The existence of a nitrosoguanidine-immutable strain which responds normally to EMS confirms the hypothesis that the mechanism of action of these two alkylating agents is different.

The most likely mechanism for the majority of nitrosoguanidine-induced mutations is misreplication. A more specific proposal suggested long ago (8) remains a viable working hypothesis: nitrosoguanidine interacts with the replication machinery by inducing it to commit errors at the time DNA replication is resumed; shortly thereafter, the defective components are replaced, confining mutations to small chromosome segments.

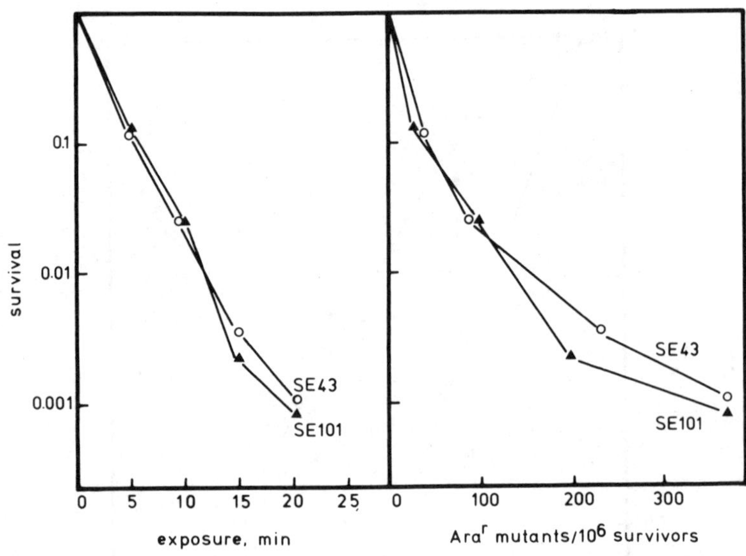

FIG. 3. *Survival and mutation to arabinose resistance after exposure of strains SE101 and SE43 to 0.1 M ethyl methane sulfonate.*

TABLE 1. *Survival and mutation of several strains (see legend to Fig. 1) after nitrosoguanidine treatment (100 μg/ml, pH 7.5, 37° C, 30 min)*

Strain	Percent survival	Auxotrophs among the survivors (%)
SE101	22	5.8
SE47	40	4.3
SE43	1.1	0.6
SE53	38	0.78

REFERENCES

1. **Androsov, V. V.** 1976. (Synchronization of the replication cycle of the chromosome of *Escherichia coli* K12 by nalidixic acid: origin, direction, symmetry, and termination of replication.) Dokl. Akad. Nauk SSSR **228:**205–208.
2. **Androsov, V. V., and V. S. Levashev.** 1974. (Point of origin of bidirectional replication of chromosome F^- of strains of *Escherichia coli* K12.) Dokl. Akad. Nauk SSSR **219:**1475–1477.
3. **Asato, Y., and C. E. Folsome.** 1970. Temporal genetic mapping of the blue-green alga, *Anacystis nidulans.* Genetics **65:**407–419.
4. **Beattie, K. L., and R. F. Kimball.** 1974. Involvement of DNA replication and repair in mutagenesis of *Haemophilus influenzae* induced by N-nitrosocarbaryl. Mutat. Res. **24:**105–115.
5. **Booker, R. J., and J. S. Loutit.** 1975. The replication of the chromosome of *Pseudomonas aeruginosa* strain 1. II. Sequential mutagenesis of synchronized cultures. Genet. Res. **25:**215–228.
6. **Burke, W., and W. L. Fangman.** 1975. Temporal order in yeast chromosome replication. Cell **5:**263–269.
7. **Cerdá-Olmedo, E., and P. C. Hanawalt.** 1967. Repair of DNA damaged by N-methyl-N′-nitro-N-nitrosoguanidine in *Escherichia coli.* Mutat. Res. **4:**369–371.
8. **Cerdá-Olmedo, E., and P. C. Hanawalt.** 1967. Macromolecular action of nitrosoguanidine in *Escherichia coli.* Biochim. Biophys. Acta **142:**450–464.
9. **Cerdá-Olmedo, E., and P. C. Hanawalt.** 1968. Diazomethane as the active agent in nitrosoguanidine mutagenesis and lethality. Mol. Gen. Genet. **101:**191–202.
10. **Cerdá-Olmedo, E., P. C. Hanawalt, and N. Guerola.** 1968. Mutagenesis of the replication point by nitrosoguanidine: map and pattern of replication of the *Escherichia coli* chromosome. J. Mol. Biol. **33:** 705–719.
11. **Coulondre, C., and J. H. Miller.** 1977. Genetic studies of the *lac* repressor. IV. Mutagenic specificity in the *lacI* gene of *Escherichia coli.* J. Mol. Biol. **117:**577–606.
12. **Dawes, I. W., and B. L. A. Carter.** 1974. Nitrosoguanidine mutagenesis during nuclear and mitochondrial gene replication. Nature (London) **250:**709–712.
13. **Dawes, I. W., D. A. Mackinnon, D. E. Ball, I. D. Hardie, D. M. Sweet, F. M. Ross, and F. Macdonald.** 1977. Identifying sites of simultaneous DNA replication in eukaryotes by N-methyl-N′-nitro-N-nitrosoguanidine multiple mutagenesis. Mol. Gen. Genet. **152:**53–57.
14. **Delaney, S. F., and N. G. Carr.** 1975. Temporal genetic mapping in blue-green alga *Anacystis nidulans* using ethyl methanesulfonate. J. Gen. Microbiol. **88:**259–268.
15. **Edlund, T., P. Gustafsson, and H. Wolf-Watz.** 1976. Effect of thymine concentration on the mode of chromosomal replication in *Escherichia coli* K12. J. Mol. Biol. **108:**295–303.
16. **Godfrey, O. W.** 1974. Directed mutation in *Streptomyces lipmanii.* Can. J. Microbiol. **20:**1479–1485.
17. **Guerola, N., and E. Cerdá-Olmedo.** 1975. Distribution of mutations induced by ethyl methanesulfonate and ultraviolet radiation in the *Escherichia coli* chromosome. Mutat. Res. **29:**145–147.
18. **Guerola, N., J. L. Ingraham, and E. Cerdá-Olmedo.** 1971. Induction of closely linked multiple mutations by nitrosoguanidine. Nature (London) **230:**122–125.
19. **Hince, T. A., and S. Neale.** 1974. A comparison of the mutagenic action of the methyl and ethyl derivatives of nitrosamides and nitrosamidines on *Escherichia coli.* Mutat. Res. **24:**383–387.
20. **Hohlfeld, R., and W. Vielmetter.** 1973. Bidirectional growth of the *Escherichia coli* chromosome. Nature (London) **242:**130–132.
21. **Ishii, Y., and S. Kondo.** 1975. Comparative analysis of deletion and base change mutabilities of *Escherichia coli* B strains differing in DNA repair capacity (wild type, *uvrA, polA, recA*) by various mutagens. Mutat. Res. **27:**27–44.
22. **Jeggo, P., M. Defais, L. Samson, and P. Schendel.** 1977. An adaptive response of *E. coli* to low levels of alkylating agent: comparison with previously characterized DNA repair pathways. Mol. Gen. Genet. **157:**1–9.
23. **Jiménez-Sánchez, A., and E. Cerdá-Olmedo.** 1975. Mutation and DNA replication in *Escherichia coli* treated with low concentrations of N-methyl-N′-nitro-N-nitrosoguanidine. Mutat. Res. **28:**337–345.
24. **Jyssum, K.** 1969. Origin and sequence of chromosome replication in *Neisseria meningitidis:* influence of a genetic factor determining competence. J. Bacteriol. **99:**757–763.
25. **Kee, S. G., and J. E. Haber.** 1975. Cell cycle-dependent induction of mutations along a yeast chromosome. Proc. Natl. Acad. Sci. U.S.A. **72:**1179–1183.
26. **Kimball, R. F., and J. K. Setlow.** 1974. Mutation fixation in MNNG-treated *Haemophilus influenzae* as determined by transformation. Mutat. Res. **22:**1–14.
27. **Koníčková-Radochová, M., and J. Koníček.** 1974. Mutagenesis by N-methyl-N′-nitro-N-nitrosoguanidine in synchronized cultures of *Mycobacterium phlei.* Folia Microbiol. (Prague) **19:**16–23.
28. **Koníčková-Radochová, M., and J. Koníček.** 1976. Mapping of the chromosome of *Mycobacterium phlei* by means of mutagenesis of the replication point. Folia Microbiol. (Prague) **21:**10–20.
29. **Koyama, A. H., C. Wada, T. Nagata, and T. Yura.** 1975. Indirect selection for plasmid mutants: isolation of ColVB *trp* mutants defective in self-maintenance in *Escherichia coli.* J. Bacteriol. **122:**73–79.
30. **Kunicki-Goldfinger, W. J. H., and R. Mycielski.** 1966. Sequential UV-induced mutations in synchronized cultures of *Escherichia coli* K12. Acta Microbiol. Polon. **15:**113–118.
31. **Lapchinskaya, O. A., T. P. Saburova, B. A. Filicheva, and N. A. Lvova.** 1976. (Successive mutagenesis in

Actinomadura carminata, a carminomycin-producing organism, under ultraviolet light effect.) Antibiotiki (Moscow) **21**:791–795.

32. **Lawley, P. D.** 1974. Some chemical aspects of dose-response relationships in alkylation mutagenesis. Mutat. Res. **23**:283–295.
33. **Lee, R. W., and R. F. Jones.** 1973. Induction of Mendelian and non-Mendelian streptomycin resistant mutants during the synchronous cell cycle of *Chlamydomonas reinhardtii.* Mol. Gen. Genet. **121**: 99–108.
34. **Mishankin, B. N., Yu. G. Suchkov, M. I. Bogdanova, E. G. Koltzova, and V. Yu. Ryzhkov.** 1973. (Nalidixic acid and the selection of auxotrophic mutants of *Yersinia pestis* and *Vibrio cholerae.*) Genetika **9**: 135–140.
35. **Miyaki, M., G. Sai, S. Katagiri, N. Akamatsu, and T. Ono.** 1977. Enhancement of DNA polymerase II activity in *E. coli* after treatment with N-methyl-N′-nitro-N-nitrosoguanidine. Biochem. Biophys. Res. Commun. **76**:136–141.
36. **Neale, S.** 1976. Mutagenicity of nitrosamides and nitrosamidines in microorganisms and plants. Mutat. Res. **32**:229–266.
37. **Newton, A., D. Masys, E. Leonardi, and D. Wygal.** 1972. Association of induced frameshift mutagenesis and DNA replication in *Escherichia coli.* Nature (London) **236**:19–22.
38. **Prakash, L., and F. Sherman.** 1973. Mutagenic specificity: reversion of iso-1-citochrome *c* mutants of yeast. J. Mol. Biol. **79**:65–82.
39. **Randazzo, R., G. Sciandrello, A. Carere, M. Bignami, A. Velcich, and G. Sermonti.** 1976. Localized mutagenesis in *Streptomyces coelicolor* A3 (2). Mutat. Res. **36**:291–302.
40. **Ruiz-Vásquez, R., C. Pueyo, and E. Cerdá-Olmedo.** 1978. A mutagen assay detecting forward mutations in an arabinose-sensitive strain of *Salmonella typhimurium.* Mutat. Res. **54**:121–129.
41. **Ryan, F. J., and S. D. Cetrullo.** 1963. Directed mutation in a synchronized bacterial population. Biochem. Biophys. Res. Commun. **12**:445–447.
42. **Salganik, R. I.** 1972. Some possibilities of mutation control concerned with local increase of DNA sensitivity to chemical mutagens. Biol. Zentralbl. **91**:49–59.
43. **Samson, L., and J. Cairns.** 1977. A new pathway for DNA repair in *Escherichia coli.* Nature (London) **267**:281–283.
44. **Schwartz, M., and A. Worcel.** 1971. Reinitiation of chromosome replication in a thermosensitive DNA mutant of *Escherichia coli.* II. Synchronization of chromosome replication after temperature shifts. J. Mol. Biol. **61**:329–342.
45. **Scudiero, D., and B. Strauss.** 1976. Increased repair in DNA growing point regions after treatment of human lymphoma cells with N-methyl-N′-nitro-N-nitrosoguanidine. Mutat. Res. **35**:311–324.
46. **Siccardi, A. G., F. A. Ferrari, G. Mazza, and A. Galizzi.** 1976. Identification of coreplicating chromosomal sectors in *Bacillus subtilis* by nitrosoguanidine-induced comutation. J. Bacteriol. **125**:755–761.
47. **Stonehill, E. H., and D. J. Hutchison.** 1966. Chromosomal mapping by means of mutational induction in synchronous populations of *Streptococcus faecalis.* J. Bacteriol. **92**:136–143.
48. **Suchkov, Y. G., and B. N. Mishankii.** 1972. (Genetic mapping of *Pasteurella pestis* chromosome by the method of induced mutations.) Genetika **8**:140–143.
49. **Sweet, D. M., and B. E. B. Moseley.** 1976. The resistance of *Micrococcus radiodurans* to killing and mutation by agents which damage DNA. Mutat. Res. **34**:175–186.
50. **Vikhanskii, Yu. D., and N. I. Zhdanova.** 1976. (Localization of some auxotrophic mutations induced by nitrosoguanidine during synchronous DNA replication in *Micrococcus glutamicus.*) Genetika **12**: 139–143.
51. **Ward, C. B., and D. A. Glaser.** 1969. Origin and direction of DNA synthesis in *E. coli* B/r. Proc. Natl. Acad. Sci. U.S.A. **62**:881–886.
52. **Ward, C. B., and D. A. Glaser.** 1969. Evidence for multiple growing points on the genome of rapidly growing *E. coli* B/r. Proc. Natl. Acad. Sci. U.S.A. **63**:800–804.
53. **Ward, C. B., and D. A. Glaser.** 1970. Control of initiation of DNA synthesis in *Escherichia coli* B/r. Proc. Natl. Acad. Sci. U.S.A. **67**:255–262.
54. **Ward, C. B., M. W. Hane, and D. A. Glaser.** 1970. Synchronous reinitiation of chromosome replication in *E. coli* B/r after nalidixic acid treatment. Proc. Natl. Acad. Sci. U.S.A. **66**:365–369.
55. **Webb, S. J., and C. C. Tai.** 1970. Differential lethal and mutagenic action of 254 nm and 320-400 nm radiation on semidried bacteria. Photochem. Photobiol. **12**:119–143.
56. **Yabe, Y., and S. Mitsuhashi.** 1971. Replication and transfer of the R factor in a synchronized culture of *Escherichia coli.* Jpn. J. Microbiol. **15**:21–27.

Mutations Affecting Antibiotic Synthesis in Fungi Producing β-Lactam Antibiotics

RICHARD P. ELANDER

Industrial Division, Bristol-Myers Co., Syracuse, New York 13201

Mutation has played an important role in the development of penicillin and cephalosporin antibiotics. The low productivity of the original strains of *Penicillium chrysogenum* and *Cephalosporium acremonium,* now classified as *Acremonium chrysogenum,* and the inherent advantages of both the natural and semisynthetic penicillin and cephalosporin antibiotics over others led to an exploitation of the genetic potential of beneficial mutations to generate superproductive commercial strains of these β-lactam–producing fungi. Mutation has proved to be important in other areas of basic antibiotic research. (i) It has resulted in the development of genetically blocked mutants which were useful in the elucidation of biosynthetic pathways of β-lactam antibiotics. (ii) It has led to the discovery of a number of new β-lactam primordial peptides and closed-ring β-lactam derivatives which have not been discovered to date to be produced by wild-type isolates. (iii) The feeding of fraudulent precursor analogs to mutants blocked in side-chain precursors has resulted in new tailor-made β-lactam molecules. This last approach may result in a variety of novel antibiotically superior molecules through mutasynthesis.

This report reviews the mutational and recombinational highlights in the development of superproductive mutants and recombinant strains of β-lactam–producing fungi, with emphasis on the application of classical mutation-selection concepts to the development of superior recombinant strains by use of parasexual genetics. Classical mutation and selection techniques, important in the development of highly productive mutants in the absence of fundamental knowledge, have now been largely replaced by more rational (directed) selection techniques grounded on more scientific bases. The application of directed selection, coupled with programs using recombinational genetics, the new techniques of protoplast fusion, and the potentials of recombinant DNA, should provide the rationale for the development of new tailor-made recombinant strains elaborating β-lactam antibiotics.

MUTATION AND ENHANCED PENICILLIN FORMATION IN *P. CHRYSOGENUM*

Large programs concerned with the induction, selection, and utilization of superior penicillin-producing variants of *P. chrysogenum* have now been proceeding for over 30 years. From the screening of hundreds of thousands of strains, a series of superior penicillin-producing mutants has been developed from strain Wisconsin Q-176, and distant relatives are now used throughout the world for the manufacture of penicillin (3, 10, 11).

The phylogeny of the Wisconsin strains of *P. chrysogenum* and two modern industrial lineages is presented in Fig. 1. Although the initial screening was based on nonmutated populations, subsequent selection failed to yield further increases. Key strains in the early ancestry of the Wisconsin series were the famous Q-176 culture, which had improved antibiotic titers, and strain BL3-D10, which failed to produce pigment chrysogenin (3). All further selections were derived from the Q-176 culture over the next decade.

Three selection lines were established from the pigmentless mutant. One line was based on a selection of natural variants, another was based on UV-radiation survivors, and a third successful line was based on treatment with nitrogen mustard. Industrial lines were later derived from the 51-20 strain by use of a variety of mutagens (10).

Throughout the programs at the University of Wisconsin and at Wyeth Laboratories (Fig. 1), it was noted that the highest mutation rate with UV radiation did not necessarily coincide with the highest rate of kill. The Wisconsin group also reported that lower doses were effective in yielding mutants with higher productivity compared to doses which would effect a maximum kill (3).

During the period of intensive selection work, attention was focused on strain characteristics which correlated with high yield. The Wisconsin group demonstrated a correlation between increased productivity and reduced sporulation and growth, characteristics which correlated with improved cephalosporin C variants (12, 47). In the Wisconsin series, the greatest change was observed in the early ancestry. Between the NRRL-1951 and Q-176 strains, growth and sporulation were reduced by 60%, with a concomitant yield increase of sixfold. Later, an additional threefold increase in antibiotic titer was associated with only an additional 10% reduction in mycelial vigor. These changes may represent a correlated response due to linkage of loci determining growth with those influencing penicillin titer. Selection can also be utilized to isolate strains with improved growth and sporulation characteristics. The latter was most important for long-term preservation and for providing vigorous vegetative development in tank fermentations (13). The correlation of strain vigor, decreased sporulation, and weak vegetative development may reflect conflicting physiological or metabolic balances of pleiotropic effects of high yield-determining genes. However, it is not possible to assess the basis of genetic and environmental interactions as related to improved yield because of the lack of relevant information. There are correlations which include strain tolerance to phenylacetic acid (19), ability to accumulate intracellular sulfate (49), ability to assimilate carbohydrate and precursor (43), sensitivity to iron (44), penicillin acylase activity (15), levels of acetohydroxy acid synthetase (21), and acyltransferase activity (45).

MUTAGENESIS AND IMPROVEMENT IN THE CEPHALOSPORIN C FERMENTATION

A strain improvement program was initiated in the 1950s to improve the low levels of cephalosporin C in Brotzu's strain of *C. acremonium.* Mutagenesis of the Brotzu isolate resulted in the selection of a mutant, M-8650, which was the progenitor strain for many industrial programs (12). The synthesis of cephalosporin C in laboratory fermentations by a series of improved UV variants is shown in Fig. 2. An improved mutant developed at Lilly Industries Ltd. (CW-19) produced 3 times more antibiotic than the Brotzu culture. When CW-19 was fermented under more favorable conditions, the culture synthesized 15 times more antibiotic than the progenitor strain. The CW-19 variant also had a significantly improved cephalosporin C to penicillin N ratio (12).

Biometric considerations of the data using normal populations of UV survivor strains showed an 11 to 1 advantage for "normal" clones versus "abnormal" clones in searching for mutants producing 20% higher antibiotic titers. The probability

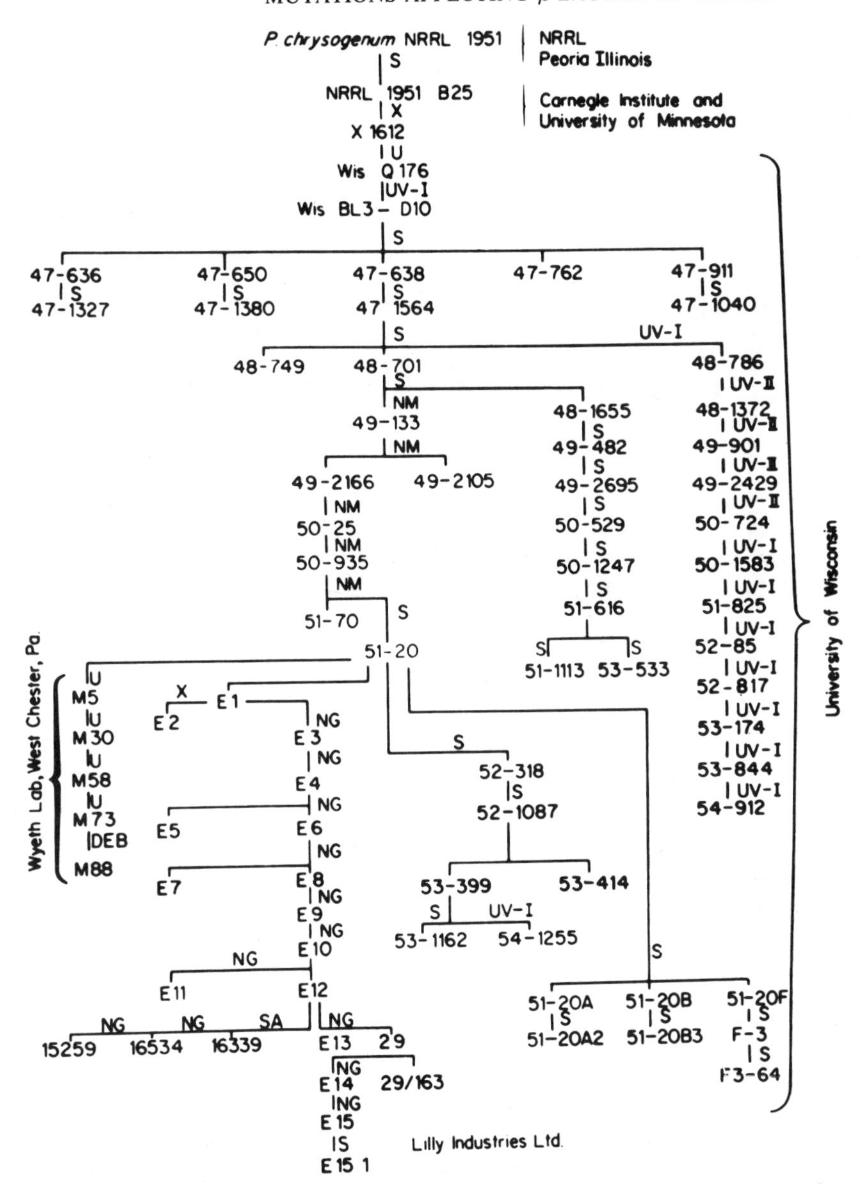

FIG. 1. *Genealogy of Wisconsin strains of* P. chrysogenum *and subsequent development of two industrial lines. Mutations brought about: spontaneously, S; by X irradiation, X; by UV irradiation, wavelength unspecified, U; 275 nm, UV-I; 253 nm, UV-II; by nitrogen mustard, NM; by nitrosoguanidine, NG; by diepoxybutane, DEB. (Elander [10].)*

statements were calculated for many survivor populations on a statistical basis (12).

The improved UV variants differed markedly from the progenitor strains in cultural and biochemical properties. Untreated populations of the improved variants showed a progressive reduction in colony diameter, decreased vegetative development, and decreased sporulation vigor, features which were also characteristic of the improved penicillin variants (11). Fasani et al. (17) reported that dimethylsulfate-

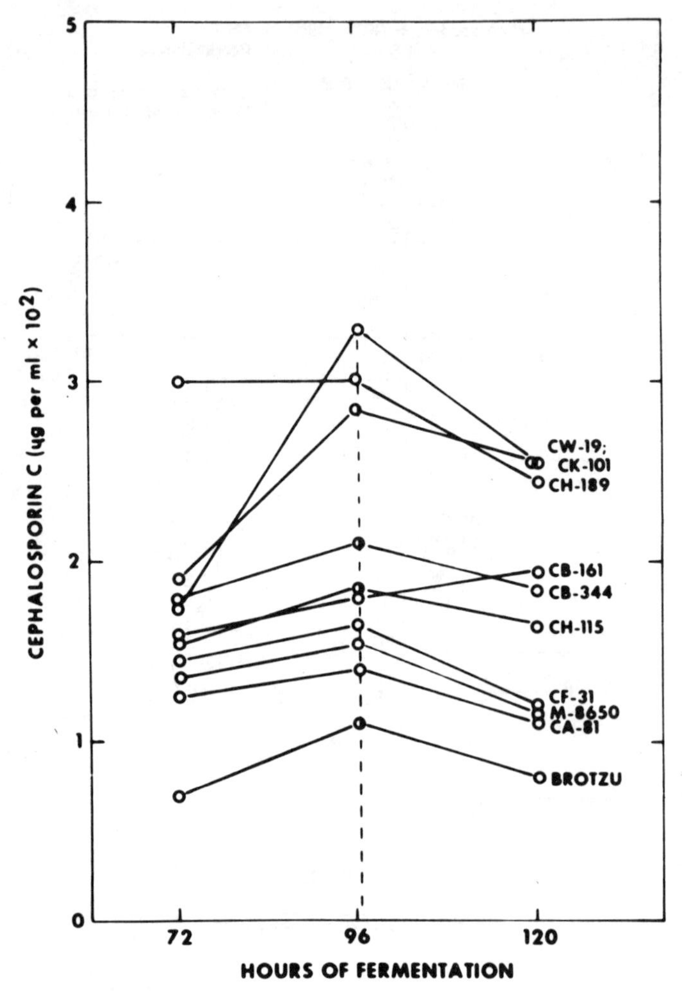

FIG. 2. *Cephalosporin production by Brotzu strain and early improved Lilly variants of* C. acremonium. *(After Elander et al. [12].)*

and phenethyl alcohol-treated populations yielded higher-producing cephalosporin C variants. The highest antibiotic producers were obtained by phenethyl alcohol treatment. Strains of *C. acremonium* or *C. polyaleurum* resistant to polyene antibiotics produced 10 g of cephalosporin C/liter (Takeda Pharmaceutical Co., Japanese Patent JA-110723, 16 January 1975). In another patent, polyploid clones were reported to be potent cephalosporin C producers. The higher ploidy clones were induced by exposure to camphor followed by selection of large cells (Takeda Pharmaceutical Co., Japanese Patent JS-109680, 16 January 1975).

Submerged cultures of *C. acremonium* form arthrospores, and the differentiation coincides with the maximal rate of cephalosporin C synthesis. In the improved variants, arthrospore formation was proportional to the increased antibiotic formation (39). Methionine supplementation enhanced the onset of differentiation, and the requirement for methionine was increased with higher-yielding mutants (39). Methionine and sulfate metabolism is important with respect to cephalosporin C synthesis, and the metabolism of methionine, its analog norleucine (9), and sulfate may trigger antibiotic synthesis through cellular differentiation.

Cysteine is the immediate donor of sulfur to cephalosporin C, but the amino acid is not stimulatory of cephalosporin C synthesis in media containing sulfate (7). Methionine stimulates antibiotic synthesis, and the stimulation is not due to sulfur donation but to an unresolved role in antibiotic regulation. Nüesch et al. (42) and Drew et al. (9) have obtained mutants with blocks between sulfate and cysteine and cystathionine and homocysteine. The mutants respond to cysteine, cystathionine, and homocysteine, but do not replace methionine with respect to cephalosporin C stimulation. An interesting Ciba *slp* mutant, blocked in the sulfate reduction pathway prior to sulfide formation, was able to assimilate more exogenous methionine and synthesized four times more antibiotic than its sulfide-proficient parent (42). Revertant strains of the mutant assimilated less methionine and synthesized low levels of cephalosporin C. Drew and Demain (8) showed a similar result with the 274-1 mutant. Nüesch and co-workers have proposed that higher levels of cephalosporin C obtained with non-sulfate-utilizing mutants were due to their inability to synthesize cysteine, an amino acid which acts as a repressor of methionine permease (42).

In a study with improved mutants at Lilly Industries Ltd., Queener et al. (47) reported that the specific activity of glutamate dehydrogenase was derepressed whereas two mutants in a low-yielding series had repressed levels of glutamate dehydrogenase. The altered regulation pattern for this enzyme may have removed a nitrogen limitation for cephalosporin C synthesis. An inverse relationship appeared to exist between vegetative development and enhanced cephalosporin C synthesis.

Mutants of *C. acremonium*, altered in sulfur metabolism and in their potential to synthesize cephalosporin C from sulfate, have been derived (41). One of the mutants, M8650-sp-1, corresponded to the *cys-3* mutant of *Neurospora crassa* in which the locus exerts coordinate control over the synthesis of sulfate permease as well as arylsulfatase. This mutant was facilitated for sulfate transport and repressed for arylsulfatase, and it utilized sulfate as effectively as methionine in providing sulfur for cephalosporin C. In this connection, the parent strain, M8650, is considered to be a derepressed mutant for arylsulfatase synthesis. The sulfatase repression in M8650-sp-1 may be related to the accumulation of sulfide which regulates sulfatase synthesis, since sulfide is believed to be a corepressor of sulfatase in fungi. Mutants of *C. acremonium* are increasingly derepressed for arylsulfatase and concomitantly exhibit increased potentials for synthesis of antibiotics from methionine (7).

Another mutant, IS-5, with enhanced potential to use sulfate for cephalosporin C production, produced two times more antibiotic than its parent, with sulfate (29). Cephalosporin C production by this mutant was sensitive to methionine, in contrast to its parent. In addition to the mutant IS-5, several other mutants with an increased potential to produce higher levels of cephalosporin C from sulfate were methionine sensitive. Therefore, the increase in productivity from sulfate and in methionine sensitivity may be metabolically related and caused by the same mutational event.

Recently, Komatsu and Kodaira (28) reported enzymatic studies on sulfate-efficient strains of *C. acremonium*. In sulfate-starved cells, norleucine showed an inhibitory effect on cephalosporin C and penicillin N formation in the presence of inorganic S sources and L-cysteine. However, antibiotic production was stimulated by methionine in the parental strain. High cysteine pools were formed in the sulfate-efficient strains. One of the cysteine biosynthetic enzymes, L-serine sulfhydrylase, was elevated twofold in the mutant, thereby rendering the improved mutant with a high pool of cysteine, an important biosynthetic intermediate of cephalosporin C.

PARASEXUAL RECOMBINATION AND ANTIBIOTIC PRODUCTION IN β-LACTAM-PRODUCING FUNGI

P. chrysogenum

Sermonti (50) carried out heterokaryon experiments between a low-producing auxotrophic strain, NRRL 1951 *pro,* and a high-producing strain, Wis. 49-133 *nic.* Homokaryotic segregants from the heterokaryon showed a clear association between the nuclear marker and yield, i.e., *pro* segregants were low producers and *nic* segregants were high yielding. Thus, penicillin yield is determined by nuclear genes, and there is evidence from both heterokaryons and parasexual recombinants that several different genes are involved. Both the *y-met* and *w ade* heterokaryon of Elander (10) and the "New Hybrid" strain of Alikhanian and Borisova (1) show significantly higher yields than their parent homokaryons, suggesting complementation between nonallelic genes.

Caglioti and Sermonti (5) attempted to map *pen-1,* a determinant for increased yield, by mitotic recombination. Its locus was distal to *pro-1* and *met-1,* two auxotrophic markers in one of three linkage groups which they tentatively identified. By haploidization analysis, Ball (4) confirmed the existence of three linkage groups of a Glaxo strain. Increased yield was induced in five separate strains, each of which carried conidial color or auxotrophic markers.

Two phenomena have been described which suggest that genes determining increased yield of penicillin are recessive and that independently induced mutations could be allelic. When a diploid strain is derived via the parasexual cycle between a low-yielding and a high-yielding strain, its yield is comparable to that of the former. Diploids obtained between strains from a common ancestor, but carrying independently induced mutations for higher yield, gave yields equivalent to the parental strain, but higher than the yield of the common ancestor (10, 13, 33). An explanation of these observations is that the strains carried a number of allelic mutant sites in common, even though the sites had been independently mutated. It appears there are a number of genetic determinants, recessive in their expression and located on more than one chromosome, which determine increased penicillin yield in *P. chrysogenum.* Their expression can be modified by genetic background, for most conidial color and auxotrophic markers reduced yield drastically. To date, the technique of mutation and selection has been the most reliable procedure for improving penicillin titers. Breeding involves the parasexual cycle, and, therefore, the peculiarities of the cycle in *P. chrysogenum* present difficulties in achieving recombination and subsequent unimpaired segregation. Strain variability and instability cause difficulties in the initiation of the parasexual cycle because heterokaryons are not easily produced with strains of *P. chrysogenum.* When heterokaryons are produced, instability of diploids and their segregants often occurs. Elander (10) described a highly stable diploid strain at Lilly Industries Ltd. derived from the haploid production strain E-15 (Fig. 3). Spontaneous variation in the diploid and haploid was 9.2% and 31.9%, respectively, as assessed by the color types which segregated. The difference was even more striking after exposure to UV irradiation, where the diploid:haploid derived variants were 11.6% and 41.1%, respectively. In this case, the diploid strain was far more stable than its haploid progenitor.

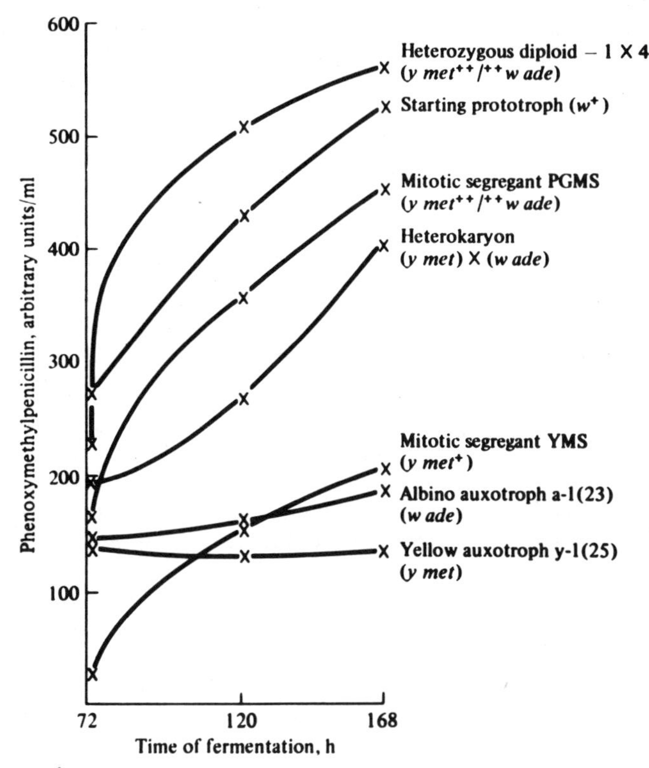

FIG. 3. *Penicillin production by mutant and recombinant strains of* P. chrysogenum *E15. (From Elander [10].)*

In contrast, Ball (4) reported that diploids show greater instability than their haploid progenitors, but within the same range. The diploid showed twice as many poorly sporulating types as the haploid, while densely sporulating types were increased 10 times. Ball (4) and Roper (48) have compared this phenomenon with that of "mitotic nonconformity" described by Nga and Roper (40) in *A. nidulans.* Mitotic nonconformity stems from the existence of duplicate segments of small fractions of the genome which probably originated through translocations. If the difficulties described in producing diploids can be overcome, they may be attractive to the fungal geneticist. There is the possibility of a heterotic effect on yield (52). Even if this does not occur, productive segregants from diploid strains may be selected. If the latter approach is used, it is recommended to start initially with strains lacking chromosomal rearrangements or to induce variants in increased yield from an existing strain without inducing chromosomal aberrations. Since spontaneous de-diploidization can always occur, systems have been proposed to reduce such effects. Macdonald (33) has suggested the use of parental haploids which grow poorly in complex fermentation media to select against the parental segregants arising during fermentation. Azevedo and Roper (2) suggested the induction of recessive lethals in the diploid so that haploid segregants would be eliminated. It has been suggested that the Lilly diploid may be a balanced lethal diploid in that it was extremely stable but, even after treatment with UV irradiation, the proportion of viable segregants it produced was hardly increased. Provided that diploid stability and unrestricted parasexual recombination can be achieved, it should be possible to make further selection progress. Elander (10) described a spontaneous segregant from the Lilly diploid which produced nearly 25% more antibiotic than its parent, which itself yielded better than the production haploid E-15. Ball (4) has also described recombinant segregants with improved yield.

The first 30 years of mutation-selection and scientific breeding have demonstrated the potentials of mutation breeding and the dangers present in utilizing mutagens which induce chromosomal aberrations (4). The next 25 years should produce more extensive application of the parasexual cycle to breeding (22, 23, 35).

C. acremonium (*A. chrysogenum*)

The commercial strains of cephalosporin-producing fungi lack true sexual reproduction but can be manipulated by parasexual recombination. Parasexuality is based on mitotic recombination during normal cell division. It consists of the formation of heterokaryons, rare nuclear fusion within the heterokaryon to form a diploid nucleus, and mitotic recombination in diploid and haploid nuclei by genetic crossing-over or loss of chromosomal segments through nondisjunction. Mitotic nonconformity is an additional genetic mechanism allowing for genetic recombination and may explain the genetic basis for culture instability of many fungal organisms. Large segments of chromosomes invert in high frequency, and extensive crossing-over within the inversion loops may generate chromosomal deletion or extensive duplication of genetic material. In this manner, mitotic nonconformity can lead to excessive genetic heterogeneity in cultures due to genetic imbalance.

Parasexual genetic studies with strains derived from Brotzu's *Cephalosporium* isolate have been disappointing (12; K. D. Macdonald, personal communication). Heterokaryons and diploids have been difficult to obtain because the cells of the fungus are typically uninucleate and nuclear migration is limited, thereby restricting heterokaryosis. However, Nüesch et al. (42) have adopted parasexual techniques for cephalosporin C biosynthetic studies and obtained a number of relatively stable diploid strains.

The elaboration of cephalosporins by selected *Emericellopsis* species (14, 18) offers fertile areas for genetic studies of fungi which have a true sexual mode of reproduction. However, the vegetative cells of *Emericellopsis* are similar to *Cephalosporium* in that they are typically uninucleate, with restricted nuclear migration. All species of *Emericellopsis* thus far investigated are self-compatible or homothallic, and meiotic or sexual recombination occurs only rarely (16).

USE OF BIOCHEMICAL MUTANTS FOR THE ELUCIDATION OF BIOCHEMICAL PATHWAYS OF SYNTHESIS OF β-LACTAM ANTIBIOTICS

Blocked mutants have been useful in the elucidation of biosynthetic pathways for β-lactam antibiotics. Blocked mutants of *C. acremonium* accumulated exclusively deacetylcephalosporin C with a D-α-aminoadipyl side chain at C-7. The finding that deacetylcephalosporin C was formed early in the fermentation in the absence of cephalosporin C was unexpected, because the deacetyl moiety was thought to be a breakdown product of preformed cephalosporin C (24). Examination of acetylhydrolase levels of both the parental strains and the mutant strains showed equivalent

Please substitute for the incomplete structures printed on page 29.

$CH_3CO\text{-}NH$

$CH\text{-}(CH_2)_3\text{-}CO\text{-}NH$

HOOC

S

N

O

CH_3

COOH

(I)

$HOOC\text{-}CH\text{-}(CH_2)_3\text{-}CO\text{-}HN$

NH_2

S

N

O

CH_2SCH_3

COOH

(II)

$HOOC\text{-}CH\text{-}(CH_2)_3\text{-}CO\text{-}HN$

NH_2

S

N

O

$CH_2S_2O_3H$

COOH

(III)

$HOOC\text{-}CH\text{-}(CH_2)_3\text{-}CO\text{-}HN$

NH_2

S

N

O

COOH

CH_3

$CH_2\text{-}S\text{-}C\text{-}CH$

CH_3

COOH

NH_2

(IV)

levels of intracellular and extracellular enzyme, suggesting that deacetyl is a precursor of cephalosporin C and is synthesized by enzymatic acetylation of deacetylcephalosporin C (18).

The accumulation of deacetoxycephalosporin C by *C. acremonium* blocked in cephalosporin C synthesis was reported independently (25, 32, 46). Kanzaki and Fujisawa (25) showed that a deacetylcephalosporin C-blocked mutant derived from a deacetyl-producing strain accumulated deacetoxycephalosporin C and only a trace quantity of the deacetyl moiety, suggesting that deacetylcephalosporin C is synthesized by enzymatic hydroxylation of deacetoxycephalosporin C.

Mutants have also been important in attempts to establish the role of penicillin N in the biosynthesis of cephalosporin C. Konomi and Demain (T. Konomi and A. L. Demain, Abstr. Annu. Meet. Am. Soc. Microbiol. 1977, O24, p. 249) presented evidence that type A mutants of *C. acremonium* were capable of synthesizing a cephalosporin antibiotic from crude penicillin N by use of cell-free extracts.

NOVEL β-LACTAM PRECURSORS OR ANTIBIOTICS FROM MUTANTS OF *C. ACREMONIUM*

Mutants of *C. acremonium* blocked in the synthesis of cephalosporin C have been reported to produce novel β-lactam molecules or precursors in fermentation beers. Traxler et al. (52) reported a new antibiotic, *N*-acetyldeacetoxycephalosporin C (I), isolated from broths of blocked mutants of *C. acremonium*. The titers of this new metabolite in three independently obtained deacetoxycephalosporin C-producing mutants varied from 0.05 to 0.2 mg/ml.

CH_3CO-NH
CH-$(CH_2)_3$-CO-NH
HOOC
O S N COOH CH_3
(I)

HOOC-CH-$(CH_2)_3$-CO-HN
NH_2
O N S COOH
CH_2SCH_3
(II)

HOOC-CH-$(CH_2)_3$-CO-HN
NH_2
O N S COOH
$CH_2S_2O_3H$
(III)

HOOC-CH-$(CH_2)_3$-CO-HN
NH_2
O N S COOH
CH_2-S-C-CH
CH_3
CH_3
COOH

NH_2
(IV)

Kitano et al. (27) isolated a family of cephem compounds containing a glutaryl group instead of the α-aminoadipyl residue at C-7. Another new cephalosporin, 3-methylthiomethyl cephem (II), was detected in broths of a potent mutant at Takeda (26). The new compound may originate from a reaction between cephalosporin C and methanethiol. Kanzaki and Fujisawa (25) described a cephem metabolite in which the *O*-acetyl group of cephalosporin C was replaced by thiosulfate (a Bunte salt) (III).

Thiosulfate may be a degradation product of methionine, an amino acid shown to be stimulatory to cephalosporin C synthesis. A new cephem compound identified as 7-(D-5-amino-5-carboxyacetamido)-3-(1,1-dimethyl-2-amino-2-carboxyethyl)-3-cephem-4-carboxylic acid (IV) was found in another Takeda mutant (27).

A deacetyl-producing mutant from Takeda was shown to be defective in acetyl coenzyme A: deacetylcephalosporin C acetyltransferase and produced a new cephem identified as D-5-amino-5-carboxyvaleramido-(5-formyl-4-carboxy-2H,3H,6H-tetrahydro-1,3-thiazinyl)glycine.

RATIONAL SELECTION FOR IMPROVED β-LACTAM PRODUCTIVITY

In efforts to improve the efficiency of large-scale strain improvement programs, rational direct selection procedures have been reported to be more efficient than random blind screening for the selection of improved variants. Many of the techniques involve the use of fermentation prescreening of mutagenized cells prior to laboratory fermentation studies. More importantly, the techniques are based on known or probable biochemical mechanisms and, therefore, remove some of the empiricism, serendipity, and labor commonly associated with random empirical screening.

Mutants Selected Directly on Agar Plates

Direct demonstration of antibiotic production by a colony growing on solidified fermentation medium can be observed by overlaying of a sensitive organism after colonial development of either *Penicillium* or *Acremonium.* The colony-plate method has advantages in that it can eliminate many of the poorly producing isolates, thereby increasing the probability of discovering superior mutants in laboratory programs. The application of the colony-plate procedure has meaning only if plate performance is correlated with submerged fermentation performance. The program has been useful in selecting superior cephalosporin producers by researchers at the University of Wisconsin (14, 48) and has recently been advocated by Trilli et al. (53). The Wisconsin workers sprayed plates containing mature fungal colonies of mutagenized spores with suspensions of *Alcaligenes faecalis.* Strains having a greater inhibition zone diameter compared to colony diameter (potency index) were examined in flask fermentations. With this procedure, approximately 60% of the isolates were discarded prior to the flask evaluation stage. Using the above technique, the Wisconsin workers obtained a strain showing a fivefold improvement over a 4-year period in a small-scale program. Trilli et al. (53) grew colonies of *C. acremonium* originating from mutagenized spores on small disks of agar medium. After 5 days of growth, the antibiotic contents of the disks were assayed with a sensitive assay organism. By varying the concentration of nitrogen in the agar, these workers were able to control the quantities of antibiotic produced. The relation of agar disk inhibition zone diameter to log shaken flask titer was linear with short incubation times, but shifted toward a higher order upon more prolonged incubation periods. Interestingly, their results suggest that the shaken flask performance test underestimated the improvement in strain productivity.

Selection of Mutants for Resistance or Sensitivity to Antibiotic Precursors (Stimulants), Biosynthetic Amino Acid Precursors, or Analogs of Precursors

Analogs of end products may act as false feedback effectors, thereby inhibiting growth of the producer organism. The side-chain precursor of penicillin G, phenylacetic acid, is a highly toxic agent to strains of *P. chrysogenum.* Fuska and Welwardov á (19) reported that a high percentage of a population of strains resistant to high concentrations of phenylacetic acid were superior penicillin producers.

Use of Biosynthetic Precursor Amino Acid Analogs for Improved Productivity

Godfrey (20) reported that analog-resistant mutants of the cephamycin-producing actinomycete *Streptomyces lipmanii* showed increased cephamycin productivity. Trifluoroleucine and 2-amino-ethyl-L-cysteine were particularly effective. Nüesch et al. (42) reported that strains of *C. acremonium* resistant to selenomethionine had impaired methionine uptake and were poor producers of cephalosporin. They used the methionine analogs DL-methionine-DL-sulfoxide, DL-norleucine, and DL-ethionine in studies on the effects of methionine, sulfate, and sulfur metabolism and their effects on cephalosporin synthesis in strains of *C. acremonium.*

Use of Mutants Resistant to Heavy Metals and Other Toxic Agents

Godfrey (20) reported the use of phenylmercuric acetate resistance in attempts to discover high-yielding strains of the cephamycin-producing organism, *S. lipmanii.* He found decreased production with phenylmercuric acetate-resistant mutants. Niss and Nash (41) reported the use of *C. acremonium* resistant to potassium chromate, a compound known to impair sulfate uptake. A mutant designated as M8650-*chr* showed severe impairment for the synthesis of cephalosporin C from sulfate. However, the chromate resistance mutation did not alter the capability of the resistant mutant to synthesize cephalosporin C from methionine. Marzluf (34), working with *Neurospora crassa,* reported that chromate-resistant mutants were defective in sulfate transport. Lemke (30) reported the effects of a variety of toxic substances on the growth of *C. acremonium.*

Use of Methionine Auxotrophic Mutants for the Improved Synthesis of Cephalosporin C

H. J. Treichler et al. (U.S. Patent 3,776,815, 4 December 1973) reported that methionineless strains of *C. acremonium* produced 5 times more cephalosporin C on the addition of 4 g of methionine/liter compared to the parental strain on the addition of optimal methionine at 2 g/liter. However, the high concentrations of methionine required exclude any commercial application.

Table 1 shows a comparison of random empirical screening procedures against a variety of directed rational screening in large screening programs with penicillin and cephalosporin strain improvement at Bristol-Myers Co. The data clearly demonstrate the superiority of rational selection procedures as expressed by the numbers of isolates retained for preservation and tertiary screening prior to small-tank evaluations. Such data are probably representative of other current industrial laboratories involved in rational screening for improved mutants.

TABLE 1. *Comparison of random selection versus directed selection procedure in strains of* P. chrysogenum *and* A. chrysogenum *(adapted from Chang and Elander [6])*

TYPE	TREATMENT	ORGANISMS EXAMINED	SELECTION PROCEDURE	NO. TESTED	NO. RETAINED	% RETAINED[a]
Random selection	UV, NG[b]	Penicillium and Acremonium	None	860	7	0.81
Directed selection	UV, NG X-ray	Penicillium and Acremonium	1. Colony-plate	438	6	1.36
			2. Auxotrophs	35	2	5.71
			3. Haploidization-inducing agents	503	8	1.59
			4. Mitotic inhibitors (R), (S)	225	3	1.33
			5. Mercury (R), (S)	162	2	1.23
			6. Amino acid analogs (R), (S)	567	8	1.41
			7. Sulfur analogs (R), (S)	452	22	3.98
		Acremonium	1. Methionine analogs (R), (S)	605	22	3.64
			2. Increased sensitivity to methionine			
			a) Growth	247	4	1.62
			b) β—lactam synthesis	52	2	3.85

[a] Superior on both primary and secondary screening tests.
[b] Nitrosoguanidine.

MUTANTS AND NEW BIOSYNTHETIC β-LACTAM MOLECULES

Mutation of microorganisms producing secondary metabolites has resulted in the selection of biochemical or blocked mutants capable of producing new modified metabolites either directly or in response to some precursor analog. The modified metabolites usually possess the basic structural features of the parent compound, but either lack or contain modified functional groups which often convey differing biological activities. The biosynthetic approach using mutants and biosynthetic analogs has been useful in the generation of new aminoglycoside antibiotics.

To date, there has been only a single report of a new biosynthetic analog produced by mutants of β-lactam-producing fungi in response to the feeding of biosynthetic precursor analogs. Troonen et al. (54) reported on a lysine auxotroph of *Acremonium chrysogenum* ATCC 20389 producing cephalosporin C and penicillin N only, in media supplemented with DL-α-aminoadipic acid. The mutant was found to incorporate a fraudulent side-chain analog, L-*S*-carboxymethylcysteine, to generate a new biosynthetic penicillin (Fig. 4). The new penicillin (RIT-2214) was identified as 6-(D)-{[(2-amino-2-carboxy)ethylthio]-acetamido}-penicillanic acid. However, no corresponding modified 7-aminocephalosporanic acid derivative was reported. Lemke and Nash (31) reported lysine-requiring strains of *C. acremonium* which were unable to synthesize either penicillin N or cephalosporin C. The mutants grew in a minimal medium supplemented with lysine but not with α-aminoadipate. The presence of exogenous DL-α-aminoadipic acid was sufficient for both growth and the production of penicillin N and cephalosporin C (38).

Incorporation of side-chain precursors appears to be nonspecific for strains of *P. chrysogenum* and has generated a variety of biosynthetic penicillins. However, in *C. acremonium,* only the L-*S*-carboxymethylcysteine analog has been reported to be incorporated into a new penicillin N-like antibiotic. It was proposed that L-*S*-carboxymethylcysteine or a derivative may interfere with an enzyme involved in an oxidative process essential for the synthesis of cephalosporin C (54). L-*S*-carboxymethylcysteine, which contains a thioether group, may be a competitive inhibitor for some oxidase that acts on the sulfur atom of a hypothetical β-lactam intermediate which in nonmutant cultures forms a sulfoxide intermediate in the conversion of penicillins to cephalosporins (37).

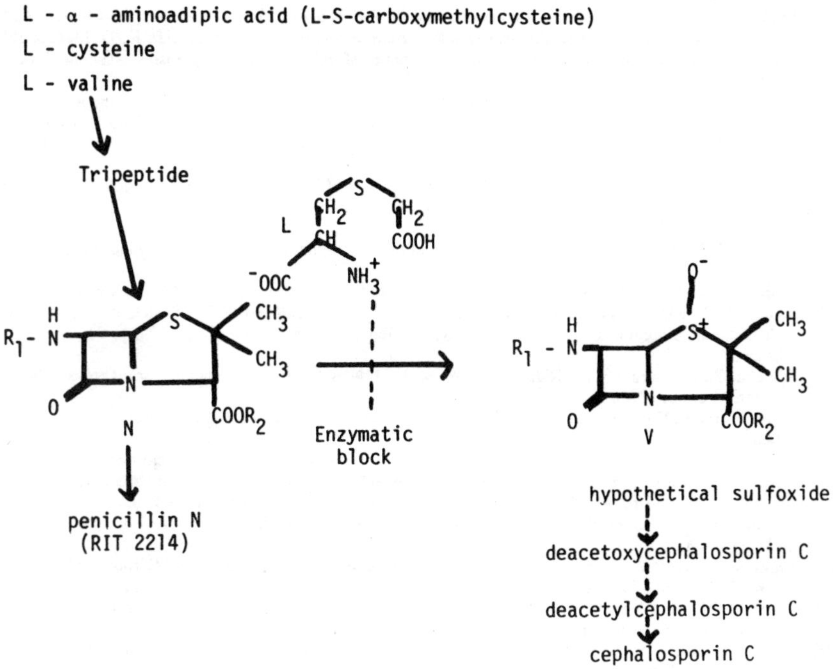
L - α - aminoadipic acid (L-S-carboxymethylcysteine)
L - cysteine
L - valine
Tripeptide
S
CH2
CH2
L
CH
COOH
-OOC
NH3+
H
R1- N
S
CH3
CH3
N
O
COOR2
N
penicillin N
(RIT 2214)
Enzymatic
block
O-
H
R1 - N
S+
CH3
CH3
N
O
COOR2
V
hypothetical sulfoxide
deacetoxycephalosporin C
deacetylcephalosporin C
cephalosporin C

FIG. 4. *Hypothetical branched pathway for the biosynthesis of penicillin N, cephalosporin C, and a new biosynthetic penicillin analog (RIT 2214) by a mutant strain of* A. chrysogenum. *(From Troonen et al. [54].)*

ACKNOWLEDGMENTS

I dedicate this paper to my former mentors at the University of Wisconsin: Professors Stauffer, Backus, Raper, Johnson, and Lederberg. The fundamental scientific contributions of these distinguished researchers and scholars form the basic concepts for both the fundamental and applied microbial genetics and biochemistry of industrial microorganisms.

REFERENCES

1. **Alikhanian, S. I., and L. N. Borisova.** 1956. Vegetative hybridization of fungi of the genus *Penicillium.* Izv. Akad. Nauk SSSR. **2:**74–85.
2. **Azevedo, J. L., and J. A. Roper.** 1967. Lethal mutations and balanced lethal systems in *Aspergillus nidulans.* J. Gen. Microbiol. **49:**149–155.
3. **Backus, M. P., and J. F. Stauffer.** 1955. The production and selection of a family of strains in *Penicillium chrysogenum.* Mycologia **47:**429–463.
4. **Ball, C.** 1973. The genetics of *Penicillium chrysogenum.* Prog. Ind. Microbiol. **12:**47–72.
5. **Caglioti, M. T., and G. Sermonti.** 1956. A study of the genetics of penicillin-producing capacity in *Penicillium chrysogenum.* J. Gen. Microbiol. **14:**38–52.
6. **Chang, L. T., and R. P. Elander.** 1979. Rational selection for improved cephalosporin C productivity in strains of *Acremonium chrysogenum* Gams. Dev. Ind. Microbiol., vol. 20 (in press).
7. **Dennen, D. W., and D. D. Carver.** 1969. Sulfatase regulation and antibiotic synthesis in *Cephalosporium acremonium.* Can. J. Microbiol. **15:**175–181.
8. **Drew, S. W., and A. L. Demain.** 1975. Production of cephalosporin C by single and double sulfur auxotrophic mutants of *Cephalosporium acremonium.* Antimicrob. Agents Chemother. **8:**5–10.
9. **Drew, S. W., D. J. Winstanley, and A. L. Demain.** 1976. Effect of norleucine on morphological differentiation in *Cephalosporium acremonium.* Appl. Microbiol. **31:**143–145.
10. **Elander, R. P.** 1967. Enhanced penicillin synthesis in mutant and recombinant strains of *Penicillium chrysogenum,* p. 403–423. *In* H. Stübbe (ed.), Induced mutations and their utilization. Akademie-Verlag, Berlin.
11. **Elander, R. P.** 1976. Mutation to increased product formation, p. 517–521. *In* D. Schlessinger (ed.), Microbiology—1976. American Society for Microbiology, Washington, D.C.

12. **Elander, R. P., C. J. Corum, H. DeValeria, and R. M. Wilgus.** 1976. Ultraviolet mutagenesis and cephalosporin synthesis in strains of *Cephalosporium acremonium,* p. 253–271. *In* K. D. Macdonald (ed.), Second international symposium on the genetics of industrial microorganisms. Academic Press, London.
13. **Elander, R. P., M. A. Espenshade, S. G. Pathak, and C. H. Pan.** 1973. The use of parasexual genetics in an industrial strain improvement program with *Penicillium chrysogenum,* p. 239–253. *In* Z. Vaněk, Z. Hošťálek, and J. Cudlín (ed.), Genetics of industrial microorganisms, vol. 2. Elsevier Publishing Co., Amsterdam.
14. **Elander, R. P., J. F. Stauffer, and M. P. Backus.** 1961. Antibiotic production by various species and varieties of *Emericellopsis* and *Cephalosporium.* Antimicrob. Agents Annu. 1960, p. 91–102. Plenum Press, New York.
15. **Erickson, R. C., and L. D. Dean.** 1966. Acylation of 6-aminopenicillanic acid in *Penicillium chrysogenum.* Appl. Microbioi. **14:**1047–1048.
16. **Fantini, A. A.** 1962. Genetics and antibiotic production of *Emericellopsis* species. Genetics **47:**161–177.
17. **Fasani, M., F. Marini, and L. Teatini.** 1974. Variability of cephalosporin C production by phenethyl alcohol and dimethylsulfate. Abstr. Int. Symp. Genet. Ind. Microorg. 2nd, 1974, p. 31. Sheffield, England.
18. **Fujisawa, Y., H. Shirafugi, M. Kida, and K. Nara.** 1973. New findings on cephalosporin C biosynthesis. Nature (London) New Biol. **246:**154–155.
19. **Fuska, J., and F. Welwardová.** 1969. Selection of productive strains of *Penicillium chrysogenum.* Biologia **24:**691–698.
20. **Godfrey, O. W.** 1973. Isolation of regulator mutants of the aspartic and pyruvic acid families and their effect on antibiotic production in *Streptomyces lipmanii.* Antimicrob. Agents Chemother. **4:**73–79.
21. **Goulden, S. A., and F. W. Chattaway.** 1969. End-product control of acetohydroxy acid synthetase by valine in *Penicillium chrysogenum* Q-176 and a high penicillin-yielding mutant. J. Gen. Microbiol. **59:** 111–118.
22. **Hopwood, D. A.** 1977. Genetic recombination and strain improvement. Dev. Ind. Microbiol. **18:**9–21.
23. **Hopwood, D. A., and M. J. Merrick.** 1977. Genetics of antibiotic production. Bacteriol. Rev. **41:**595–635.
24. **Huber, F. M., R. H. Baltz, and P. G. Caltrider.** 1968. Formation of deacetylcephalosporin C in the cephalosporin C fermentation. Appl. Microbiol. **16:**1011–1014.
25. **Kanzaki, T., and Y. Fujisawa.** 1976. Biosynthesis of cephalosporins. Adv. Appl. Microbiol. **20:**159–201.
26. **Kanzaki, T., T. Fukita, H. Shirafuji, Y. Fujisawa, and K. Kitano.** 1974. Occurrence of a 3-methylthiomethylcephem derivative in a culture broth of a *Cephalosporium* mutant. J. Antibiot. **27:**361–362.
27. **Kitano, K., K. Kintaka, S. Suzuki, K. Katamoto, K. Nara, and Y. Nakao.** 1975. Screening of microorganisms capable of producing β-lactam antibiotics. J. Ferment. Technol. **53:**327–338.
28. **Komatsu, K., and R. Kodaira.** 1977. Sulfur metabolism of a mutant of *Cephalosporium acremonium* with enhanced potential to utilize sulfate for cephalosporin C production. J. Antibiot. **30:**226–233.
29. **Komatsu, K., M. Mizumo, and R. Kodaira.** 1975. Effect of methionine on cephalosporin C and penicillin N production by a mutant of *Cephalosporium acremonium.* J. Antibiot. **28:**881–888.
30. **Lemke, P. A.** 1969. A century of compounds and their effect on fungi. Mycopathol. Mycol. Appl. **38:** 49–59.
31. **Lemke, P. A., and C. H. Nash.** 1972. Mutations that affect antibiotic synthesis by *Cephalosporium acremonium.* Can. J. Microbiol. **18:**255–259.
32. **Liersch, M., J. Nüesch, and H. J. Treichler.** 1976. Final steps in the biosynthesis of cephalosporin C, p. 179–195. *In* K. D. Macdonald (ed.), Second international symposium on the genetics of industrial microorganisms. Academic Press, London.
33. **Macdonald, K. D.** 1964. Preservation of the heterozygous diploid condition in industrial microorganisms. Nature (London) **204:**404–405.
34. **Marzluf, G. A.** 1970. Genetic and metabolic controls for sulfate metabolism in *Neurospora crassa:* isolation and study of chromate-resistant and sulfate-transport-negative mutants. J. Bacteriol. **102:** 716–721.
35. **Merrick, M. J.** 1976. Hybridization and selection for penicillin production in *Aspergillus nidulans*—a biometrical approach to strain improvement, p. 229–242. *In* K. D. Macdonald (ed.), Second international symposium on the genetics of industrial microorganisms. Academic Press, London.
36. **Metzenberg, R. L., and J. W. Parson.** 1966. Altered repression of some enzymes for sulfur utilization in a temperature-conditional lethal mutant of *Neurospora.* Proc. Natl. Acad. Sci. U.S.A. **55:**629–635.
37. **Morin, R. B., R. G. Jackson, R. A. Mueller, E. R. Lavagnino, W. B. Scanlon, and S. L. Andrews.** 1963. Chemistry of cephalosporin antibiotics. III. Chemical correlation of penicillin and cephalosporin antibiotics. J. Am. Chem. Soc. **85:**1896–1897.
38. **Nash, C. H., N. De La Higuera, N. Nuess, and P. A. Lemke.** 1974. Application of biochemical genetics to the biosynthesis of β-lactam antibiotics. Dev. Ind. Microbiol. **15:**114–132.
39. **Nash, C. H., and F. M. Huber.** 1971. Antibiotic synthesis and morphological differentiation of *Cephalosporium acremonium.* Appl. Microbiol. **22:**6–10.
40. **Nga, B. H., and J. A. Roper.** 1969. A system generating spontaneous intrachromosomal changes at mitosis in *Aspergillus nidulans.* Genet. Res. **14:**63–70.
41. **Niss, H. F., and C. H. Nash III.** 1973. Synthesis of cephalosporin C from sulfate by mutants of *Cephalosporium acremonium.* Antimicrob. Agents Chemother. **4:**474–478.

42. **Nüesch, J., H. J. Treichler, and M. Liersch.** 1973. Biosynthesis of cephalosporin C, p. 309–344. *In* Z. Vanek, Z. Hošťálek, and J. Cudlín (ed.), Genetics of industrial microorganisms, vol. 2. Elsevier Publishing Co., Amsterdam.
43. **Pan, C. H., L. Hepler, and R. P. Elander.** 1972. Control of pH and carbohydrate addition in the penicillin fermentation. Dev. Ind. Microbiol. **13:**103–112.
44. **Pan, C. H., L. Hepler, and R. P. Elander.** 1975. The effect of iron on a high-yielding industrial strain of *Penicillium chryosgenum* and production levels of penicillin G. J. Ferment. Technol. **53:**854–861.
45. **Preuss, D. L., and M. J. Johnson.** 1967. Penicillin acyltransferase in *Penicillium chrysogenum.* J. Bacteriol. **94:**1502–1508.
46. **Queener, S. W., J. J. Capone, A. B. Radue, and R. Nagarajan.** 1974. Synthesis of deacetoxycephalosporin C by a mutant of *Cephalosporium acremonium.* Antimicrob. Agents Chemother. **6:**334–337.
47. **Queener, S. W., J. McDermott, and A. B. Radue.** 1975. Glutamate dehydrogenase-specific activity and cephalosporin C synthesis in the M8650 series of *Cephalosporium acremonium* mutants. Antimicrob. Agents Chemother. **7:**646–651.
48. **Roper, J. A.** 1973. Mitotic recombination and mitotic nonconformity in fungi, p. 81–88. *In* Z. Vaněk, Z. Hošťálek, and J. Cudlín (ed.), Genetics of industrial microorganisms, vol. 2. Elsevier Publishing Co., Amsterdam.
49. **Segel, I. H., and M. J. Johnson.** 1961. Accumulation of intracellular inorganic sulfate by *Penicillium chrysogenum.* J. Bacteriol. **81:**91–98.
50. **Sermonti, G.** 1959. Genetics of penicillin production. Ann. N.Y. Acad. Sci. **81:**950–972.
51. **Stauffer, J. F., L. J. Schwartz, and C. W. Brady.** 1966. Problems and progress in a strain selection program with cephalosporin-producing fungi. Dev. Ind. Microbiol. **7:**104–113.
52. **Traxler, P., H. J. Treichler, and J. Nüesch.** 1975. Synthesis of N-acetyl-deacetoxycephalosporin C by a mutant of *Cephalosporium acremonium.* J. Antibiot. **28:**605–606.
53. **Trilli, A., V. Michelini, V. Mantovani, and S. J. Pirt.** 1978. Development of the agar disk method for the rapid selection of cephalosporin producers with improved yields. Antimicrob. Agents Chemother. **13:** 7–13.
54. **Troonen, H., P. Roelants, and B. Boon.** 1976. RIT 2214, a new biosynthetic penicillin produced by a mutant of *Cephalosporium acremonium.* J. Antibiot. **29:**1258–1266.

Yeast Transformation: a New Approach for the Cloning of Eucaryotic Genes

ALBERT HINNEN, JAMES B. HICKS, CHRISTINE ILGEN, AND GERALD R. FINK

Section of Botany, Genetics, and Development, Cornell University, Ithaca, New York 14853

Transformation of cells by DNA molecules provides a powerful tool for genetic exchange and manipulation. Until recently, however, transformation was possible in only a few bacteria. These bacterial transformation systems have played a key role in our understanding of gene expression by permitting the analysis of defined DNA sequences in vivo as well as in vitro. The recent development of recombinant DNA technologies has made this analysis even more sophisticated by allowing the joining of virtually any DNA molecules.

Reports of lower eucaryotic transformation systems have appeared sporadically during the past 20 years. However, these systems had little credibility because it was impossible, with the available technologies, to rule out trivial explanations such as reversion and DNA-induced mutations (5, 6).

We have used cloned yeast genes to develop a transformation system for *Saccharomyces cerevisiae* (3). The use of pYeleu10, an *Escherichia coli*-yeast hybrid plasmid, containing the *LEU2*$^+$ gene from yeast inserted into the ColE1 plasmid DNA from *E. coli* (7) provided us with a highly enriched source of *LEU2*$^+$ genes from yeast and enabled us to verify the transformation event at the molecular level by the identification of the ColE1 sequences in the transformants.

The ability to introduce new DNA sequences into the yeast genome by transformation adds a new dimension to the molecular biology of this organism. Yeast has been well characterized genetically and biochemically, but its life cycle offered no mechanism for the transfer of defined DNA sequences into the cells. Transformation fills this void in the genetic manipulation by providing the means for the introduction of novel molecules into the yeast genome. Our studies show that the DNA which enters during transformation can be inserted into the chromosomes or, if it carries a portion of the yeast 2μ plasmid, it can remain unintegrated.

YEAST TRANSFORMATION BY USE OF CLONED YEAST DNA

The recombinant plasmids pYeleu10 and pYehis1 were constructed by Ratzkin and Carbon (7). They cloned randomly sheared yeast DNA into the *Eco*RI restriction site of the *E. coli* plasmid ColE1 by use of the polydeoxyadenylic acid-polydeoxythymidylic acid "connector" method. The plasmids pYeleu10 and pYehis1 have been isolated from a collection of such hybrids by transforming *leuB* and *hisB* auxotrophs of *E. coli*. The *leuB* gene codes for the leucine biosynthetic enzyme β-isopropyl malate dehydrogenase and corresponds to the *leu2* gene in yeast; similarly, *hisB* codes for imidazole glycerol phosphate dehydratase and corresponds to the *his3* gene in yeast.

To eliminate the ambiguities caused by revertants, we have constructed stable recipients by recombining within the *leu2* gene (and *his3* gene, respectively) two stable point mutations, thereby creating double mutants which are extremely stable and revert at a frequency of $<10^{-10}$. Transformation of these stable auxotrophs with the circular plasmids pYeleu10 and pYehis1 occurs at a frequency of 10^{-6} per regenerated spheroplast. Spheroplasts of the auxotrophs are treated with Ca^{2+}, mixed with the DNA and polyethylene glycol, and regenerated in 3% agar on minimal medium plates. About 10% of the spheroplasts plated give rise to large colonies which appear after 2 to 4 days of incubation (3).

INTEGRATION TYPES OF pYeleu10 TRANSFORMANTS

Using biochemical and genetic techniques, we have analyzed the pYeleu10 transformants in more detail. We have been able to distinguish three different types of transformants (types I, II, and III; Fig. 1). Type I and II transformants are "addition transformants" because transformation occurs by a recombination event which adds the plasmid to the yeast genome. Type III transformants are "substitution transformants" which have the mutated *leu2⁻* region replaced by the *LEU2⁺* region of the pYeleu10 plasmid.

Type I transformants have the entire pYeleu10 plasmid integrated at the *leu2* region on chromosome III, which results in a tight linkage of the *leu2⁻* genes in the resident chromosome to the incoming *LEU2⁺* genes of the plasmid. This close linkage relationship can be verified by the Southern hybridization technique (9), as diagrammed schematically on the right in Fig. 1. Type I transformants have a duplicated *leu2* region which is genetically unstable and leads to Leu⁻ segregants at a frequency of about 1%. In type II transformants the *LEU2⁺* gene segregates independently from the resident *LEU2⁺* gene and leads to a DNA restriction fragment which can be separated from the DNA fragment carrying the *leu2⁻* genes. Type III transformants can be explained if a double crossover event is postulated as depicted in Fig. 1. The result of this integration event is genetically and biochemically indistinguishable from a revertant; however, we have never seen any revertants of the double mutation at the *leu2* locus. Type II and type III transformants display a stable Leu⁺ phenotype.

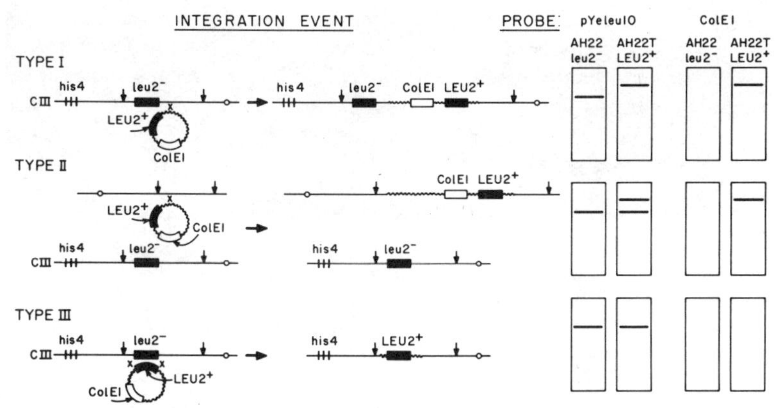

FIG. 1. *Schematic interpretation of the integration events proposed for transformant types I, II, and III. Each type of integration event (left) gives rise to a unique chromosome structure (center) that can be visualized by hybridization of pYeleu10 and ColE1 DNA to* HindIII *restriction digests (right). The arrows (↓) represent* HindIII *restriction sites. Type I: integration of plasmid pYeleu10 into chromosome III at a sequence which is complementary to a yeast sequence carried by the plasmid. Type II: integration of plasmid pYeleu10 into a chromosomal location genetically unlinked to the* leu2 *region of chromosome III. Type III: integration of yeast DNA sequences of plasmid pYeleu10 into the* leu2 *region by a double crossover event. These different integration events lead to predictable patterns when* HindIII *restriction digests of these strains are hybridized with pYeleu10 or ColE1 DNA. These hypothetical hybridization patterns are in agreement with the actual patterns.*

We have developed a yeast colony hybridization procedure (3) which has enabled us to identify and characterize the yeast transformants easily. Since pYeleu10 carries ColE1 sequences, completely absent in our untransformed yeast strains, ^{32}P-labeled ColE1 DNA can be used to determine the presence of these sequences in the transformants. The results of such an analysis reveal that 80 to 90% of all pYeleu10 transformants have ColE1 sequences stably integrated into their genome. Standard genetic techniques have allowed us to establish the linkage relationship between the *LEU2*$^+$ genes and the ColE1 sequences in the transformants. Figure 2 shows such an

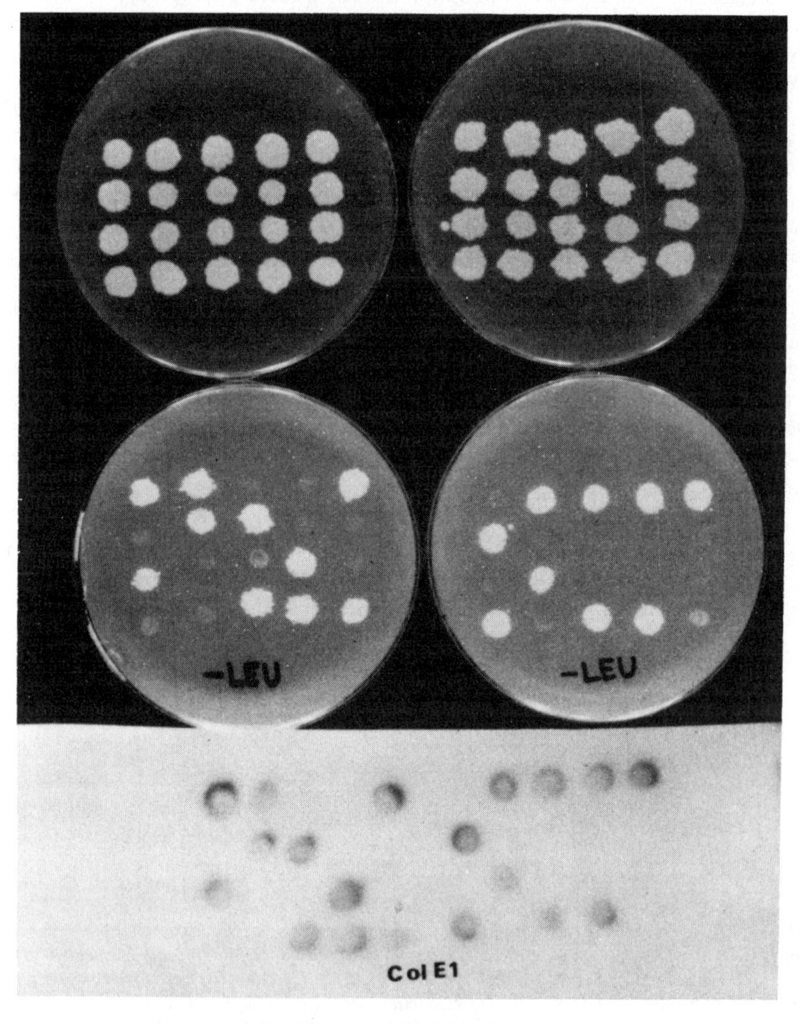

FIG. 2. *Analysis of transformants by a yeast colony hybridization procedure. A type I transformant was crossed with a* leu2$^-$ *tester strain, and the diploid cells were sporulated. After meiosis, 10 yeast tetrads were dissected into the four individual spores and placed on two complete medium plates (top two plates). These plates were replicated on two* leu$^-$ *plates to determine the segregation of the Leu*$^+$ *phenotype (bottom two plates). The same spores were analyzed by colony hybridization with* ^{32}*P-labeled ColE1 DNA as radioactive probe (lower panel).*

analysis with a type I transformant in a cross with an *leu2*$^-$ tester strain. In all of the 10 tetrads the segregation of the Leu$^+$ phenotype is in the parental arrangement with the ColE1 sequence. Therefore, *LEU2*$^+$ and ColE1 are tightly linked.

The above analysis shows that: (i) yeast transformation using hybrid plasmids such as pYeleu10 involves integration of the added DNA into yeast chromosomes, (ii) integration does not always occur at the site where the cloned yeast genes originate but also at a number of other chromosomal locations, (iii) bacterial sequences on the plasmid (e.g., ColE1) become stably associated with the yeast chromosome and are tightly linked to the *leu2* gene.

SEQUENCE HOMOLOGY DIRECTS INTEGRATION

About 80% of the pYeleu10 transformants have the *leu2* gene integrated at the *leu2* region on chromosome III. This close linkage of the plasmid sequences and the resident sequences suggests that sequence homology plays a role in the integration event. Nevertheless, in about 20% of the transformants the *LEU2*$^+$ genes segregate independently of the *leu2*$^-$ genes on chromosome III. We have analyzed six of these isolates genetically and by colony hybridization. In all six, the integration sites for pYeleu10 are unlinked to each other and map on other chromosomes. This means that there are at least six other locations on the yeast genome besides *leu2* where pYeleu10 can integrate. If homology is necessary for the integration event, as our model in Fig. 1 suggests, we must assume that pYeleu10 has sequences which are repeated all over the yeast genome.

To test the possibility that some of the pYeleu10 sequences are reiterated, we have carried out hybridization analysis under conditions which allow the detection of small sequence homologies. Such an analysis revealed that pYeleu10 has sequences which are repeated at least 20 times in the yeast genome. We have re-cloned the restriction fragments of pYeleu10 made with *Eco*RI on bacteriophage λ. This allowed us to use, independently, different segments of pYeleu10 as probes for the further localization of these sequences.

The hypothesis that these repeated sequences might be the reason for the apparently random integration of pYeleu10 is supported by the different behavior of pYehis1 in yeast transformation experiments. pYehis1, which carries the *his3* genes of yeast, integrates only into the *his3* region on chromosome XV. Furthermore, pYehis1 does not show any of the repetitive sequences which are present in pYeleu10.

To determine further whether sequence homology is recognized in the integration process, we introduced DNA sequences into the yeast genome which are not naturally present in this organism. This can be done by transforming a *his3 leu2* double mutant with pYehis1 to His$^+$, thereby introducing ColE1 sequences at the *his3* locus on chromosome XV of yeast. Such a transformant was then used in a subsequent transformation experiment using pYeleu10. If the ColE1 sequences in the pYeleu10 plasmid recognize the ColE1 sequence at *his3,* it should be possible to isolate transformants which have the *leu2* genes transposed near to the *his3* locus on chromosome XV. Genetic analysis of these doubly transformed strains identified such transpositions of the *leu2* genes (Fig. 3). A further analysis of these strains with *LEU2*$^+$ transposed into chromosome XV revealed a characteristic instability during mitotic growth. About 1% of the vegetative haploid cells are Leu$^-$ His$^+$ and 0.2% are Leu$^-$ His3$^-$. Thus, the *LEU2*$^+$ region is lost alone, but the *HIS3*$^+$ region is lost only in concert with the *LEU2*$^+$ region. We assume that this mitotic segregation behavior is a consequence of intrachromosomal crossovers which depend upon exchanges

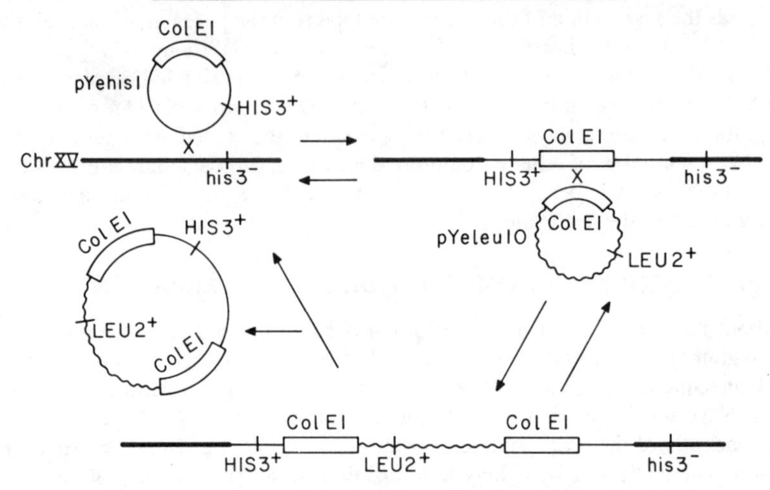

FIG. 3. *Schematic interpretation of the integration of pYeleu10 into the ColE1 sequence on chromosome XV. At the top, the integration of pYehis1 into the* his3 *region on chromosome XV is shown. Upon retransformation with pYeleu10, a chromosomal rearrangement is formed, as diagrammed at the bottom. We assume that the transposition occurred by homology between the ColE1 sequences. Because of the twofold duplication present in this structure, the transformant is unstable and can give rise to two types of segregants. It can lose the pYeleu10 segment and segregate a His$^+$ Leu$^-$ clone by virtue of a crossover event between the two ColE1 regions or lose both plasmids at once using the homology of the two* his3 *regions and segregate a His3$^-$ Leu2$^-$ double mutant clone.*

between homologous sequences. A schematic interpretation of this behavior is shown in Fig. 3.

CLONING OF GENES ON THE YEAST 2μ PLASMID

Plasmids are defined as self-replicating extrachromosomal elements. Bacterial plasmids which can confer resistance to a variety of antibiotics are widely used as cloning vehicles in *E. coli.* Yeast contains a plasmid which has been shown to be a closed circular DNA molecule with a circumference of 2 μm, and is referred to as "2μ DNA" (8). There are about 50 to 100 copies of this 2μ DNA per haploid genome. The 2μ DNA does not seem to be associated with either nuclear DNA or the mitochondrial genome. Although the 2μ DNA is well characterized physically, there is little information available as to the biological significance of these yeast plasmids (2, 4). Resistance to a number of compounds has been associated with 2μ DNA; however, because of the lack of a yeast transformation system, this assignment could not be verified (2).

We connected yeast chromosomal genes to the yeast 2μ DNA in order to determine the effect of the extrachromosomal element on transformation. The 2μ DNA can be cut twice with the restriction endonuclease *Eco*RI, thereby creating two linear DNA fragments. We cloned each of these two restriction fragments into the *Eco*RI site of pYehisl (Fig. 4). These newly constructed circular hybrids, called phis2μR1.4 and phis2μR2.6, which consist of ColE1 DNA, nuclear yeast genes, and 2μ sequences were used to transform a yeast *his3*$^-$ mutant to His$^+$. The presence of 2μ DNA in pYehisl had two effects: (i) the transformation frequency was increased 100-fold and

(ii) the transformants became extremely unstable; i.e., about 40% of the vegetative cells segregated His$^-$ clones.

We analyzed the phis2μ transformants genetically by crossing them with a *his3*$^-$ tester strain and following the segregation of the His$^+$ phenotype in tetrads. Figure 5 shows that the *HIS3*$^+$ genes segregate either $4^+:0^-$ or $4^-:0^+$, which is indicative of non-Mendelian inheritance. We assume that the $4^-:0^+$ segregation is due to the loss

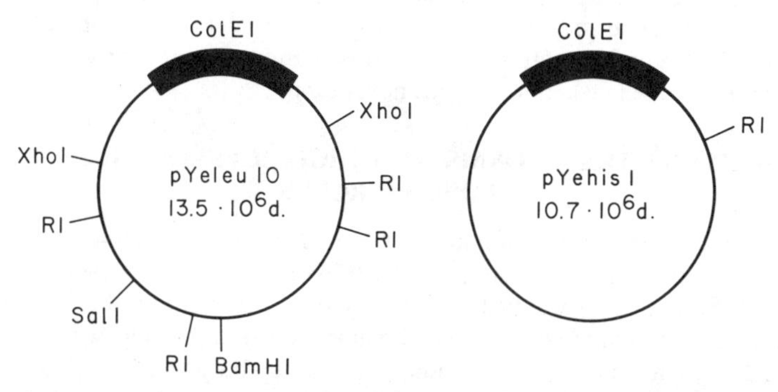

FIG. 4. *Restriction map of pYeleu10 and pYehis1 (for pYehis1 only the* Eco*RI site is given).*

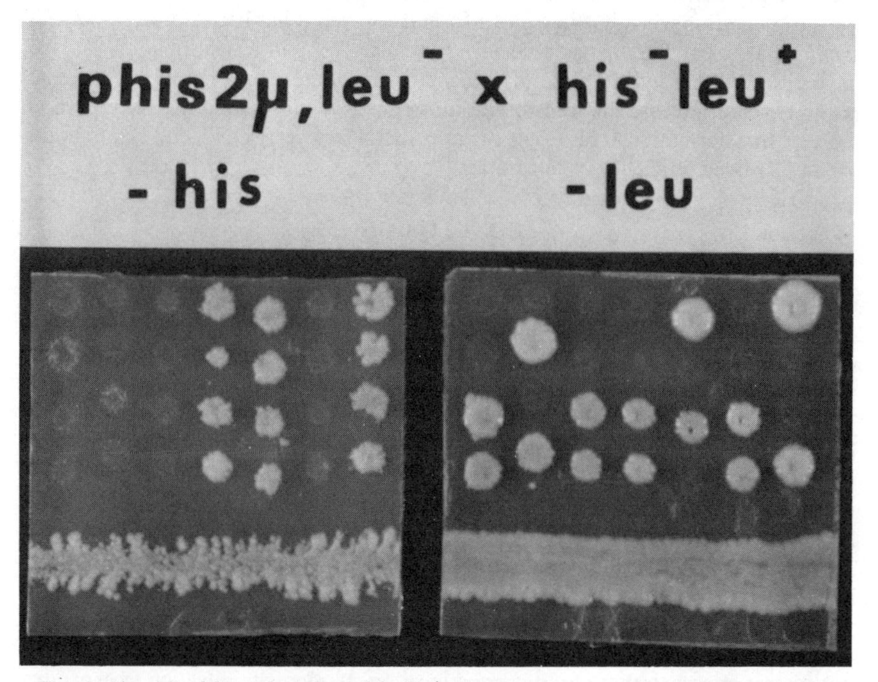

FIG. 5. *Non-Mendelian inheritance of the His$^+$ phenotype in a cross of a phis2µR1.4 transformant with a* his3$^-$ *tester strain. There are two genes segregating in this cross: a* LEU2$^+$ *gene and a* HIS3$^+$ *gene. The* LEU2$^+$ *genes show the normal $2^+:2^-$ segregation as expected for a chromosomal marker. The* HIS3$^+$ *genes, however, segregate either $4^+:0^-$ or $4^-:0^+$, which is indicative of a nonchromosomal location of the* HIS3$^+$ *genes. We interpret the non-Mendelian segregation of four His$^+$ spores in three of the tetrads shown as a random distribution of the phis2µR1.4 plasmid at meiosis. The presence of tetrads with four His$^-$ spores is probably due to the loss of phis2µR1.4 before meiosis.*

of the *HIS3*$^+$ genes prior to meiosis and that the $4^+:0^-$ segregation is due to distribution of the phis2μR1.4 plasmid to all four meiotic progeny independent of the nuclear chromosomal segregation. This result is dramatic since the pYehisl plasmid without the 2μ insert gives transformation only by integration at the *his3* region on chromosome XV.

Our results suggest that the *HIS3*$^+$ gene, once it becomes associated with 2μ DNA, is transmitted independently of the chromosomes. In a formal sense the *HIS3*$^+$ gene behaves like a plasmid in yeast, located either in the nucleus or in the cytoplasm. Unfortunately, the pattern of transmission gives no indication of the site of transcription of the *HIS3*$^+$ gene when it is in this novel form.

LINEAR DNA TRANSFORMS AT A HIGHER FREQUENCY THAN CLOSED CIRCLES

Cutting circular plasmids into linear DNA fragments can increase the yeast transformation frequency. Digesting pYeleu10 with restriction endonucleases *Sal*I, *Xho*I or *Bam*HI (Fig. 4) resulted in a 5- to 20-fold increase in the efficiency of transformation as compared with that obtained by transforming with the uncut plasmid. This increase in transformation frequency by linear plasmids is surprising. Conventional models predict that a single crossover event is sufficient to integrate the plasmid when it is in a circular form, whereas a double crossover event is required for the integration when it is in a linear form. However, it has been shown in *Bacillus subtilis* that linear DNA is as efficient as circular DNA in transformation (1). We have observed that hybrid plasmids which are made linear by a cut in the bacterial portion do not increase the transformation frequency; only cuts in the yeast portion increase transformation. This result suggests that a free yeast DNA end is important for an increased efficiency of integration.

CONCLUSION

The yeast transformation system has made possible a spectrum of novel approaches to yeast molecular biology and genetics. It provides the means to test in vitro-engineered DNA molecules for their in vivo function. Yeast-*E. coli* hybrid molecules which have been constructed by conventional cloning techniques can be tested by transformation to determine whether they carry the desired yeast structural genes. The ability to test the expression in yeast of yeast sequences cloned in *E. coli* is important in cases where yeast genes have been obtained by complementing corresponding mutations in *E. coli*, and it is absolutely essential in those instances where particular yeast genes have no genetic counterpart in *E. coli* and cannot be assayed by complementation.

We have shown that the presence of unique bacterial sequences, stably integrated into the yeast genome, allows the directed transposition of yeast genes. In other words, once bacterial sequences have been inserted in the yeast chromosome by the methods we have described here, any yeast gene cloned on the same bacterial plasmid can be transposed to that chromosomal location. If adequate selection systems can be found, the recognition of ColE1 sequences within the yeast genome can be used to create well-defined regions of chromosomes containing a small or large segment of DNA whose sequence and functions are completely known. For example, if the expression of foreign genes requires a strong promoter, this technology would permit the insertion of the foreign sequences adjacent to a known genetic environment containing a strong promoter.

Our results with 2μ DNA open up another avenue for the cloning of yeast genes. These yeast plasmids are potential vehicles for the development of a yeast host-vector system which would complement the already existing procaryotic cloning systems.

Yeast has been extensively studied for the past 50 years and is probably the best understood lower eucaryote. The ease of genetic and biochemical analyses coupled with the availability of a transformation system now makes yeast ideal as a host for the genes of higher eucaryotes. Yeast has all the features of a typical eucaryotic organism (nuclear membrane, organelles, transcription and translation machinery, etc.) and is very likely to have regulatory properties which are similar to those in higher eucaryotic organisms. Thus, higher eucaryotic genes may be expressed when inserted into yeast by transformation.

ACKNOWLEDGMENTS

This work was supported by National Science Foundation grant PCM76-11667 and Public Health Service grant GM15408 from the National Institute of General Medical Sciences to G.R.F. and by a National Institutes of Health postdoctoral fellowship to J.B.H. All recombinant DNA work was carried out under P2 conditions as approved by the local biohazard committee, the National Science Foundation, and the National Institutes of Health.

REFERENCES

1. **Duncan, C. H., G. A. Wilson, and F. E. Young.** 1977. Transformation of *Bacillus subtilis* and *Escherichia coli* by a hybrid plasmid pCD1. Gene **1:**153–167.
2. **Guerineau, M., C. Grandchamp, and P. P. Slonimski.** 1976. Circular DNA of a yeast episome with two inverted repeats: structural analysis by a restriction enzyme and electron microscopy. Proc. Natl. Acad. Sci. U.S.A. **73:**3030–3034.
3. **Hinnen, A., J. B. Hicks, and G. R. Fink.** 1978. Transformation of yeast. Proc. Natl. Acad. Sci. U.S.A. **75:** 1929–1933.
4. **Hollenberg, C. P., A. Degelmann, B. Kustermann-Kuhn, and H. D. Royer.** 1976. Characterization of 2-μm DNA of *Saccharomyces cerevisiae* by restriction fragment analysis and integration in an *Escherichia coli* plasmid. Proc. Natl. Acad. Sci. U.S.A. **73:**2072–2076.
5. **Khan, N. C., and S. P. Sen.** 1974. Genetic transformation in yeasts. J. Gen. Microbiol. **83:**237–250.
6. **Mishra, N. C., and E. L. Tatum.** 1973. Non-Mendelian inheritance of DNA-induced inositol independence in *Neurospora.* Proc. Natl. Acad. Sci. U.S.A. **70:**3875–3879.
7. **Ratzkin, B., and J. Carbon.** 1977. Functional expression of cloned yeast DNA in *Escherichia coli.* Proc. Natl. Acad. Sci. U.S.A. **74:**487–491.
8. **Sinclair, J. H., B. J. Stevens, P. Sanghari, and M. Rabinowitz.** 1967. Mitochondrial-sattelite and circular DNA filaments in yeast. Science **156:**1234–1237.
9. **Southern, E. M.** 1975. Detection of specific sequences among DNA fragments separated by gel electrophoresis. J. Mol. Biol. **98:**503–517.

Regulation of Primary Metabolism

R. HÜTTER

Mikrobiologisches Institut, Eidgenössische Technische Hochschule, CH-8092 Zurich, Switzerland

The metabolic and catabolic pathways of primary metabolism are involved in the supply of materials essential for growth of the organism, e.g., amino acids for protein synthesis, nucleotides for nucleic acids, fatty acids for lipids, and sugars for carbohydrates. The biosynthetic and catabolic pathways as such are in many cases identical or at least very similar over broad groups of microorganisms, but the regulations imposed on the pathways show considerable diversity.

In addition to the supply of materials for growth, primary metabolism also furnishes the precursors for secondary metabolism, and it follows that knowledge of primary metabolism and its regulation is crucial for industrial microbial processes in a triple sense. First, we have to grow the organisms, and we need a well-functioning primary metabolism for this. Second, some industrial fermentation processes are directed toward the production of compounds of the primary metabolism, e.g., amino acids, nucleotides, vitamins, and, of course, ethanol. Third, primary metabolism has to supply the necessary precursors for the synthesis of secondary metabolites in sufficient amount and often in well-adjusted ratios, as in the case of antibiotics or ergot alkaloids (see also 13, 23).

Studies on the regulation of primary metabolism have mainly been performed in suitable model systems such as the eubacteria *Escherichia coli, Salmonella typhimurium, Bacillus subtilis, Pseudomonas aeruginosa,* and *P. putida.* The fungi *Neurospora crassa* and *Aspergillus nidulans* and the yeast *Saccharomyces cerevisiae* have served in the same role. From the studies with these organisms, rather precise molecular models have evolved which describe specific controls for enzyme synthesis and enzyme activity for particular pathways as well as models for generalized control patterns. However, as the systems analyzed are usually not used for the industrial production of microbial metabolites, we can only derive *principles* of regulation from them.

In the following discussion, I will try to summarize some of the principles of regulatory mechanisms in the primary metabolism of microorganisms. The discussion will be divided into three sections: gene expression, enzyme expression, and pools and flux.

GENE EXPRESSION

To illustrate some principal points, the regulation of the tryptophan pathway in different microbial species will be discussed. Although the sequence of tryptophan biosynthesis is identical in all cases investigated, there is great variability in the pattern of gene distribution (9; Fig. 1). In *E. coli* and *Salmonella,* the tryptophan genes are organized in an operon, as seems also the case for *Serratia* and *Aeromonas.* In other bacteria, the situation is more complicated, as diverse gene arrangements occur. In fungi, no gene clustering is observed, but the structural genes are dispersed over different chromosomes. A great variability in the number of genes and their coding content has also evolved.

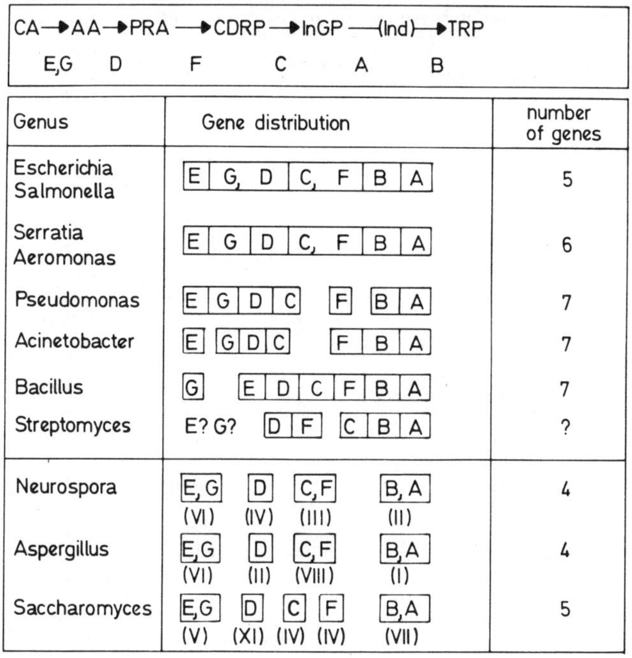

CA → AA → PRA → CDRP → InGP —(Ind)→ TRP

E,G D F C A B

Genus	Gene distribution	number of genes
Escherichia Salmonella	[E G, D C, F B A]	5
Serratia Aeromonas	[E G D C, F B A]	6
Pseudomonas	[E G D C] [F] [B A]	7
Acinetobacter	[E] [G D C] [F B A]	7
Bacillus	[G] [E D C F B A]	7
Streptomyces	E? G? [D F] [C B A]	?
Neurospora	[E,G] (VI) [D] (IV) [C,F] (III) [B,A] (II)	4
Aspergillus	[E,G] (VI) [D] (II) [C,F] (VIII) [B,A] (I)	4
Saccharomyces	[E,G] (V) [D] (XI) [C] (IV) [F] (IV) [B,A] (VII)	5

FIG. 1. *Tryptophan gene distribution in microorganisms. CA, Chorismate; AA, anthranilate; PRA, N-(5'-phosphoribosyl)-anthranilate; CDRP, 1-O-carboxypenylamino)-1-deoxyribulose phosphate; InGP, indole glycerolphosphate; Ind, indole; TRP, L-tryptophan. The gene designations are taken from Crawford (9), and the data are compiled from the same reference. Arrangement of genes in one block designates physical vicinity (operon or gene cluster). Letters in one square designate the enzymatic functions coded by this gene and the letters refer to: A, tryptophan synthase α-chain; B, tryptophan synthase β-chain (EC 4.2.1.20); C, indole glycerolphosphate synthase (EC 4.1.1.48); D, anthranilate phosphoribosyltransferase (EC 2.4.2.18); F, phosphoribosyl-anthranilate isomerase; E, anthranilate synthase large subunit; and G, anthranilate synthase small, glutamine amido-transferase subunit (EC 4.1.3.27). A question mark following a gene designation indicates doubt about its true location. The Roman numbers in parentheses below the genes in fungi designate the chromosomes on which these genes are located.*

An even greater diversity is observed with respect to the regulation of enzyme synthesis (9, 15, 26; Table 1). In thc bcst-studied system of *E. coli,* a double control is observed: a repressor-operator system and an attenuator system (51). If the basal levels of tryptophan biosynthetic enzymes are arbitrarily taken as 1, relief of repression can lead to a 100-fold increase in enzyme levels; if attenuation is eliminated in addition to repression, a further 8- to 10-fold increase in enzyme levels is observed. Whereas the repressor-operator system controls enzyme synthesis through an inhibition of transcription initiations, the attenuator system works through premature termination of transcription (4, 32, 35). In the closely related *S. typhimurium,* regulation seems very similar to that in *E. coli.* In *P. aeruginosa* and in *P. putida,* the tryptophan biosynthetic enzymes occur in three groups. No common regulation has been found for the three gene groups, but each chromosomal location seems associated with a unique regulatory mechanism. In *Bacillus,* a repressor system seems to occur which influences the tryptophan operon as well as the independently located *trpG* locus.

TABLE 1. *Regulatory patterns for tryptophan biosynthesis (examples)*

Genus	Gene organization[a]	Repressor system	Attenuation	"General control system"	Inducibility by InGP
Escherichia, Salmonella	Operon	+[b]	+	−	−
Pseudomonas	E, G, D, C	+	?	−	−
	F	−	−	−	−
	BA	−	−	−	+
Bacillus	Operon	+	−	−	−
	G	+	−	−	−
Neurospora	E, G	−	−	+	−
	D	−	−	+	−
	C, F	−	−	+	−
	B, A	−	−	+	+
Saccharomyces	E, G	−	−	+	−
	D	−	−	+	−
	C	−	−	+	−
	F	−	−	−	−
	B, A	−	−	+	−

[a] For gene organization, gene designation, and abbreviations, see legend to Fig. 1.
[b] +, This regulation system has been clearly demonstrated; −, absence of the system or ignorance about it; ?, controversial data.

In *N. crassa* and *S. cerevisiae,* the genes are separate. No *trpR*-type mutants have been found, but two types of regulation seem to occur. In *N. crassa,* tryptophan synthase is inducible by indole glycerolphosphate (InGP), as has been found in *Pseudomonas,* but no such inducibility is present in *S. cerevisiae.* Both organisms react, however, to tryptophan starvation by a two- to threefold increase in their enzyme levels. This increase can be stimulated not only by tryptophan starvation but also by starvation for other amino acids such as histidine, lysine, arginine, or leucine. In the same instances, biosynthetic enzymes for all these amino acids are derepressed. This generalized control system is completely different from the enterobacterial pattern, in which a strong pathway-specific control occurs. The fungal system has been named "general control of amino acid biosynthesis," but its mechanism is not yet clear (11, 34, 42, 50).

Even though the list is far from complete, it is obvious that the synthesis of particular enzymes in homologous pathways is subject to diverse regulatory patterns.

Specific regulatory systems for particular enzymes have also been found in fungi. The best-investigated systems are the arginine-regulatory mutants in *S. cerevisiae* described by Wiame and his group (49) and the *cis*-dominant regulatory mutants described in different systems of *A. nidulans* (see, e.g., 1, 2, 27). Other specific regulatory systems (such as autoregulation) also exist (6). This presentation will not discuss different pathways which are subject to different types of control in a given organism and will also omit the field of general controls, such as carbon catabolite repression, nitrogen catabolite repression, or the phosphate effect. The reader is referred to the reviews by Gots and Benson (17), Goldberger et al. (16), and Zimmermann (52), or in this volume to the articles by J. Martín and by Y. Aharonowitz.

ENZYME ACTIVITY

Four types of regulatory systems are preponderant: (i) enzyme modifications through, for example, phosphorylation and dephosphorylation, (ii) proteolytic enzyme activation and inactivation, (iii) feedback inhibition of enzyme activity, and (iv) activation through noncovalent binding of small co-activator molecules. A few examples will illustrate this large and complex field.

Enzyme modifications through phosphorylation and dephosphorylation play an important role in higher organisms and have been investigated in mammalian systems, e.g., glycogen synthase or mammalian pyruvate dehydrogenase (7, 18, 40, 44). Uy and Wold (48) discussed other post-translational covalent modifications of proteins. As examples of microbial systems, the extensive studies on glutamine synthase of *E. coli* (45) or on alterations and modifications of DNA-dependent RNA polymerase after phage T4 infection of *E. coli* can be mentioned (for review, see 12).

Similarly, proteolytic enzyme activation and inactivation are well known for higher organisms, e.g., the activation of zymogens to enzymes or of pro-peptide hormones to peptide hormones. For a general survey, the paper by Neurath and Walsh (39) is a good basis, and Holzer (22) has provided a good review of possible mechanisms for a selective control of proteinase action.

Feedback inhibition of enzyme activity has been well studied in a variety of microbial systems. Feedback inhibition of the first (or second) enzyme of amino acid biosynthesis plays a crucial role for regulation of the flux through the pathway. Different types of the feedback effect have been found operative in different systems and in different organisms: multiple isoenzymes for some reactions in the common stem of branched pathways, sequential feedback inhibition, cumulative or multivalent (concerted) inhibition of a common enzyme in a branched pathway, or reversal of inhibition (8, 24, 25). For the industrial microbiologist, it may be of primary importance to know which system is present in the organism in question. Lysine biosynthesis in *E. coli* and *Corynebacterium glutamicum* will be discussed and compared as an illustration (Fig. 2).

Three critical points exist in *E. coli* (8, 43). (i) The conversion of aspartate to aspartyl phosphate is catalyzed by three isozymes, and each is subject to regulation by one of the end products lysine, methionine, and threonine. (ii) Two isozymes occur for homoserine dehydrogenase, one methionine specific and one threonine specific. (iii) In the lysine branch, the first enzyme (dihydrodipicolinate synthase) is subject to feedback control by lysine. To maximize the flow from aspartate to lysine, regulation of all three aspartokinases has to be alleviated, the activities of both homoserine dehydrogenases have to be reduced, and feedback inhibition of dihydrodipicolinate synthase has to be eliminated.

The situation in *C. glutamicum* is much simpler (10, 36–38, 41). Only one enzyme (aspartokinase) subject to multivalent inhibition by threonine and lysine and only one homoserine dehydrogenase are present, and dihydrodipicolinate synthase is not subject to feedback inhibition by lysine. Three methods have been used to obtain high lysine production. First, mutants with defective (either completely blocked or leaky) homoserine dehydrogenase are used. This allows a reduction of drainage of aspartate semialdehyde to homoserine and, through a low threonine pool, alleviates multivalent feedback inhibition of aspartokinase. The same goal is achieved through the use of a mutant with a homoserine dehydrogenase supersensitive to feedback inhibition by threonine. A third approach is the use of a mutant with an aspartokinase insensitive to feedback inhibition (use of *S*-2-amino-ethyl-L-cysteine). In every case,

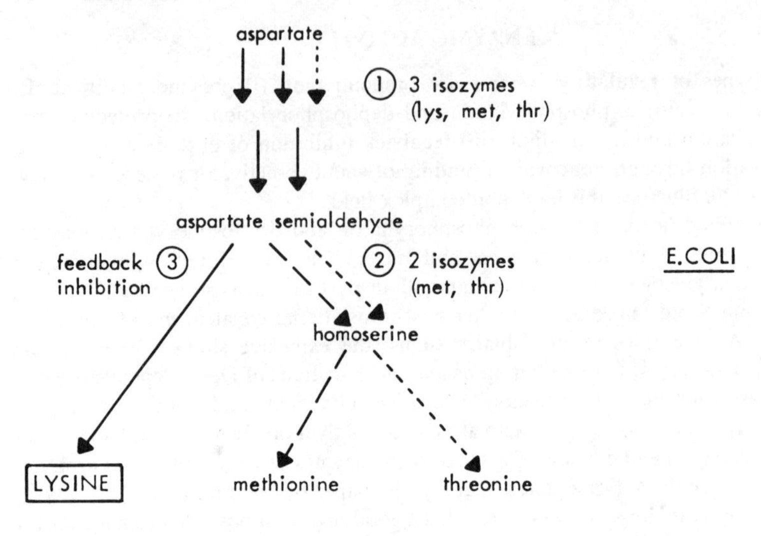

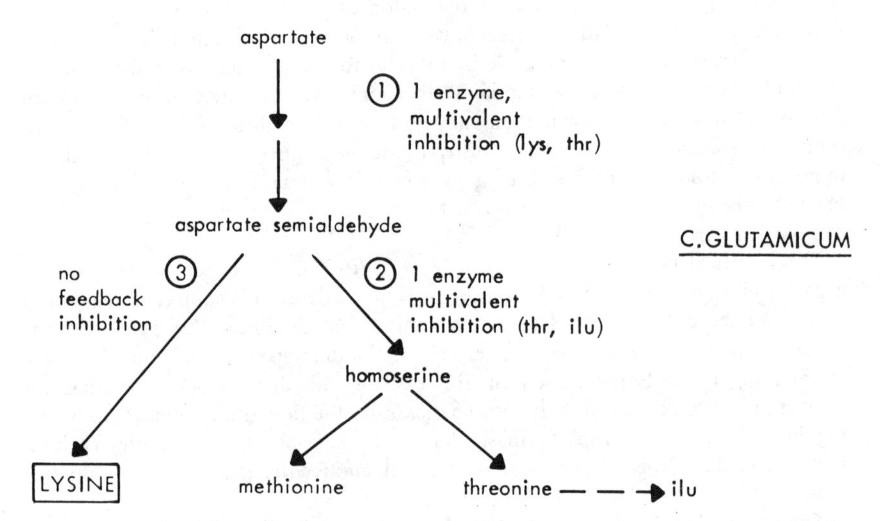

FIG. 2. *Lysine biosynthesis in* E. coli *and* C. glutamicum. *Abbreviations: ilu, L-isoleucine; lys, L-lysine; met, L-methionine; thr, L-threonine. See text for other explanations.*

a single mutational step leads to good lysine production. We can thus see that the occurrence of a suitable regulatory pattern considerably facilitates the task of the "microbial strain developer."

A fourth kind of regulation of enzyme activity is enzyme activation by noncovalent binding of small molecules. For example, chorismate mutase (EC 5.4.99.5) catalyzes the first step of tyrosine and phenylalanine biosynthesis from chorismate. It is inhibited in *S. cerevisiae* by L-tyrosine, but the inhibition is reversed by L-tryptophan.

In crude extracts and partially purified enzyme preparations, a direct stimulation of chorismate mutase activity by L-tryptophan may be observed. The activity of chorismate mutase may increase more than 10 times. The activation puts a twofold strain on tryptophan biosynthesis: it reduces the availability of chorismate for tryptophan biosynthesis, and it enhances the feedback sensitivity of anthranilate synthase for tryptophan, as tryptophan inhibition is competitive to chorismate. The relative activity of chorismate mutase plays an important role in tryptophan synthesis since mutants with increased activities of this enzyme have a reduced flux through the tryptophan pathway (30). Similar observations have been made with different wild-type isolates of *B. subtilis* (21, 33).

As in the case of enzyme synthesis, the regulation of enzyme activity is diverse even in homologous steps and sequences. Different techniques have been used to circumvent, alleviate, or eliminate steps critical for the excessive production of the desired metabolite.

POOL AND FLUX

Except when the industrial process leads directly to the production of a specific enzyme (e.g., amylase or protease), the industrial microbiologist is not primarily interested in the regulation of enzyme synthesis and enzyme activity. In most cases his main interest lies in maximizing the flux through the pathways leading to the desired product.

The "pool" is the "total quantity of low molecular weight compounds that may be extracted under conditions such that the macromolecules are not degraded into low-molecular-weight subunits" (5). Its concentration controls, directly or indirectly, enzyme synthesis and enzyme activity. Compartmentation of the enzyme pool is usually not taken into account although structural heterogeneity of the cells would seem to impose heterogeneities in the distribution of enzymes and their substrates (for discussion, see 46). The simple measurement of enzyme activities in cell-free extracts and the determination of kinetic parameters of these enzymes are usually of limited help in the estimation of the actual performance of a given pathway in vivo. To judge the situation within the cell, actual flux measurements have to be made.

Again, the tryptophan pathway will serve as an illustration. In the synthesis of aromatic amino acids (tryptophan, tyrosine, and phenylalanine), chorismic acid is the branch compound. Tryptophan synthesis proceeds through anthranilic acid, its phosphoribosyl derivative (PRA), 1-(*o*-carboxyphenylamino)-1-deoxyribulose phosphate (CDRP), InGP, and indole (see Fig. 1). PRA is a very unstable compound, and CDRP and InGP are easily dephosphorylated. On the other hand, anthranilic acid and indole are rather stable and can be readily analyzed. We have carried out a number of experiments (34) with mutants of *S. cerevisiae* blocked in anthranilate phosphoribosyl transferase (step D).

The actual flux capacity (enzyme levels) was first determined. By comparing the specific enzyme activities in crude extracts or in cells treated with Triton X-100, it was shown that anthranilate synthase within the cell functions with an activity comparable to that determined in vitro (Table 2). The observed enzyme activities of the wild-type strain would allow the synthesis of about 1 nmol of tryptophan per min per mg of protein. With 140-min doubling time and with a tryptophan content of protein of about 1.2% (by weight), the necessary flux through the pathway for protein synthesis was calculated to be 0.35 nmol per min per mg of protein. We assumed in the further calculations that all tryptophan is used for protein synthesis and that its

TABLE 2. *Enzyme activities and accumulation rates in the tryptophan pathway of* S. cerevisiae

Enzymatic step	Specific enzyme activity[a]	Rate of accumulation[a]
E,G (anthranilate synthase)	1.5	
		→ 1.5 (anthranilate in D-negative mutant strain)
D (anthranilate phosphoribosyltransferase)	1.1	
F (phosphoribosylanthranilate isomerase)	1.4	
C (indole glycerolphosphate synthase)	1.6	
		→ around 1 (indole in mutant strain with partially defective tryptophan synthase)
A, B (tryptophan synthase)	4.3	
		→ 0.001 (pool) (+ 0.35 for protein synthesis)

[a] Expressed in nanomoles per minute per milligram of protein.

use for the synthesis of nicotinic acid can be neglected. If the cell did not regulate the flux, it would overproduce approximately 0.7 nmol of tryptophan per min per mg of protein and would accumulate a pool of about 80 μmol per g (dry weight). We actually found a tryptophan pool of about 0.07 μmol/g (dry weight) in minimal medium (14). This pool corresponds to an internal concentration of 2×10^{-5} to 3×10^{-5} M, a concentration which inhibited anthranilate synthase in vitro by about 75%.

It would therefore seem that feedback inhibition of anthranilate synthase activity by tryptophan is the main regulatory mechanism responsible in the wild-type strain for keeping a low tryptophan pool. As pointed out by Atkinson (3), the restriction of the pool size not only is advantageous because of the savings it could contribute to the cellular economy but also could be necessary for maintaining a "solvent phase" in the living cell.

These data also suggest that the second enzyme, anthranilate phosphoribosyl transferase, is present in a very low yield and that it determines the maximal flux capacity. All other enzymes seem to be present in approximately four times or more the amounts that are needed for protein synthesis, as has also been verified by a tetraploid gene dosage series (34).

To increase the capacity of the pathway, we can introduce into our wild-type strain a mutation leading to elevated enzyme synthesis (Table 3). However, the introduction of such a mutation enhances the maximum flux capacity only to a small degree, as the third enzyme, phosphoribosyl-anthranilate isomerase, does not react to this mutation. Nevertheless, the pool increases slightly, but this increase in the pool size has to be attributed mainly to the slower growth rate of this mutant (200-min doubling time on minimal medium). More directly, a mutation that eliminates feedback sensitivity of anthranilate synthase leads to a considerable improvement in the actual flux and to an increase in the rate of tryptophan accumulation, which, however, is only 3 μmol whereas the theoretically expected pool size amounts to 80 μmol per g (dry weight). The total tryptophan accumulation rate of 0.17 shows, on the other hand, that only about one-third of the total tryptophan produced is retained

TABLE 3. *Tryptophan pool and tryptophan accumulation in regulatory mutants of* S. cerevisiae

Strain[a]	Anthranilate synthase activity		Tryptophan pool[d]	Tryptophan accumulation rate[b]
	Amt[b]	K_i[c]		
X 2180-1A, wild type	1.4	2.4×10^{-5}	0.07	0.001
RH 558, "constitutively derepressed" (*cdr*)	4.9	2.1×10^{-5}	0.17	0.002
RH 511, feedback-insensitive anthranilate synthase (*fbr*)	1.4	$\gg 10^{-3}$	2.8	0.17
RH 511 · 558, *cdr* · *fbr*	4.5	$\gg 10^{-3}$	2.9	0.7

[a] Specific activity of phosphoribosyl-anthranilate isomerase in all strains approximately 1.1 nmol per min per mg of protein.

[b] Expressed in nanomoles per minute per milligram of protein.

[c] Concentration of tryptophan (molar) leading to 50% inhibition of anthranilate synthase activity in vitro.

[d] Expressed in nanomoles per milligram (dry weight).

within the cell and that the remainder is excreted into the medium. The situation becomes more favorable in a double-mutant strain carrying a mutation for "constitutivity" as well as a mutation for feedback insensitivity to anthranilate synthase. The pool is not higher, but the rate of tryptophan accumulation increases to 0.71 nmol per min per mg of protein. The excess tryptophan produced cannot be retained in the cell under the conditions applied and is excreted into the medium. The fact that we could reach only approximately 60% of the theoretical value can be partially attributed to the competition for chorismic acid by chorismate mutase and, more importantly, to tryptophan degradation, which seems to become considerable as soon as high internal tryptophan concentrations have been reached.

How does *S. cerevisiae* compare to other microbial strains in which tryptophan accumulation has been analyzed and to the industrial processes for tryptophan production by *Hansenula anomala, Corynebacterium glutamicum,* or the alkaloid-producing *Claviceps* strains? Anthranilate is the starting material for the production of tryptophan by *Hansenula* (47). In this way, two critical steps are avoided, namely, feedback inhibition of anthranilate synthase and the availability of chorismic acid. In the literature, the average amount of tryptophan produced per gram of cells per hour is given as 0.1, which corresponds to 0.15 nmol per min per mg of protein, with maxima around 0.5 nmol per min per mg of protein. *C. glutamicum* was described as producing up to 12 g of L-tryptophan per liter (19). The strains used are derived from the wild-type strain by 10 mutational steps. The capacity of the processes is estimated to be around 1.0 to 1.5 nmol of L-tryptophan per min per mg of protein. The production capacity of 1 to 2 mg of tryptophan equivalents per h per g (dry weight) of *Claviceps* cells can be estimated (31; own unpublished data), which corresponds to 0.2 nmol per min per mg of protein. Thus, in all tryptophan processes studied, the actual realized or realizable flux is in the same range, around 0.2 to 1.5 nmol per min per mg of protein.

The studies of the tryptophan system may be compared to the more intensive study by Kacser and co-workers on the flux of arginine biosynthesis and its control in *N. crassa* (28, 29) as well as to other studies (21, 46). In nature, we usually do not encounter "rate-limiting" reaction steps, but should identify and quantitate "rate-controlling" steps. As the "rate-controlling" mechanisms in partially developed industrial strains may be eliminated by mutations, the removal of the "rate-limiting" steps becomes increasingly important.

REFERENCES

1. **Arst, H. N., and D. W. MacDonald.** 1975. A gene cluster in *Aspergillus nidulans* with an internally located cis-acting regulatory region. Nature (London) **254**:26–31.
2. **Arst, H. N., and C. Scazzocchio.** 1975. Initiator constitutive mutation with an "up-promoter" effect in *Aspergillus nidulans.* Nature (London) **254**:31–34.
3. **Atkinson, D. E.** 1969. Limitation of metabolic concentrations and the conservation of solvent capacity in the living cell. Curr. Top. Cell. Regul. **1**:29–43.
4. **Bennett, G. M., M. E. Schweingruber, K. D. Brown, C. Squires, and C. Yanofsky.** 1976. Nucleotide sequence of region preceding tryptophan mRNA initiation site and its role in promoter and operator function. Proc. Natl. Acad. Sci. U.S.A. **73**:2351–2355.
5. **Britten, R. J., and F. T. McClure.** 1962. The amino acid pool in *Escherichia coli.* Bacteriol. Rev. **26**:292–335.
6. **Calhoun, D. H., and G. W. Hatfield.** 1975. Autoregulation of gene expression. Annu. Rev. Microbiol. **29**:275–299.
7. **Chock, P. B., and E. R. Stadtman.** 1977. Superiority of interconvertible enzyme cascades in metabolic regulation. Analysis of multicyclic systems. Proc. Natl. Acad. Sci. U.S.A. **74**:2766–2770.
8. **Cohen, G.** 1967. 1971. Le métabolisme cellulaire et sa régulation. Hermann, Paris.
9. **Crawford, I. P.** 1975. Gene rearrangements in the evolution of the tryptophan synthetic pathway. Bacteriol. Rev. **39**:87–120.
10. **Daoust, D. R. and T. H. Stoudt.** 1966. The biosynthesis of L-lysine in a strain of *Micrococcus glutamicus.* Dev. Ind. Microbiol. **7**:22–34.
11. **Delforge, J., F. Messenguy, and J. M. Wiame.** 1975. The regulation of arginine biosynthesis in *Saccharomyces cerevisiae.* The specificity of argR-mutation and the general control of amino acid biosynthesis. Eur. J. Biochem. **59**:231–239.
12. **Doi, R. H.** 1977. Role of ribonucleic acid polymerase in gene selection in procaryotes. Bacteriol. Rev. **41**:568–594.
13. **Drew, S. W., and A. L. Demain.** 1977. Effect of primary metabolites on secondary metabolism. Annu. Rev. Microbiol. **31**:343–356.
14. **Fantes, P. A., L. M. Roberts, and R. Hütter.** 1976. Free tryptophan pool and tryptophan biosynthetic enzymes in *Saccharomyces cerevisiae.* Arch. Microbiol. **107**:207–214.
15. **Gibson, F., and J. Pittard.** 1968. Pathways of biosynthesis of aromatic amino acids and vitamins and their control in microorganisms. Bacteriol. Rev. **32**:465–492.
16. **Goldberger, R. F., R. G. Deeley, and K. P. Mullinix.** 1976. Regulation of gene expression in prokaryotic organisms. Adv. Genet. **18**:1–67.
17. **Gots, J. S., and C. E. Benson.** 1974. Biochemical genetics of bacteria. Annu. Rev. Genet. **8**:79–101.
18. **Greengard, P.** 1978. Phosphorylated proteins as physiological effectors. Science **199**:146–152.
19. **Hagino, H., and K. Nakayama.** 1975. L-tryptophan production by analog-resistant mutants derived from a phenylalanine and tyrosine double auxotroph of *Corynebacterium glutamicum.* Agric. Biol. Chem. **39**:343–349.
20. **Higgins, J.** 1965. Dynamics and control in cellular reactions, p. 13–46. *In* B. Chance, R. W. Estabrook, and J. R. Williamson (ed.), Control of enzyme metabolism. Academic Press Inc., New York.
21. **Hoch, S. O., C. W. Roth, I. P. Crawford, and E. W. Nester.** 1971. Control of tryptophan biosynthesis by the methyltryptophan resistance gene in *Bacillus subtilis.* J. Bacteriol. **105**:38–45.
22. **Holzer, H.** 1974. Possible mechanism for selective control of proteinase action. Adv. Enzyme Regul. **12**:1–10.
23. **Hopwood, D. A., and M. J. Merrick.** 1977. Genetics of antibiotic production. Bacteriol. Rev. **41**:595–635.
24. **Hütter, R.** 1969. Genetik industrieller Mikroorganismen. Pathol. Microbiol. **34**:195–212.
25. **Hütter, R.** 1971. Regulation of amino-acid biosynthesis and industrial production of amino acids, p. 169–179. *In* Proc. Symp. Radiation and Radioisotopes for Industrial Microorganisms. International Atomic Energy Agency, Vienna.
26. **Hütter, R.** 1973. Regulation of the tryptophan biosynthetic enzymes in fungi, p. 109–124. *In* Z. Vaněk, Z. Hošťálek and J. Cudlín (ed.), International symposium on the genetics of industrial microorganisms, vol. 2. Academia, Prague.
27. **Hynes, M. J.** 1975. A cis-dominant regulatory mutation affecting enzyme induction in the eukaryote *Aspergillus nidulans.* Nature (London) **253**:210–211.
28. **Kacser, H.** 1963. The kinetic structure of organisms, p. 25–40. *In* R. J. C. Harris (ed.), Biological organization at the cellular and supracellular level. Academic Press Inc., New York.
29. **Kacser, H., and J. A. Burns.** 1973. The control of flux, p. 65–104. *In* D. D. Davies (ed.), Rate of control of biological processes. Cambridge University Press, London.
30. **Kradolfer, P., J. Zeyer, G. Miozzari, and R. Hütter.** 1977. Dominant regulatory mutants in chorismate mutase of *Saccharomyces cerevisiae.* FEMS Microbiol. Lett. **2**:211–216.
31. **Krupinsky, V. M., J. E. Robbers, and H. G. Floss.** 1976. Physiological study of ergot: induction of alkaloid synthesis by tryptophan at the enzymatic level. J. Bacteriol. **125**:158–165.
32. **Lee, F., and C. Yanofsky.** 1977. Transcription termination at the tryptophan operon attenuators of *Escherichia coli* and *Salmonella typhimurium:* RNA secondary structure and regulation of termination. Proc. Natl. Acad. Sci. U.S.A. **74**:4365–4369.
33. **Lorence, J. H., and E. W. Nester.** 1967. Multiple forms of chorismate mutase. Biochemistry **6**:1541–1553.

34. **Miozzari, G., P. Niederberger, and R. Hütter.** 1978. Tryptophan biosynthesis in *Saccharomyces cerevisiae:* control of the flux through the pathway. J. Bacteriol. **134:**48–59.
35. **Miozzari, G. F., and C. Yanofsky.** 1978. Translation of the leader region of the *Escherichia coli* tryptophan operon. J. Bacteriol. **133:**1457–1466.
36. **Miyajima, R., S. I. Otsuka, and I. Shiio.** 1968. Regulation of aspartate family amino-acid biosynthesis in *Brevibacterium flavum.* I. Inhibition by amino-acids of the enzymes in threonine biosynthesis. J. Biochem. (Tokyo) **63:**139–148.
37. **Miyajima, R., and I. Shiio.** 1970. Regulation of asparatate family amino acid biosynthesis in *Brevibacterium flavum.* III. Properties of homoserine dehydrogenase. J. Biochem. (Tokyo) **68:**311–319.
38. **Nakayama, K., K. Tanaka, H. Ogino, and S. Kinoshita.** 1966. Studies on lysine fermentation. V. Concerted feedback inhibition of aspartokinase and the absence of lysine inhibition on aspartic semialdehyde-pyruvate condensation in *Micrococcus glutamicus.* Agric. Biol. Chem. **30:**611–616.
39. **Neurath, H., and K. A. Walsh.** 1976. Role of proteolytic enzymes in biological regulation. Proc. Natl. Acad. Sci. U.S.A. **73:**3825–3832.
40. **Rubin, C. S., and O. M. Rosen.** 1975. Protein phosphorylation. Annu. Rev. Biochem. **44:**831–887.
41. **Sano, K., and I. Shiio.** 1970. Microbial production of L-lysine. III. Production by mutants resistant to S-(2-aminoethyl)-L-cysteine. J. Gen. Appl. Microbiol. **16:**373–391.
42. **Schürch, A., J. Miozzari, and R. Hütter.** 1974. Regulation of tryptophan biosynthesis in *Saccharomyces cerevisiae*: mode of action of 5-methyl-tryptophan and 5-methyl-tryptophan-sensitive mutants. J. Bacteriol. **117:**1131–1140.
43. **Stadtman, E. R.** 1963. Symposium on multiple forms of enzymes and control mechanisms. II. Enzyme multiplicity and function in the regulation of divergent metabolic pathways. Bacteriol. Rev. **27:**170–190.
44. **Stadtman, E. R., and P. D. Chock.** 1977. Superiority of interconvertible enzyme cascades in metabolic regulation. Analysis of monocyclic systems. Proc. Natl. Acad. Sci. U.S.A. **74:**2761–2765.
45. **Stadtman, E. R., and A. Ginsburg.** 1974. The glutamine synthetase of *Escherichia coli.* Structure and control, p. 755–807. *In* P. D. Boyer (ed.), The enzymes, vol. 10, 3rd ed. Academic Press, Inc., New York.
46. **Stebbing, N.** 1974. Precursor pools and endogenous control of enzyme synthesis and activity in biosynthetic pathways. Bacteriol. Rev. **38:**1–28.
47. **Terui, G.** 1972. Tryptophan, p. 515–531. *In* K. Yamada, S. Kinoshita, T. Tsunoda, and K. Yamada (ed.), The microbial production of amino acids. Kodansha Ltd., Tokyo.
48. **Uy, R., and F. Wold.** 1977. Posttranslational covalent modification of proteins. Science **198:**890–896.
49. **Wiame, J. M., and E. L. Dubois.** 1976. The regulation of enzyme synthesis in arginine metabolism of *Saccharomyces cerevisiae,* p. 391–406. *In* K. D. Macdonald (ed.), Second international symposium on the genetics of industrial microorganisms. Academic Press Inc., New York.
50. **Wolfner, M., D. Yep, F. Messenguy, and G. R. Fink.** 1975. Integration of amino acid biosynthesis into the cell cycle of *Saccharomyces cerevisiae.* J. Mol. Biol. **90:**273–290.
51. **Yanofsky, C.** 1976. Regulation of transcription initiation and termination in the control of expression of the tryptophan operon of *E. coli,* p. 75–87. *In* D. P. Nierlich et al. (ed.), Molecular mechanisms in the control of gene expression. ICN-UCLA Symposium on Molecular and Cell Biology V. Academic Press Inc., New York.
52. **Zimmermann, F. K.** 1975. Genetic regulatory mechanisms in fungi. Prog. Bot. **37:**247–258.

Control of Lysine Metabolism in the Petroleum Yeast *Saccharomycopsis lipolytica*

H. HESLOT, C. GAILLARDIN, J. M. BECKERICH, AND P. FOURNIER

Service de Génétique, Institut National Agronomique, Paris, France

Saccharomycopsis lipolytica is an ascogenous, heterothallic yeast which has received special attention because of its ability to metabolize hydrocarbons. Inbreeding programs resulted in the isolation of genetically manageable strains (2, 6), which made possible genetic and physiological studies of various pathways: protoporphyrine biosynthesis (1), alkane utilization (2), and production of protease (10). Our objective was to obtain lysine-excreting strains of this organism, and the work involved the study of the regulation of the biosynthetic pathway, the storage and catabolism of this amino acid, and the excretion process itself.

The biosynthesis of lysine by yeasts was elucidated some 25 years ago. It involves a pathway different from that found in plants and molds, but relatively little is known about the regulation of this pathway. The first step appears to be feedback sensitive to lysine, but the physiological significance of this effect has been questioned (11). A coordinated repression of the first three enzymes has been reported (3), but, as judged from in vivo experiments, the main regulatory effect of lysine seems to be exerted at an unidentified step in the late part of the pathway (11). The gene-enzyme relationships are not completely understood in *Saccharomyces cerevisiae* since at least 14 different loci (and possibly more; see 5) have been defined for the 11 biosynthetic steps: there is no mutant for the first enzyme, for instance, whereas up to three loci are associated with a defect in aminoadipic semialdehyde reductase.

The catabolism of lysine in yeast is also poorly understood, and the pathway is generally assumed to be similar to that found in other fungi.

LYSINE BIOSYNTHETIC PATHWAY AND ITS REGULATION

Structural genes. There are 11 steps leading from acetyl coenzyme A (CoA) and α-ketoglutarate to L-lysine (Fig. 1). Our approach was to isolate a number of lysine auxotrophs after mutagenic treatment. These mutants were tested for their capacity to grow on intermediates of the pathway, but only α-aminoadipate could be used by some of them. They were classified as utilizers (class 1) and nonutilizers (class 2) of α-aminoadipate.

Complementation tests were performed between mutants of each class. We found seven complementation groups in class 1 and four in class 2. As there is no overlap, the total number of complementation groups is 11, defining a priori 11 genes (or loci). These genes are not organized into an operon-like structure, as most of them are unlinked. However, *lys2* and *lys3* are closely linked (2% recombination), and all mutants of group 1 are very closely linked and probably belong to a single locus, *lys1*. Measurement of the enzymatic activities of our lysineless mutants allowed the location of some blocks: *lys1* and *lys11* lack activity 1; *lys6* and *lys7* have no homoaconitase; *lys9* and *lys10* have no homoisocitrate dehydrogenase; *lys2* and *lys3* are affected in steps 7, 8, and 9; *lys4* is blocked in step 10; and *lys5* has no saccharopine dehydrogenase. There is some complementation, most probably intragenic, between *lys1* mutants. It implicates an oligomeric structure for the corresponding enzyme, homocitrate synthetase. This is not surprising in view of the regulatory properties of this enzyme, as described below.

Regulation of the pathway. We have studied the effectors of the first enzyme of the pathway, i.e., homocitrate synthetase catalyzing the condensation of acetyl CoA and α-ketoglutarate (8). Inhibition of homocitrate synthetase by lysine is highly specific, and only very close analogs such as cyclohexylalanine and transdehydrolysine are inhibitory.

In view of a possible branching of the L-lysine pathway to another (as yet unidentified) product, we made a search for other inhibitory compounds (Table 1). Several arguments indicate that sites involved in the fixation of aminoadipate, pipecolate, and dipicolinate are distinct from those of lysine, suggesting a lysine-independent control of the first part of the pathway. No additional feedback-sensitive enzyme has been detected.

In *S. cerevisiae,* the two first enzymes of the pathway are repressed by lysine. We have investigated whether there are any variations under these conditions in the specific activities of homocitrate synthase and homoaconitase. The latter decreases very slightly (about 1.5-fold) when the wild type is grown in minimal media with increasing lysine concentrations (up to 10 mM), but there is no detectable variation

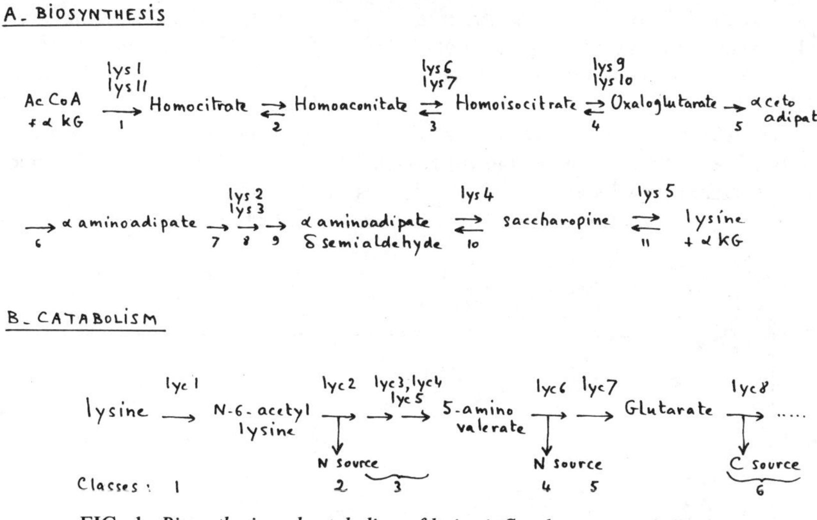

A. BIOSYNTHESIS

Ac CoA + α kG → (1; lys 1, lys 11) Homocitrate ⇄ (2) Homoaconitate ⇄ (3; lys 6, lys 7) Homoisocitrate ⇄ (4; lys 9, lys 10) Oxaloglutarate → (5) α ceto adipate

→ (6) α aminoadipate → (7; lys 2, lys 3) → (8) → (9) α aminoadipate δ semialdehyde ⇄ (10; lys 4) saccharopine ⇄ (11; lys 5) lysine + α kG

B. CATABOLISM

lysine → (lyc 1) N-6-acetyl lysine → (lyc 2) → (lyc 3, lyc 4, lyc 5) → 5-amino valerate → (lyc 6) → (lyc 7) Glutarate → (lyc 8) .....

N source; N source; C source

Classes: 1; 2, 3; 4, 5; 6

FIG. 1. *Biosynthesis and catabolism of lysine in* Saccharomycopsis lipolytica.

TABLE 1. *Properties of homocitrate synthase from the wild strain of* Saccharomycopsis lipolytica

Ligands	Type of inhibition observed with variable substrate		Effect on cooperativity of fixation sites for	
	α-Ketoglutarate	Acetyl CoA	α-Ketoglutarate	Acetyl CoA
Lysine	Competitive	Mixed	+	−
Aminoadipate	Competitive	Mixed	−	+
Pipecolate	Competitive	Mixed	−	+
Dipicolinate	Competitive	Competitive	+	+
CoASH	No inhibition	Competitive	−	+

of homocitrate synthase levels. There is also no effect of lysine on the levels of homoisocitric dehydrogenase, aminoadipic reductase, and aminoadipyl semialdehyde reductase. Consequently, it is probable that, if any one of these enzymes is indeed repressed by lysine, repression is already at or near its maximum.

The last enzyme of the pathway (saccharopine dehydrogenase, coded for by the *lys5* locus) is apparently repressed twofold by L-lysine in the wild type (Fig. 2). This repression disappears completely in mutants (*lyc1, lyc2, lyc3, lyc4, lyc5*) which are unable to carry the first step of the lysine catabolic pathway (see below), i.e., lysine → *N*-6-acetyllysine (Fig. 1). *N*-6-acetyllysine itself did not show any repressive effect. Derepression of homocitrate synthase, homoaconitase, homoisocitric dehydrogenase, aminoadipyl semialdehyde reductase, and saccharopine dehydrogenase (but not aminoadipate reductase) has been repeatedly observed in slow-growing strains (*ade1, his1, arg1, lys2,* and *lys3* mutants). Whether or not this derepression has anything to do with specific lysine controls or with a general control of amino acid biosynthesis (such as has been described in *S. cerevisiae* for the tryptophan, arginine, histidine, lysine, etc., pathways) remains to be investigated.

Regulation mutants. Inhibition of homocitrate synthase by two analogs of L-lysine, namely, cyclohexylalanine and transdehydrolysine, has been described. These two compounds, as well as others (such as aminoethylcysteine), are also growth inhibitors.

We isolated 32 mutants resistant to transdehydrolysine from the wild strain W 29 and measured the following parameters: capacity for growth on lysine as carbon source, as a test of the integrity of the transport system; lysine pool; and percent inhibition of homocitrate synthase by 1 mM lysine. The data in Table 2 show a good correlation between: (i) absence of growth on lysine together with a normal pool and sensitive homocitrate synthase, and (ii) growth on lysine together with a higher pool and less sensitive homocitrate synthase. Furthermore, two main types of mutations which result in resistance to transdehydrolysine were identified: alteration of the L-

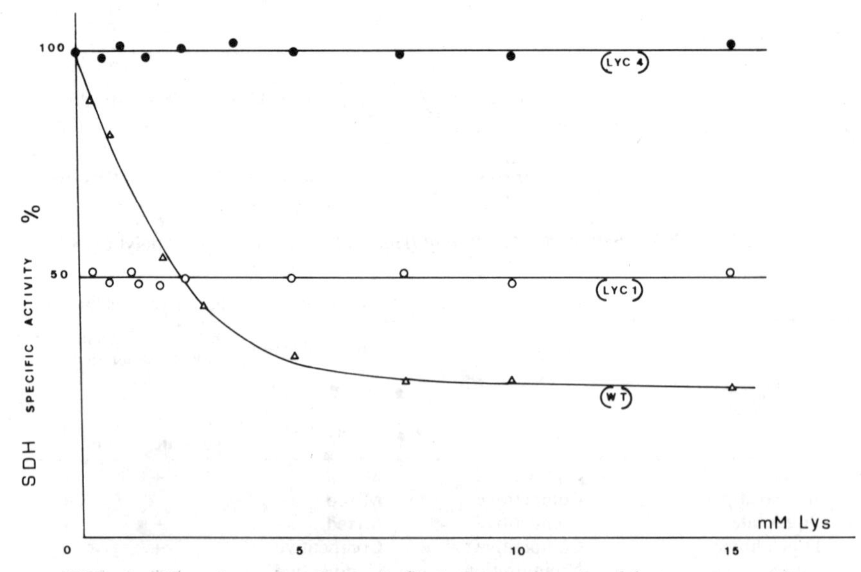

FIG. 2. *Effect of exogenously supplied lysine on saccharopine dehydrogenase activity.*

TABLE 2. *Properties of transdehydrolysine-resistant strains derived from the wild type*

Strains	Growth on lysine as carbon source	Lysine pool (μmol/g, dry wt)	Inhibition of homocitrate synthase by 1 mM lysine (%)
Wild type	+	20	70
mg-1	−	22	69
mg-2	−	19	75
mg-3	+	82	42
mg-4	+	87	38
mg-5	+	106	20
mg-7	+/−	52	56
mg-9	+	65	48
mg-10	+/−	18	75
mg-11	+	95	21
mg-15	−	25	
mg-16	+	68	37
mg-17	+	47	41
mg-18	+	53	48
mg-19	+	82	35
mg-20	+	66	47
mg-21	−	19	
mg-22	−	21	
mg-23	+	62	46
mg-24	+	53	43
mg-25	+/−	17	76
mg-26	+	52	42
mg-28	−	25	
mg-30	+	69	45
mg-32	−	21	76

lysine transport system and desensitization of homocitrate synthase to feedback by L-lysine. Mutant mg-5 is very closely linked to locus *lys1* and is most probably altered in the homocitrate synthase structural gene. The lysine pool in mg-5 is increased five times.

LYSINE CATABOLISM

The wild type W 29 of *S. lipolytica* can use lysine as a carbon or nitrogen source. Its growth response to different compounds as carbon or nitrogen source was investigated. Among the intermediates tested, only *N*-6-acetyllysine and 5-aminovalerate were able to mimic lysine as a source of carbon or nitrogen (7). Labeling experiments confirmed that *N*-6-acetyllysine is indeed an intermediate of the catabolic pathway. When [^{14}C]lysine was added to a cell culture, *N*-6-[^{14}C]acetyllysine was detected in the cellular pool by chromatography and autoradiography. However, 5-[^{14}C]aminovalerate was found in very small amounts, indicating that it is not accumulated.

Five phenotypic classes were found among mutants defective in lysine catabolism, and each mutant was shown to be the result of a single mutation. Eight complementation groups were detected. According to their growth responses on different carbon sources, the mutants can be located as indicated in Fig. 1.

The inability of class 2 mutants to use lysine or *N*-6-acetyllysine as a nitrogen source can be explained by the sequence of lysine catabolism in Fig. 1. However, all mutants of class 3 (with at least three different loci) are unable to use lysine although they grow on *N*-6-acetyllysine as nitrogen source. From their responses to different carbon sources, they have been located before 5-aminovaleric acid (Fig. 1). These mutants complement with class 1 mutants. They produce *N*-6-acetyllysine only in the presence of 5-aminovalerate and, consequently, do not lack the information to catalyze the conversion of lysine to *N*-6-acetyllysine. It appears that 5-aminovaleric acid is probably the inducer or activator of this enzyme. Mutants of classes 1, 2, and 3 accumulate more lysine when grown in the presence of this amino acid then does the wild strain W 29 (Fig. 3). Combining a mutant of class 1 and a mutant lacking feedback control (mg-5) in a single strain results in a 20-fold increase of the free-lysine pool.

ACTIVE TRANSPORT OF LYSINE

As the highest accumulating strain did not excrete lysine, we investigated the L-lysine transport systems in order to obtain lysine-excreting mutants (4). The cells were incubated for 2 min in a minimal growth medium (pH 5.0) containing [^{14}C]-lysine. When the concentration of the latter varies between 10 and 200 μM, the initial velocity follows Michaelis kinetics, with an apparent K_m of 1.9×10^{-5} M.

Inhibitors of energetic metabolism such as 2,4-dinitrophenol and sodium azide

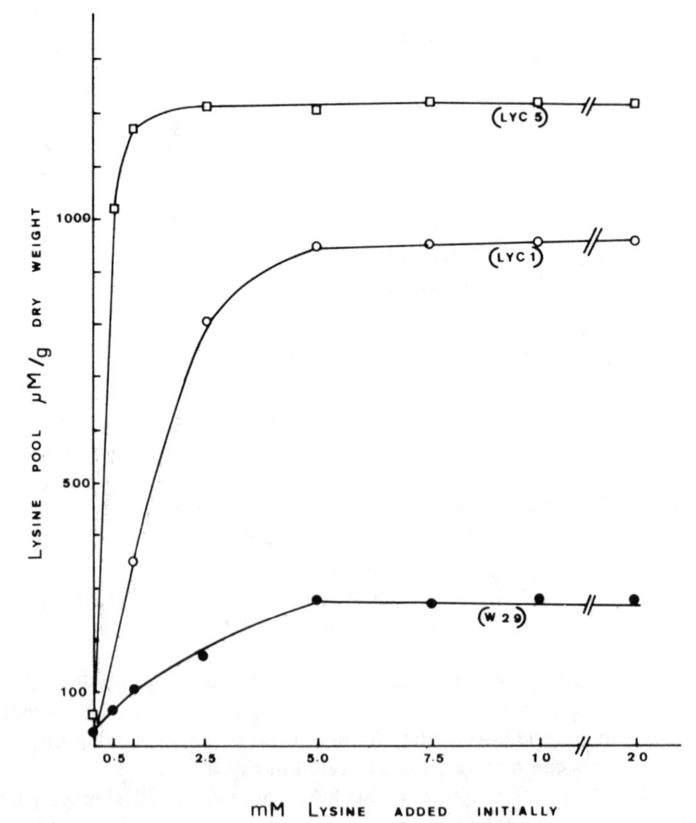

FIG. 3. *Variations of lysine pool of cells grown on minimal medium supplemented with various amounts of lysine.*

suppress up to 95% of the lysine uptake. To detect other types of inhibitors, we measured uptake of lysine (50 μM) in the presence of various compounds at concentrations of 500 to 5,000 μM. Three groups can be recognized: (i) strong inhibitors such as L-arginine, L-canavanine, L-ornithine, L-aminoethylcysteine, 4,5-transdehydrolysine, and ε-*N*-methyllysine; (ii) some amino acids allowing a residual activity of about 70% at 500 μM; and (iii) inhibitors of low effectiveness.

A comparison of these results with the structural formulas of the compounds indicates that the lysine transport system has a marked specificity for L-lysine and close analogs. Kinetic studies revealed that the following compounds are competitive inhibitors: L-arginine, L-ornithine, L-*S*-aminoethylcysteine, and L-4,5-transdehydrolysine. Since L-arginine has a K_i of 1.8×10^{-5} M, and is therefore of the same order of magnitude as L-lysine (K_m, 1.9×10^{-5} M), we are probably dealing with a permease common to L-lysine and L-arginine. With the exception of proline, all other amino acids tested were somewhat inhibitory, with noncompetitive kinetics. When the uptake was investigated at higher L-lysine concentrations (1 to 10 mM), a second transport system was detected which is also sensitive to sodium azide and has Michaelis kinetics, with a K_m of 4×10^{-4} M.

Mutants that were affected in the lysine transport systems were isolated by plating mutagenized cells on media containing two or all three of the following compounds, which are known growth inhibitors and also competitive inhibitors of lysine uptake at low L-lysine concentrations (50 μM): L-4,5-transdehydrolysine, L-canavanine, and L-5-aminoethylcysteine. Some of these mutants were tested for their uptake capacity at 2.5 mM and 50 μM L-lysine. They were found to be affected at one or both of these concentrations.

When the uptake mutants were combined with the lysine-accumulating strains, only a low-level excretion of lysine was obtained.

CONCLUSION

Yeast strains bearing two mutations (alteration of feedback control of homocitrate synthetase by lysine and a block of L-lysine catabolism) showed a 20-fold increased pool of free lysine, but no excretion of this amino acid in the growth medium took place. A further mutation inactivating the transport system had only a small beneficial effect.

A specific search for excreting clones, after UV mutagenesis, led to positive results. However, most of the clones were unstable, and those that could be stabilized excreted only small amounts of lysine (9). A better knowledge of the mechanisms involved would help in designing a rational approach to the selection of lysine-excreting strains.

REFERENCES

1. **Bassel, J., P. Hambright, R. Mortimer, and A. J. Bearden.** 1975. Mutant of the yeast *Saccharomycopsis lipolytica* that accumulates and excretes protoporphyrin IX. J. Bacteriol. **123:**118–122.
2. **Bassell, J., and R. Mortimer.** 1973. Genetic analysis of mating type and alkane utilization in *Saccharomycopsis lipolytica.* J. Bacteriol. **114:**894–896.
3. **Battacharjee, J. K.** 1970. Leaky mutations and coordinate regulation of the accumulation of lysine precursors in *Saccharomyces.* Can. J. Genet. Cytol. **12:**85–789.
4. **Beckerich, J. M., and H. Heslot.** 1978. Physiology of lysine permeases in *Saccharomycopsis lipolytica.* J. Bacteriol. **133:**492–498.
5. **Biswas, G. O., and J. K. Battacharjee.** 1974. Isolation and complementation of lysine auxotrophs in *Saccharomyces.* Antonie van Leeuwenhoek J. Microbiol. Serol. **40:**221–231.

6. **Gaillardin, C. M., V. Charoy, and H. Heslot.** 1973. A study of copulation and meiotic segregation in *Candida lipolytica.* Arch. Mikrobiol. **22:**69–73.
7. **Gaillardin, C. P., Fournier, G. Sylvestre, and H. Heslot.** 1976. Mutants of *Saccharomycopsis lipolytica* defective in lysine catabolism. J. Bacteriol. **125:**48–57.
8. **Gaillardin, C. M., L. Poirier, and H. Heslot.** 1974. Regulation of the L-lysine pathway in the yeast *Saccharomycopsis (Candida) lipolytica.* Mutant affected at the level of homocitrate synthase. Abstr. Int. Symp. Genet. Ind. Microorganisms, 2nd, Sheffield.
9. **Gaillardin, C. M., G. Sylvestre, and H. Heslot.** 1975. Studies on an unstable phenotype induced by UV irradiation. The lysine excreting (lex^-) phenotype of the yeast *Saccharomycopsis lipolytica.* Arch. Microbiol. **104:**89–94.
10. **Ogrydziak, D., and R. Mortimer.** 1977. Genetics of extracellular protease production in *Saccharomycopsis lipolytica.* Genetics **87:**621–632.
11. **Tucci, A. F., and L. N. Ceci.** 1972. Control of lysine biosynthesis in yeast. Arch. Biochem. Biophys. **153:** 751–754.

Turnover of RNA and Proteins in *Escherichia coli*

DAVID APIRION

Department of Microbiology and Immunology, Division of Biochemistry and Biomedical Sciences, Washington University Medical School, St. Louis, Missouri 63110

The ability to use single-cell proteins as a nutritional source for humans is becoming increasingly important. As the need to find cheap sources of protein becomes exceedingly urgent, single-cell cultures that can utilize for growth a variety of substrates which would otherwise be nonedible become more and more attractive.

One serious problem in increased utilization of microorganisms as a source of protein is their high content of nucleic acids compared to other foods such as cereals. The level of tolerance of mammals to nucleic acids in food is limited, and continuous feeding on a diet which is rich in nucleic acids leads to gout, because of inability to secret high levels of uric acid (24, 26). Therefore, it is useful to treat single-cell proteins in a way that would decrease the level of the nucleic acids without reducing the level of protein in the preparation.

The largest source of nucleic acids in all cells is rRNA, which constitutes up to 80% of the total nucleic acids in many unicellular organisms. Therefore, knowledge about the turnover of RNA in the cell is of major relevance to this problem. In recent years, my colleagues and I have studied rather extensively the decay of ribosomes and especially rRNA in *Escherichia coli*. Some of those studies will be summarized here.

rRNA as well as other nucleic acids contains phosphate, but other molecules such as phospholipids and lipopolysaccharides also contain phosphate. Therefore, we developed simple methods for simultaneously monitoring DNA, RNA (mRNA, tRNA, and rRNA separately), lipopolysaccharides, and phospholipids in whole cells. I shall describe here some of these techniques, present a model for the decay of the ribosome, show that it is quite feasible to degrade most of the RNA of the cell without losing much of the cellular protein, and describe a new RNase, which is in all likelihood responsible for the decay of rRNA in the cell.

Newly synthesized RNA in *E. coli* contains a number of species, including mRNA, rRNA, and tRNA. mRNA molecules are relatively unstable and are degraded soon after synthesis, whereas others (e.g., rRNA and tRNA) are relatively stable and persist in the cell for some time. If, however, cultures are starved for a carbon source, for instance, then the RNA molecules that were stable during exponential growth would start to decay (for further details and considerations of these problems, see 1, 2). To study the mechanism of rRNA and ribosome decay during carbon starvation, we used four strains (see 12–14; Fig. 1). They include a wild-type strain, a strain lacking the enzyme RNase I, another lacking RNase I and having a modified polynucleotide phosphorylase, and one which, in addition, has a thermolabile RNase II. During carbon starvation at 37°C only minor differences among the strains were detected.

Since RNase I is the only enzyme completely lacking in some of these strains (6, 11), whereas all the known RNase II and polynucleotide phosphorylase mutants contain modified forms of these enzymes (21, 22, 25), and since it has been shown that the modified enzymes become limiting only at elevated temperatures (15, 16, 22, 25), degradation of RNA was measured at 45 and 50°C (45°C is near the highest temperature [46°C] at which *E. coli* strains can grow exponentially and form colonies from single cells). The results in Fig. 1 justified the expectations. After 4 h of carbon starvation at 50°C (and 45°C; not shown), clear differences in the decay of RNA could be observed among all these four strains. However, protein degradation was less than 15% after 5 h. No differences among the strains were observed with regard to the decay of protein, whereas the wild-type strain lost more than 95% of its RNA. In Fig. 1, the results for protein turnover are shown for two strains; the other two behaved similarly. Although the differences between protein and RNA degradation are quite dramatic, the manner in which these experiments were carried out allows only one conclusion: that the degradation of RNA to acid-soluble material is largely independent of degradation of protein to acid-soluble material. These experiments thus show that in 4 h it is possible to degrade almost all the RNA in the cell without much affecting its protein content, a significant step towards utilizing single-cell protein for food.

The experiment in Fig. 1 suggested the involvement of some RNases in degradation of "stable" RNA during carbon starvation, but such experiments did not determine whether or not other enzymes might participate in "stable" RNA degradation.

To probe further the mechanism by which ribosomes and rRNA are degraded, we developed methods which allow a fast and reliable quantitation of all the phosphate-containing macromolecules. These consisted of labeling growing cultures with $^{32}P_i$, opening cells in a hot (95°C) sodium dodecyl sulfate-containing buffer, and subjecting the whole-cell extract to electrophoresis (10) in two types of thin-slab polyacrylamide gels (5, 7). The large nucleic acids, DNA and rRNA, can be displayed in a 1.5% polyacrylamide gel containing 0.5% agarose (Fig. 2), and the small molecules can be displayed in a 5%/10% tandem polyacrylamide gel (Fig. 3). In a 1.5% polyacrylamide gel a slow-migrating ^{32}P-labeled material, referred to as the 40S band, appeared (Fig.

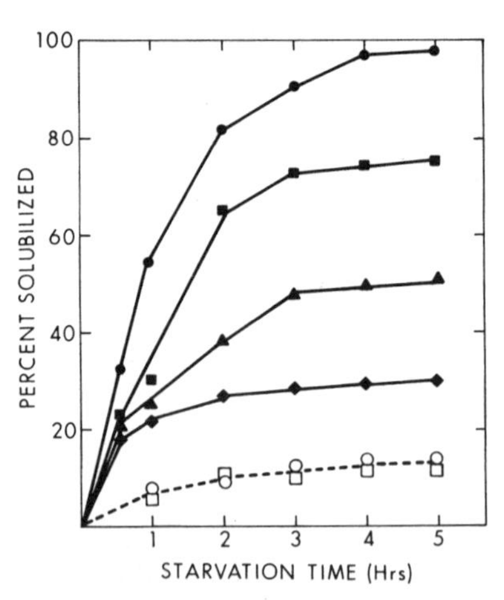

FIG. 1. *Degradation of long term-labeled RNA and protein during carbon starvation at 50°C. Strains AB301 (wild-type; ●), A19 (RNase I⁻; ■), Q13 (RNase I⁻ polynucleotide phosphorylase⁻ [PNPase⁻]; ▲), and N7060 (RNase I⁻ PNPase⁻ RNase II⁻; ◆) were grown at 37°C and starved at 50°C. The solubilization of RNA and protein was measured (RNA, closed symbols and solid lines; protein, open symbols and broken lines). Values of 100% represent 20,000 to 25,000 cpm for RNA and about 3,000 cpm for protein. (The values of protein decay for the two strains not shown here were similar to that of the strains shown.)*

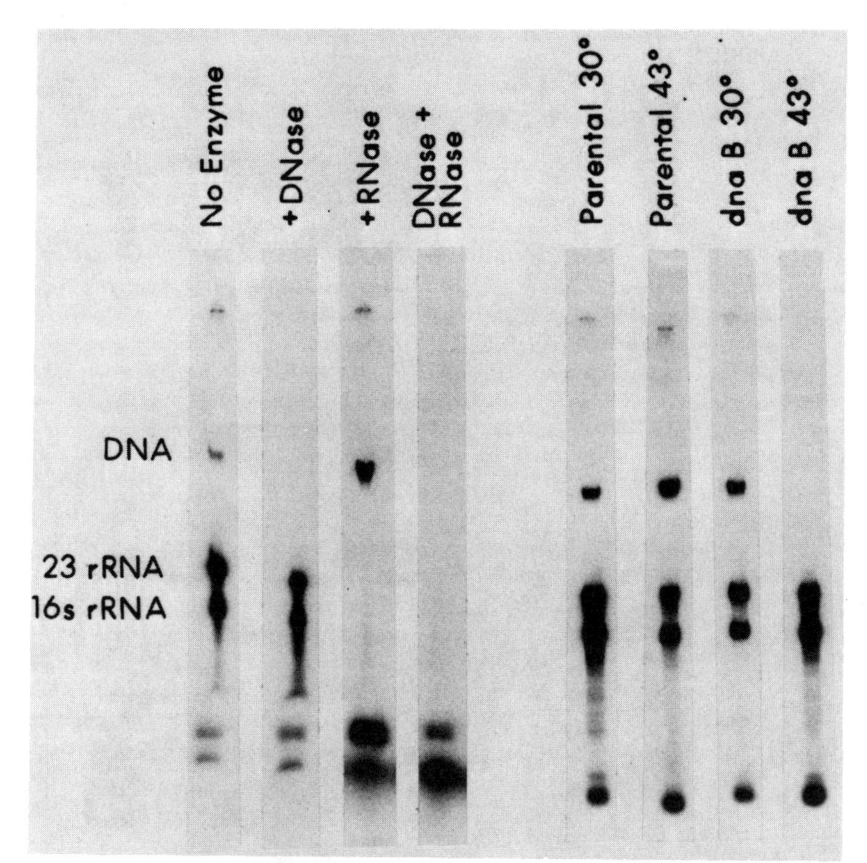

FIG. 2. *Display of DNA and large rRNA in a gel. A 30-ml culture of* E. coli *D10 (Met⁻RNase I⁻) was grown in low-phosphate Tris-base medium (13) and labeled with 3 μCi of $^{32}P_i$/ml for 2 h at 37°C with vigorous aeration. At an absorbancy at 560 nm of about 0.5, cells were collected in a volume/volume equivalence of killing buffer (80 μg of nalidixic acid/ml, 80 μg of KCN/ml, 80% ethanol, and 0.2% diethylpyrocarbonate). The samples were kept on an ice-water bath until all samples were collected. Samples were spun for 10 min at about 6,000 × g at 3°C. The supernatant fluid was poured off, and the pellets were resuspended in lysis buffer at a desired volume (0.5% sodium dodecyl sulfate, 10 mM Na_2EDTA, 20% sucrose, and 0.05% bromophenol blue; pH was adjusted to 7.0 prior to the addition of sodium dodecyl sulfate). Samples were heated in lysis buffer for 3 to 4 min in a boiling-water bath and then quenched on ice water (for enzymatic analysis the cracking buffer did not contain sucrose and bromophenol blue). To the opened cells, NaCl was added to 0.4 M, and the samples remained in the ice-bath for 15 min for protein precipitation. Extracts were spun at 10,000 × g in the cold for 15 min. The supernatant fluid was removed, 2.1 volumes of 95% ethanol were added to it, and it was stored overnight at −20°C and spun at 15,000 × g in the cold for 15 min. The pellet was resuspended in the dialysis buffer and dialyzed (1:1,000) against Tris-Mg buffer (Tris-hydrochloride, 0.01 M, pH 7.5; $MgCl_2$, 0.01 M). Three samples, each comprising one-quarter of the whole-cell lysate, were treated with DNase, RNase, and both DNase and RNase. The enzymatic reactions contained 200 μg of RNase (pancreatic) or DNase I per ml (both from Worthington Biochemicals Corp.) and were incubated at 37°C for 90 min. The samples for lanes 5–8 were derived from the labeling of a* dnaB *mutant and its parental strain (a non-temperature-sensitive strain). They both were labeled in 2-ml cultures, in a low-phosphate medium with 5 μCi of $^{32}P_i$/ml. The cells were labeled for 2 h at each temperature. For labeling at 43°C, cells were grown at 30°C, transferred to 43°C, and labeled 1 h later for 2 h. The cells were lysed as described. Samples were analyzed in a 1.5% polyacrylamide-0.5% agarose gel, containing 0.2% sodium dodecyl sulfate. Electrophoresis was carried out for 2.5 h at 80 V. The gel was dried and autoradiographed for 24 h (10, 13). The details for the gel are exactly as described by Gegenheimer and Apirion (9) for the 2% polyacrylamide-0.5% agarose gel, with the exception that 1.5 rather than 2% polyacrylamide was used. (Occasionally the bottom centimeter of the gel contained a retaining layer consisting of 3.5% polyacryalmide-0.5% agarose.)*

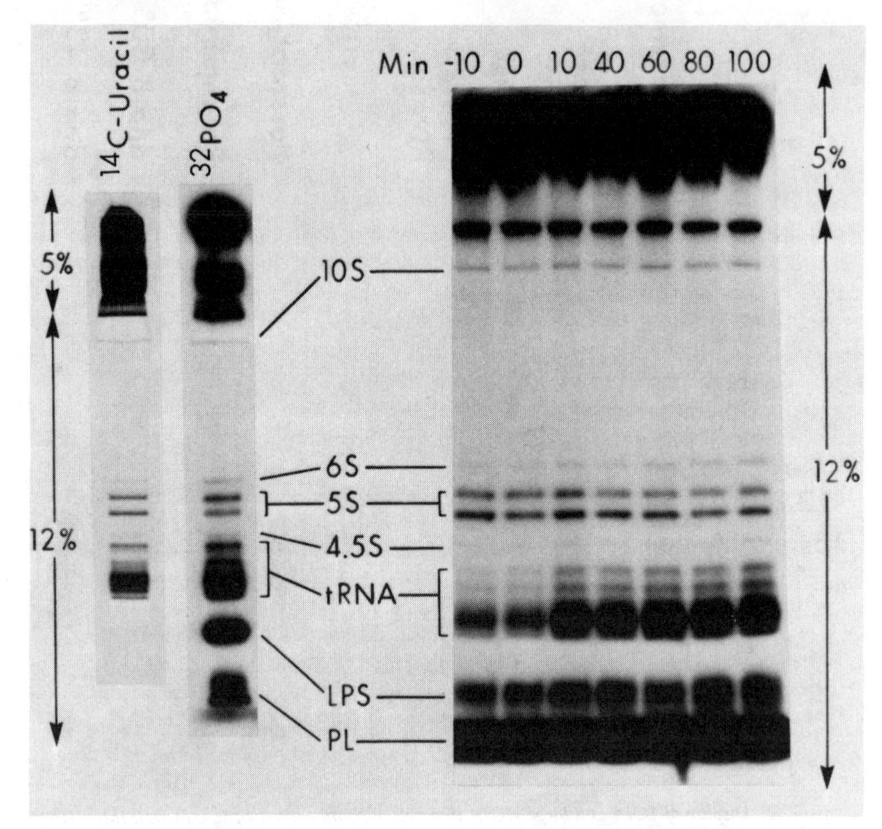

FIG. 3. *Display of low-molecular-weight RNA, lipopolysaccharides, and phospholipids in a gel. The two left lanes show [^{14}C]uracil and long-term (7 h) ^{32}P-labeled cell lysates separated in a tandem 5%/12% gel; 10S, 6S, 5S, 4.5S, and tRNA bands appear in both lanes. Lipopolysaccharides (LPS) and phospholipids (PL) are labeled with ^{32}P only (5). The seven lanes on the right represent samples collected during phosphate starvation in a Tris medium containing peptone (17) separated in a 5%/12% gel. Cells were labeled with $^{32}P_i$ for 2 h prior to starvation. Time 0 indicates the beginning of starvation. The 6S band became darker as cells were starved for phosphate as a result of the relative stability of 6S compared with 5S rRNA.*

2; ref. 7). A variety of experiments (7) showed that this band, as well as the material which remained at the origin of the gel, is DNA. For instance, it can be seen in Fig. 2 that this material is sensitive to DNase but not to RNase, and it is not synthesized at the elevated temperature in a temperature-sensitive mutant defective in DNA synthesis.

To ascertain that all or most of the cellular DNA appears either in the origin of the gel or in the 40S band, we labeled cells with [^{14}C]thymidine and subjected them to a similar type of analysis. The labeled thymidine appeared at the origin and in the 40S band. Such experiments substantiated the notion that most of the cellular DNA is either retained in the origin or migrates in a single sharp band after electrophoresis.

By subjecting whole-cell extracts to electrophoresis in 5%/10% or 5%/12% tandem thin-slab polyacrylamide gels, tRNA is separated from 5S rRNA and other small RNA molecules as well as the phospholipids and the lipopolysaccharides of the cell (see Fig. 3; ref. 5, 17). Quantitation of both types of gels revealed, for instance, that, when strain D10 (a standard *E. coli* K-12 laboratory strain) is grown in the regular medium for phosphate labeling, its phosphate-containing macromolecules are distributed as follows: DNA, 13%; rRNA, 53.3%; tRNA, 8.5%; lipopolysaccharides, 7.9%; phospholipids, 13.2%; mRNA, 3.3% (5, 8). From similar experiments, the level of mRNA can also be deduced (8).

By use of these and other techniques, it was possible to deduce a model for the decay of ribosomes in the cells (13; Fig. 4). Surprisingly, we had to conclude that the degradation starts with an endonucleolytic attack on the rRNA in the intact ribosomal subunits. This model, which resulted from a series of studies on the turnover of ribosomes in various mutant strains of *E. coli* during deprivation of various nutrients essential for growth (13, 14), suggests that a common mechanism operates during the turnover of rRNA under various starvation conditions. These studies implied that the turnover process is initiated by an endonucleolytic attack on rRNA in ribosomal subunits. It is interesting to note that, although decay of ribosomes is an adaptive response to nutrient limitation, it is rather unlikely that the enzymes involved in this adaptation are induced by the starvation. The process of ribosome decay is very efficient and economical in the sense that when 20% of the total RNA is degraded 20% of all the ribosomes are degraded rather than 20% of each rRNA molecule. Thus, during starvation one observes either complete or fully degraded ribosomes, and the cell is able to recover efficiently from starvation when nutrients are available again.

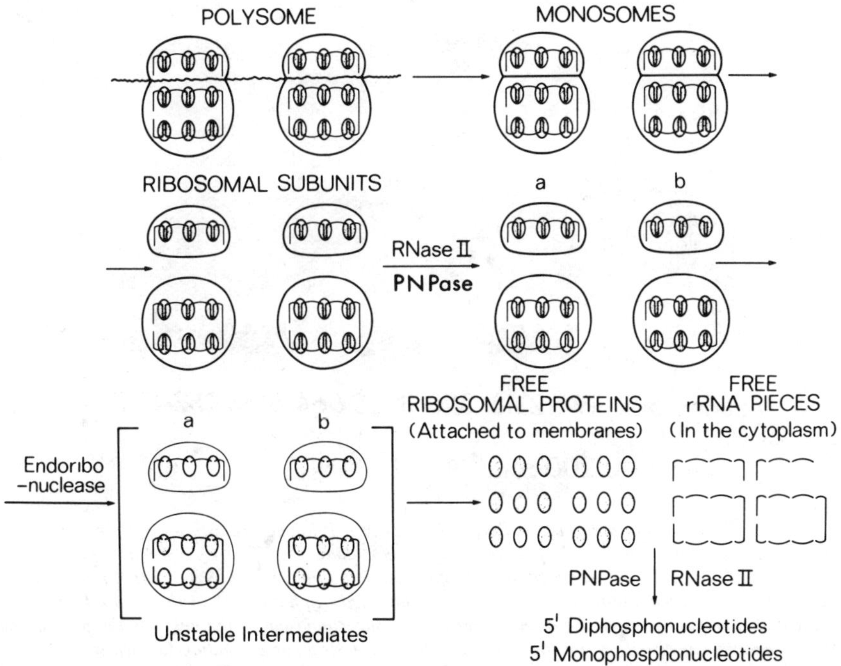

FIG. 4. *Model for disintegration of ribosomes during carbon starvation. Note that pathways a and b are identical, except that in b the 3′ end is missing from the 16S rRNA; this could result from an attack on such ends by either RNase II or polynucleotide phosphorylase. The unstable intermediates resulting from the endonucleolytic attack on the ribosomal subunits are shown in brackets since they were not observed. The small ellipsoids designate ribosomal proteins bound to double-stranded regions of the rRNA.*

Some detailed analyses of the process (13) and a review of current literature about RNases in *E. coli* (2, 20) indicated that the enzyme RNase III, which endonucleolytically degrades double-stranded RNA (23), could have been the enzyme which initiates the endonucleolytic attack on the rRNA in the ribosome during starvation. However, when isogenic RNase III$^+$ and RNase III$^-$ strains were compared, it was found that degradation of rRNA during carbon starvation was not reduced in RNase III$^-$ strains (3, 4). To understand the lacuna of the process, we started to search for a new RNase(s) in cell extracts which could endonucleolytically attack rRNA,

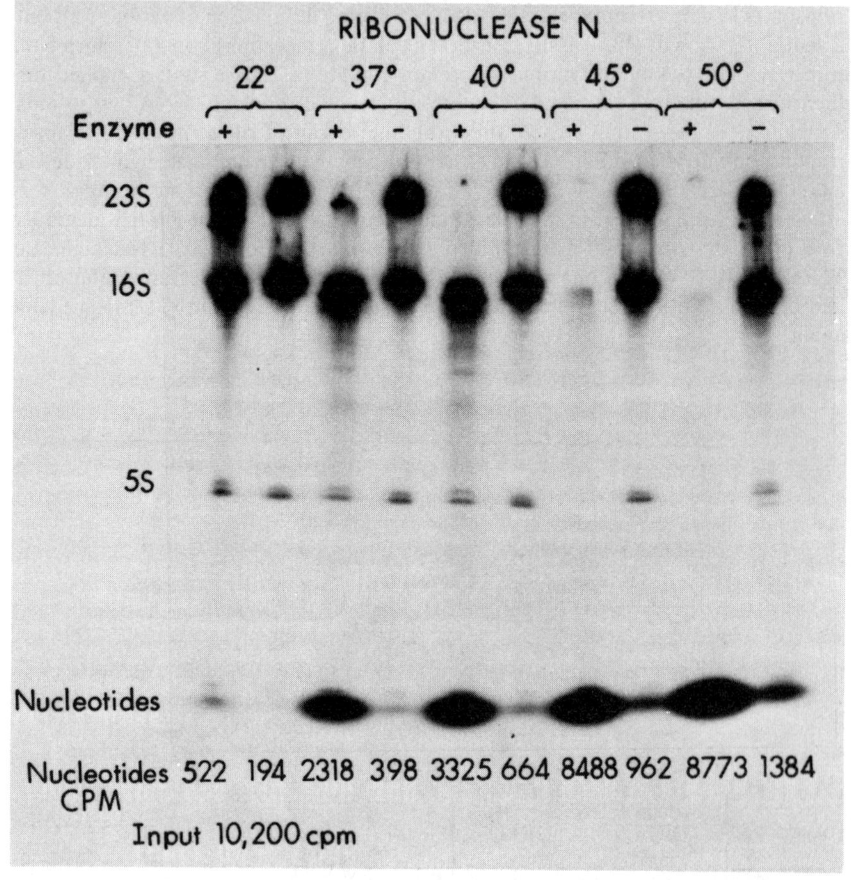

FIG. 5. *Temperature optimum of RNase N activity. ^{32}P-labeled ribosomes were incubated by themselves or with an RNase N preparation. Sodium dodecyl sulfate (0.2%) was added to the mixture, and samples were subjected to electrophoresis in a tandem 5%/15% (5% top, 15% bottom) gel. The gel was dried, autoradiographed, and quantitated. A photograph of the autoradiograph is presented. For quantiattion, the autoradiogram was superimposed on the gel, and the position of the nucleotides, as visualized on the film, was marked. A strip corresponding to each slot was cut, and the areas containing nucleotides, as well as the rest of the area in each lane, were cut and counted in a toluene-based scintillation fluid. Background values, i.e., counts, in similar gel areas in which no radioactive compounds have been separated, were subtracted. ^{32}P-labeled ribosome substrate (20 µg of RNA) was incubated by itself or with enzyme (2 µg) in 0.01 M potassium phosphate buffer, pH 7.5, containing 0.1 M KCl and 1 mM magnesium acetate. The reaction was terminated after 90 min.*

preferentially in ribosomes. We found that at least one and perhaps two such enzymes are present in *E. coli* (19). One cuts rRNA in a restricted number of sites and the other, RNase N, degrades rRNA to 5′-mononucleotides (18, 19). RNase N activity has been purified and freed from other known RNases. It is a nonspecific endoribonuclease capable of degrading various single-stranded and double-stranded RNA molecules.

When the activity of RNase N against rRNA in ribosomes was tested in vitro, the ribosomes were indeed degraded much more efficiently by RNase N at 45 and 50°C than at 22, 37, or 40°C (Fig. 5). This suggests that RNase N can attack the ribosomes during starvation in vivo at an elevated temperature. Thus, the action of RNase N on ribosomes shows parallels to the decay of the ribosome in vivo, where the process is very much accelerated at 45 and 50°C as compared to 30 and 37°C (12–14).

Studies on the localization of RNase N showed that it is an intracellular, rather than a periplasmic, enzyme. This was determined by comparing the distribution in the cell of RNase N with the bona fide intracellular β-galactosidase and the periplasmic alkaline phosphatase (18). Since our studies suggested that the enzyme is an endoribonuclease (18, 19), it was compared to RNase III, the only other known intracellular endorinbonuclease in *E. coli.* Using homopolymers and copolymers, we determined that, whereas RNase III could digest only double-stranded RNA in the presence of Mg^{2+}, RNase N digested single-stranded and double-stranded RNA with similar efficiency in the presence or absence of Mg^{2+} (Table 1). Moreover, RNase N was active against rRNA in ribosomes, but RNase III was not. Furthermore, all RNAs used, natural as well as synthetic, were substrates for RNase N. Using 5S rRNA as well as other substrates, we confirmed that the enzyme is indeed an endonuclease. The final products of the reaction of this enzyme are 5′-mononucleotides. The molecular weight of the enzyme is about 120,000, and it seems to contain two subunits which are similar in size. These properties thus differentiate this enzyme from all other known *E. coli* RNases (2, 20).

These studies show that RNase N is the most likely candidate to be the enzyme

TABLE 1. *Comparison of RNase N and RNase III activities*[a]

Enzyme	Mg^{2+} (mM)	Amt (cpm) disappearing from the stacking gel			Amt (cpm) appearing in the bottom 1-cm portion of the gel		
		[^{3}H]-poly(C)·poly(I)	[^{3}H]-poly(C)	[^{32}P]RNA in ribosomes	[^{3}H]-poly(C)·poly(I)	[^{3}H]-poly(C)	[^{32}P]RNA in ribosomes
RNase III	5	5,056	28	151	1,240	14	47
RNase N	5	6,028	7,645	7,169	2,196	3,633	1,919
RNase III	0	42	11	63	117	7	23
RNase N	0	6,061	9,149	7,502	2,400	4,093	4,902

[a] The substrates, [^{3}H]polycytidylic-polyinosinic acid (1.5:3 μg), [^{3}H]polycytidylic acid (1 μg), and ^{32}P-labeled rRNA (20 μg) in ribosomes, were incubated with 0.2 μg of RNase III and 2 μg of RNase N for 60 min at 37°C in 0.01 M Tris-hydrochloride (pH 7.5), 0.1 M NaCl, 0.08 M NH_4Cl, 0.005 M $MgCl_2$, 0.1 mM EDTA, and 0.1 mM dithioerythritol in a total volume of 50 μl in one set of experiments (first two lines). In the other set of experiments $MgCl_2$ and EDTA were excluded (last two lines). After termination of reactions, samples were applied (20,000 cpm) to a 5%/12% tandem thin-slab (1.5 mm) polyacrylamide gel, and electrophoresis was carried out. After electrophoresis, the gel was dried, the lanes were cut with tracking dye used as the marker, and the gel strips were sliced into 0.5-cm pieces, digested with hydrogen peroxide, and counted. Values were corrected for background (no enzyme incubation), and the totals were normalized.

which starts the decay of ribosomes when the cells stop growing exponentially. Therefore, this kind of enzyme, which probably exists in every type of cell, can help in degrading the RNA in cells whose protein has to be used for feeding. To ensure that the degradation is efficient, it is necessary to grow the cultures on limiting nutrients, so that the cells are starved when the culture attains a high level of density. After starvation has been reached, the temperature of the culture can be elevated, to incur an efficient decay of the RNA. Once the nucleotides are autodigested (see Fig. 1), the protein can be precipitated and thereby easily separated from the nucleotides. The nucleotides can be also used for other purposes and, by using appropriate mutants, could be shunted in a desired direction, such as the production of a particular nucleotide.

Moreover, the construction of strains lacking specific RNases could be very helpful for the expression of genes carried on recombinant DNAs, especially when they came from other procaryotic and simple eucaryotic organisms where the problem of intervening DNA sequences does not exist.

Basic studies on RNases and physiological processes of the ribosome and RNA turnover can therefore help to solve some practical problems concerning the elimination of nucleic acids from single-cell proteins.

ACKNOWLEDGMENTS

This work was supported by National Science Foundation grant PCM-76-81665.

I am much indebted to previous and present colleagues, in particular to Ruth Kaplan and Tapan K. Mirsa, who labored on the various aspects of the research summarized here.

REFERENCES

1. **Apirion, D.** 1973. Degradation of RNA in *Escherichia coli:* a hypothesis. Mol. Gen. Genet. **122:**313–322.
2. **Apirion, D.** 1974. The fate of mRNA and rRNA in *Escherichia coli. In* J. J. Dunn (ed.), Processing of RNA. Brookhaven Symp. Biol. **26:**286–306.
3. **Apirion, D., J. Neil, and N. Watson.** 1976. Consequences of losing ribonuclease III on the *Escherichia coli* cell. Mol. Gen. Genet. **144:**185–190.
4. **Apirion, D., and N. Watson.** 1974. Analysis of an *Escherichia coli* strain carrying physiologically compensating mutations one of which causes an altered ribonuclease III. Mol. Gen. Genet. **132:**89–104.
5. **Bailey, S. C., and D. Apirion.** 1977. Identification of lipopolysaccharides and phospholipids of *Escherichia coli* in polyacrylamide gels. J. Bacteriol. **131:**347–355.
6. **Cammack, K. A., and H. G. Wade.** 1965. Sedimentation behaviour of ribonuclease active and inactive ribosomes from bacteria. Biochem. J. **96:**671–680.
7. **Caras, M. A., S. C. Bailey, and D. Apirion.** 1977. The DNA of *Escherichia coli* can be displayed in a single band in polyacrylamide gels. FEBS Lett. **74:**283–286.
8. **Doerr, T. D., and D. Apirion.** 1978. Noncoordinate control of RNA, lipopolysaccharide, and phospholipid syntheses during amino acid starvation in stringent and relaxed strains of *Escherichia coli.* J. Bacteriol. **135:**274–277.
9. **Gegenheimer, P., and D. Apirion.** 1975. *Escherichia coli* ribosomal RNAs are not cut from an intact precursor molecule. J. Biol. Chem. **250:**2407–2409.
10. **Gegenheimer, P., N. Watson, and D. Apirion.** 1977. Multiple pathways for primary processing of ribosomal RNA in *Escherichia coli.* J. Biol. Chem. **252:**3064–3073.
11. **Gesteland, R. F.** 1966. Isolation and characterization of ribonuclease I mutants of *Escherichia coli.* J. Mol. Biol. **16:**67–84.
12. **Kaplan, R., and D. Apirion.** 1974. The involvement of ribonuclease I, ribonuclease II and polynucleotide phosphorylase in the degradation of stable RNA during carbon starvation in *E. coli.* J. Biol. Chem. **249:**149–151.
13. **Kaplan, R., and D. Apirion.** 1975. The fate of ribosomes in *Escherichia coli* cells starved for a carbon source. J. Biol. Chem. **250:**1854–1863.
14. **Kaplan, R., and D. Apirion.** 1975. Decay of ribosomal RNA in *Escherichia coli* cells starved for various nutrients. J. Biol. Chem. **250:**3174–3178.
15. **Krishna, R. V., and D. Apirion.** 1973. Polynucleotide phosphorylase has a role in growth of *Escherichia coli.* J. Bacteriol. **113:**1235–1239.

16. **Krishna, R. V., L. Rosen, and D. Apirion.** 1973. Increased inactivation and degradation of messenger RNA in an *Escherichia coli* strain containing a thermolabile polynucleotide phosphorylase. Nature (London) New Biol. **242:**18–20.
17. **Lee, S. Y., S. C. Bailey, and D. Apirion.** 1978. Small stable RNAs from *Escherichia coli*: evidence for the existence of new molecules and for a new ribonucleoprotein particle containing 6S RNA. J. Bacteriol. **133:**1015–1023.
18. **Misra, T. K., and D. Apirion.** 1978. Characterization of an endoribonuclease, RNase N, from *Escherichia coli.* J. Biol. Chem., **253:**5594–5599.
19. **Misra, T. K., S. Rhee, and D. Apirion.** 1976. A new endoribonuclease from *Escherichia coli*: ribonuclease N. J. Biol. Chem. **251:**7669–7674.
20. **Niyogi, S. K., and A. K. Datta.** 1975. A novel oligoribonuclease of *Escherichia coli.* I. Isolation and properties. J. Biol. Chem. **250:**7307–7312.
21. **Reiner, A. M.** 1969. Isolation and mapping of polynucleotide phosphorylase mutants of *Escherichia coli.* J. Bacteriol. **97:**1431–1436.
22. **Reiner, A. M.** 1969. Characterization of polynucleotide phosphorylase mutants of *Escherichia coli.* J. Bacteriol. **97:**1437–1443.
23. **Robertson, H. D., R. E. Webster, and N. D. Zinder.** 1968. Purification and properties of ribonuclease III from *Escherichia coli.* J. Biol. Chem. **243:**82–91.
24. **Talbot, J. H., and T. F. Yu.** 1976. Gout and uric acid metabolism. Stratton Intercontinental Medical Book Corp., New York,
25. **Weatherford, S. C., L. Rosen, L. Gorelic, and D. Apirion.** 1972. *Escherichia coli* strains with thermolabile ribonuclease II activity. J. Biol. Chem. **247:**5404–5408.
26. **Wyngaarden, J. B., and W. N. Kelley.** 1976. Gout and hyperuricemia. Grune and Stratton, New York.

Energy Production, Growth, and Product Formation by Microorganisms

A. H. STOUTHAMER

Department of Microbiology, Biological Laboratory, Free University, 1007 MC Amsterdam, The Netherlands

During growth of microorganisms, the carbon source in the medium may be used for three purposes: energy production, growth, and product formation. The energy produced is used for growth and maintenance purposes. Product formation may be associated with either energy utilization or production. A mathematical model giving the relation among these various processes has been presented (2). Most of our knowledge on the relation between growth and energy production arose from studies in which product formation did not occur (13, 14, 17).

THEORETICAL STUDIES ON THE ATP REQUIREMENT FOR BIOMASS FORMATION

In theoretical calculations of the ATP requirement for the formation of cell material, the macromolecular composition of the cells is taken as a base. Such calculations show that the amount of ATP required for the formation of cell material is dependent on the nature of the carbon source (12, 14), the complexity of the medium (12, 14), the nature of the nitrogen source (14, 15), and the effect of the assimilation pathway of the substrate (5, 12). The formation of cell material in a minimal medium requires 34.8, 73.8, 64.4, 99.5, and 153.8 mmol of ATP/g of cell material when glucose, lactate, malate, acetate, and carbon dioxide are the carbon sources, respectively. The high ATP requirements for biomass formation during growth with simple substrates are due to a larger requirement for monomer formation and for transport processes than during growth on glucose. In most cases, there is a large discrepancy between the theoretical and the experimental ATP requirements for the formation of biomass (14); in general, the amount of biomass formed per mole of ATP is only half of the amount calculated theoretically.

MAINTENANCE ENERGY AND ENERGETIC COUPLING

Growing bacteria, like all living organisms, require a certain amount of energy for maintenance processes. Maintenance energy is required for the turnover of cellular constituents, the preservation of the right ionic composition and intracellular pH of the cell, and the maintenance of a pool of intracellular metabolites against a concentration gradient. When it is assumed that during growth the consumption of the energy source is partly growth dependent and partly growth independent, a relation between molar growth yield and specific growth rate can be derived (8):

$$\frac{1}{Y_{sub}} = \frac{m_s}{\mu} + \frac{1}{Y_{sub}^{MAX}}$$

where Y_{sub} is the molar growth yield for a substrate (grams/mole), m_s is the maintenance coefficient (millimoles of substrate per gram of dry weight per hour),

μ is the specific growth rate, and Y_{sub}^{MAX} is the molar growth yield for substrate corrected for energy of maintenance. Similar equations can be derived for ATP and O_2:

$$\frac{1}{Y_{O_2}} = \frac{m_o}{\mu} + \frac{1}{Y_{O_2}^{MAX}}$$

$$\frac{1}{Y_{ATP}} = \frac{m_e}{\mu} + \frac{1}{Y_{ATP}^{MAX}}$$

These parameters are normally determined by studying the influence of the specific growth rate on the molar growth yields in continuous cultures. Under anaerobic conditions Y_{ATP}^{MAX} and m_e can be determined, since the catabolic pathways for anaerobic breakdown of substrates are known and the ATP yields can be calculated. Anaerobic experiments are therefore ideally suited to identify the environmental factors which influence these parameters. Generally, Y_{ATP}^{MAX} and m_e cannot be calculated from aerobic experiments, since the P/O ratio in growing microorganisms is not known with certainty. Methods to determine the P/O ratio in growing cultures are discussed in a recent review (14).

Under a number of conditions, the growth yields are much lower than expected on basis of the ATP yield. The term "uncoupled growth" has been used to denote such growth conditions (10). Uncoupled growth occurs (14): (i) in minimal media, (ii) in the presence of inhibitory compounds, (iii) in media which contain suboptimal amounts of an essential growth factor, (iv) at temperatures above the optimum, and (v) with compounds as carbon and energy sources in which growth is not energylimited. Under the first four conditions, the molar growth yield, Y_{ATP}, and the specific growth rate decrease. Consequently, a larger proportion of the energy source is used for maintenance purposes. Furthermore, the actual magnitude of m_e can also be altered. Indeed, it has been demonstrated that m_e is increased by growth at temperatures above the optimum for various organisms (4, 6, 7, 18, 19). Under these conditions, the rate of ATP production is in excess of the rate at which ATP is required for growth, and energy-spilling mechanisms occur (10, 14).

CHARACTERIZATION OF THE PRODUCTION PROCESS OF VARIOUS CLASSES OF MICROBIAL PRODUCTS

Various classes of low-molecular-weight fermentation products (Table 1) can be distinguished.

TABLE 1. *Characterization of low-molecular-weight fermentation products and their production process*

Class of metabolite	Examples	Coupling of growth to energy production
End products of energy metabolism	Ethanol, acetic acid, gluconic acid, lactic acid, acetone, butanol, etc.	Coupled
Intermediates of primary metabolism	Amino acids, citric acid, nucleotides	Uncoupled
Products of secondary metabolism	Antibiotics, e.g., penicillin	Coupled

Formation of end products of energy-producing pathways. End products of energy-producing pathways are thought to be formed mainly during growth which is characterized by strong coupling between growth and energy production. Part of these processes are carried out under anaerobic conditions with which the molar growth yields are low, or in rich media, in which the carbohydrate is mainly used as an energy source (13, 14). The production of acetic and gluconic acids is carried out under aerobic conditions by acetic acid bacteria, which are characterized by a low efficiency of energy conversion (11). It is evident that for the formation of these products no free choice of the carbon source as starting material is possible. The yield of product can be increased by a further decrease in the growth yield. This can be achieved by performing the production process at a temperature above the optimum, which causes an increase in m_e.

Formation of products of intermediary metabolism. The formation of large amounts of products of intermediary metabolism occurs mostly with organisms which carry mutations or which grow under conditions of nutritional stress. Both approaches lead to abnormal regulation of metabolic pathways and result in an overproduction of the desired metabolite. As an example, the formation of lysine is treated in this fashion in Fig. 1. Four moles of reduced nicotinamide adenine dinucleotide are formed whereas 1 mol of ATP is consumed per mol of lysine formed. It is thus evident that lysine formation is associated with a net production of ATP. With two or three phosphorylation sites this net ATP production amounts to 7 or 11 mol, respectively, per mol of lysine formed. In the example in Fig. 1, it is assumed that half of the glucose is converted into lysine and the other half is used for biomass formation. The biomass formed from this 0.5 mol of glucose will be higher than in the absence of lysine formation, since the ATP production associated with lysine formation may be used for biomass formation. This is illustrated by the fact that the Y_{glu} value for the case when lysine is formed is more than half the value when no lysine is formed (Fig. 1). The influence of lysine formation on the Y_{O_2} values is very small, which emphasizes that the combination of Y_{glu} and Y_{O_2} must always be considered (15).

Mass balance for lysine formation:

$$15C_6H_{12}O_6 + 2NH_3 \rightarrow C_6H_{14}O_2N_2 + 3CO_2 + H_2O + 4\text{“}H_2\text{”}$$

ATP consumption, 1 mol/mol of lysine

Mass balance for biomass formation:

$$C_6H_{12}O_6 + 1.4NH_3 + 0.25\text{“}H_2\text{”} \rightarrow C_6H_{10.84}N_{1.4}O_{3.07} + 2.93H_2O$$

	Two sites		Three sites	
	Y_{gly}	Y_{O_2}	Y_{glu}	Y_{O2}
Lysine formation	64.9	45.1	73.5	66.9
No lysine formation	100.9	52.2	112.3	76.1

FIG. 1. *Calculation of growth parameters for a lysine-producing organism, growing on glucose. It is assumed that 50% of the glucose is converted into lysine and that the remaining glucose is used for growth. The "molecular weight" of microbial cells is 152. Formation or utilization of reducing equivalents is denoted by "H_2." The calculations are given for the experimental value for Y_{ATP} of 12(3). Furthermore, the presence of two or three phosphorylation sites has been assumed. The values for the control (no lysine formation) are given for comparison and are cited from reference 7, which also gives more details about the calculations.*

In these production processes sugar concentrations as high as 1 M are used. For an experimental Y_{ATP} value of 12 for coupled growth, the biomass concentration would be about 65 g/liter for two phosphorylation sites. These concentrations are much higher than the actual ones. This example shows that a very strong uncoupling between growth and energy production must occur under these circumstances. The example given above is not an isolated case, but represents the general case for the production of metabolites of intermediary metabolism. A very important conclusion is, therefore, that such production processes are made possible by the occurrence of energy-spilling mechanisms. In these processes the carbon source can be chosen more freely than in the case of end products of energy-generating pathways. The only essential condition is that the conversion of the carbon source to the precursors of the desired metabolite is sufficiently rapid.

Formation of products of secondary metabolism. The formation of secondary metabolites has not been thought to be associated with growth. In general, the production process is divided into a phase of rapid growth (trophophase), in which no secondary metabolites are formed, followed by a subsequent production stage (idiophase) (3), but it is questionable whether this division is always that sharp. As an example, penicillin fermentation will be considered (16).

The formation of penicillin has been studied in continuous cultures (9). It was found that the specific rate of its production (q_{pen} = units per milligram per hour) is independent of the growth rate above $\mu = 0.014$/h. Below this value of μ, a decrease of penicillin production was observed (9). In the following calculations a decrease in q_{pen} of 1.5% per h was assumed below $\mu = 0.014$/h. A Y_{glu}^{MAX} of 81 g/mol and an m_s value of 0.1222 mmol of glucose/g per h were used (9). In the earlier publications a q_{pen} between 1.5 and 2.0 was observed (9), but recently a q_{pen} of about 12 was mentioned (1). This difference is most probably due to the effects of strain improvement.

Penicillin production is carried out mostly in batch cultures. A constant feed of a mol of glucose/liter per h is assumed. For the amount of glucose consumed for penicillin production, the following mass balance reaction is used:

$$1.5C_6H_{12}O_6 + 2NH_3 + H_2SO_4 + \text{“}H_2\text{”} \rightarrow C_8H_{12}O_3N_2S \text{ (aminopenicillanic acid)} + CO_2 + 8H_2O$$

It is reasoned that the side chain of the penicillin is completely formed from the precursor added to the fermentation medium. The glucose balance is thus given by the equation

$$ds/dt = (ds/dt)_{growth} + (ds/dt)_{maintenance} + (ds/dt)_{penicillin} = a = (ds/dt)_{growth} + mx + 1.5q_p^1 \cdot x$$

in which q_p^1 is the specific rate of penicillin production in moles of penicillin per gram per hour and x is the dry weight of the organism (grams per liter). Consequently, $(ds/dt)_{growth} = a - mx - 1.5q_p^1 \cdot x$. Therefore, growth is given by the equation

$$dx/dt = (a - mx - 1.5q_p^1 \cdot x)Y_{glu}^{MAX}$$

The results of a calculation of the biomass, the rate of penicillin production, and the penicillin titer are shown in Fig. 2, and from these data the amount of sugar used for biomass formation, maintenance, and penicillin formation can be calculated. The result shown in Fig. 3 demonstrates that the amounts of sugar utilized for maintenance purposes are large, amounting to about 70% at the end of the period considered. The impression that the penicillin process can be divided into a phase of rapid growth

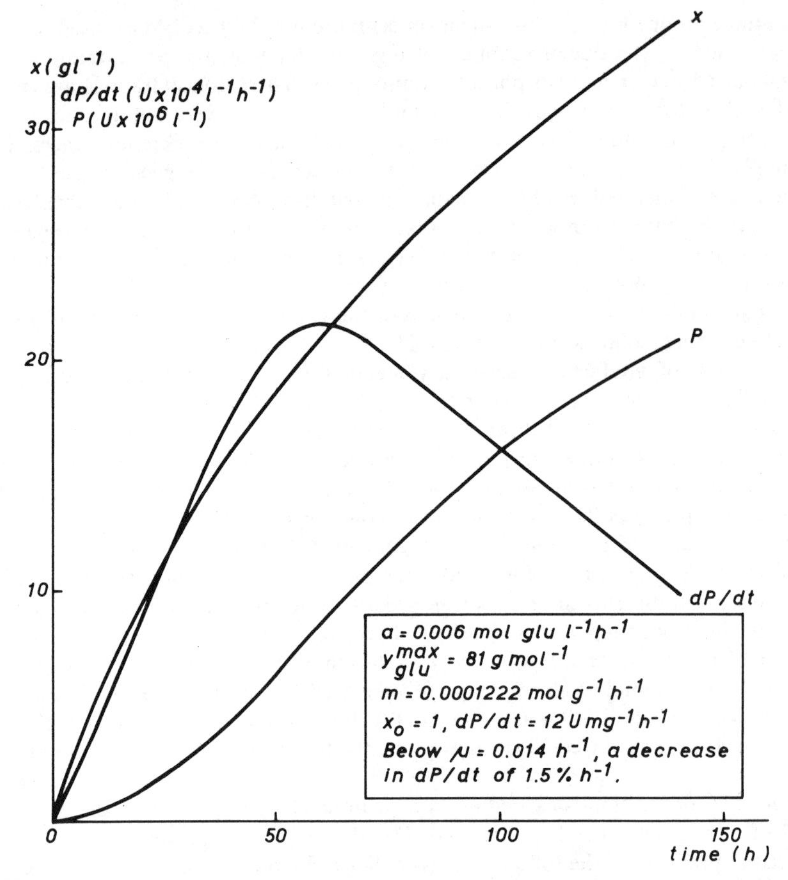

FIG. 2. *Penicillin production in a batch culture. The mycelium concentration* (x), *the rate of penicillin production* (dP/dt), *and the penicillin titer*, P, *expressed in units per liter, are given as a function of time. The constants used for these calculations are given in the inset.*

and a phase of penicillin production is just a consequence of this fact. Therefore, I conclude that the production of penicillin is carried out by growing organisms and that the growth process is characterized by a strong coupling between growth and energy production. Whereas the yield of an end product of energy-producing pathways will increase when the maintenance coefficient increases, the yield of penicillin will decrease with a higher maintenance coefficient. The formation of penicillin is also different from the formation of other types of fermentation products with respect to the influence of the carbon source. Carbohydrates are the preferred carbon sources for the formation of penicillin. The formation of penicillin from simpler substrates requires too much energy and will therefore result in low yields.

CONCLUSIONS

Considerations on the relation among energy production, growth, and product formation lead to a more accurate description of the characteristics of the production processes of various classes of fermentation products. The procedure using material

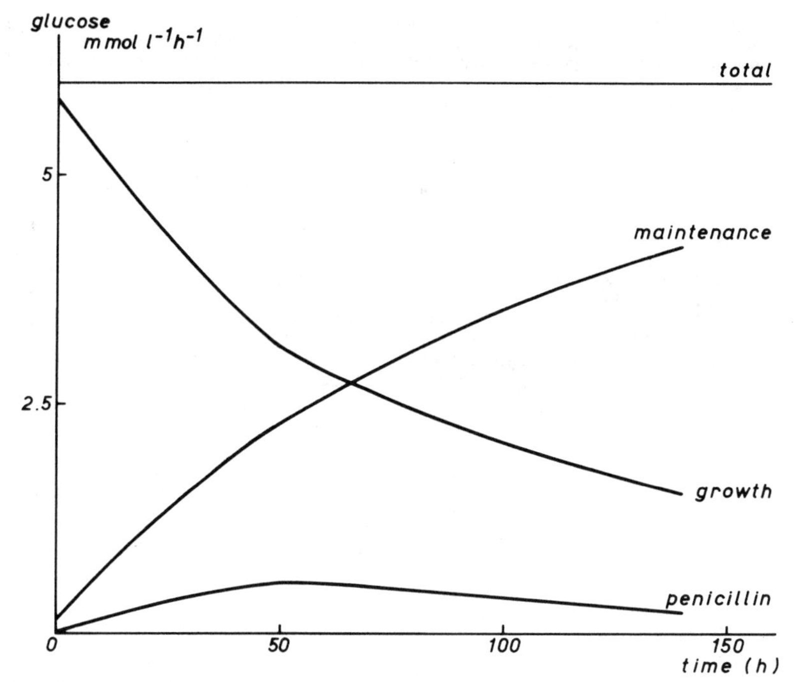

FIG. 3. *Glucose utilization in penicillin fermentation. Division of glucose over growth, maintenance, and penicillin formation in a batch culture. Calculated from the data in Fig. 2.*

balances and mathematical modeling gives indications about: (i) the way to search for improvements in production processes, (ii) the processes which are changed in genetic variants with a higher production, and (iii) situations in which no improvements can be achieved by mutagenesis.

REFERENCES

1. **Calam, C. T., and D. W. Russell.** 1973. Microbial aspects of fermentation process development. J. Appl. Chem. Biotechnol. **23:**225–237.
2. **de Kwaadsteniet, J. W., J. C. Jager, and A. H. Stouthamer.** 1976. A quantitative description of heterotrophic growth in micro-organisms. J. Theor. Biol. **57:**103–120.
3. **Demain, A. L.** 1971. Overproduction of microbial metabolites and enzymes due to alteration of regulation. Adv. Biochem. Eng. **1:**113–142.
4. **Farmer, J. S., and C. W. Jones.** 1976. The effect of temperature on the molar growth yield and maintenance requirement of *Escherichia coli* W during aerobic growth in continuous culture. FEBS Lett. **67:**359–363.
5. **Harder, W., and J. P. van Dijken.** 1976. Theoretical considerations on the relation between energy production and growth of methane-utilizing bacteria, p. 403–418. *In* H. G. Schlegel, N. Pfennig, and G. Gottschalk (ed.), Microbial production and utilization of gases (H_2, CH_4, CO). E. Goltze Verlag, Göttingen.
6. **Mainzer, S. E., and W. P. Hempfling.** 1976. Effects of growth temperature on yield and maintenance during glucose-limited continuous culture of *Escherichia coli.* J. Bacteriol. **126:**251–256.
7. **Palumbo, S. A., and L. D. Witter.** 1969. Influence of temperature on glucose utilization by *Pseudomonas fluorescens.* Appl. Microbiol. **18:**137–141.
8. **Pirt, S. J.** 1965. The maintenance energy of bacteria in growing cultures. Proc. R. Soc. London Ser. B **163:**224–231.
9. **Pirt, S. J., and R. C. Righelato.** 1967. Effect of growth rate on the synthesis of penicillin by *Penicillium chrysogenum* in batch and chemostat cultures. Appl. Microbiol. **15:**1284–1290.
10. **Senez, J. C.** 1962. Some considerations on the energetics of bacterial growth. Bacteriol. Rev. **26:**95–107.

11. **Stouthamer, A. H.** 1969. Determination and significance of molar growth yields, p. 629–663. *In* J. R. Norris and D. W. Ribbons (ed.), Methods in microbiology, vol. 1. Academic Press, London.
12. **Stouthamer, A. H.** 1973. A theoretical study on the amount of ATP required for synthesis of microbial cell material. Antonie van Leeuwenhoek J. Microbiol. Serol. **39:**545–565.
13. **Stouthamer, A. H.** 1976. Yield studies in micro-organisms. Meadowfield Press Ltd., Durham, England.
14. **Stouthamer, A. H.** 1977. Energetic aspects of the growth of micro-organisms. Symp. Soc. Gen. Microbiol. **27:** 285–315.
15. **Stouthamer, A. H.** 1977. Theoretical calculations on the influence of the inorganic nitrogen source on parameters for aerobic growth of micro-organisms. Antonie van Leeuwenhoek J. Microbiol. Serol. **43:** 351–367.
16. **Stouthamer, A. H.** 1977. Penicilline na ongeveer 50 jaar onderzoek. Versl. Afd. Natuurk. Kon. Ned. Ac. v. Wet. **86:**134–138.
17. **Thauer, R. K., K. Jungermann, and K. Decker.** 1977. Energy conservation in chemotrophic anaerobic bacteria. Bacteriol. Rev. **41:**100–180.
18. **Topiwala, H., and C. G. Sinclair.** 1971. Temperature relationship in continuous culture. Biotechnol. Bioeng. **13:**795–813.
19. **Van Uden, N., and A. Madeira-Lopes.** 1976. Yield and maintenance relations of yeast growth in the chemostat at superoptimal temperatures. Biotechnol. Bioeng. **18:**791–804.

Contribution of Genetics to the Biosynthesis of Antibiotics

J. NÜESCH

Pharmaceutical Division, Ciba-Geigy Ltd., Basel, Switzerland

The biosynthesis of antibiotics is characterized by a complicated, balanced polygenic determination. Figure 1 illustrates the main biochemical and physiological factors essential for the formation of these substances. The majority of antibiotic-producing microorganisms are haploids. Very few have complete sexual cycles. Elaborate genetic systems are not available for any of the important antibiotic-producing strains. From the beginning of the antibiotic era, research on antibiotic-producing microorganisms has been largely concentrated in industry. The main effort has been, and still is, of an applied nature and is governed largely by economic considerations. The industrial microbiologist has to solve the problem of finding high-producing strains which perform under the most favorable technical and economic conditions. The strains he discovers are frequently replaced by new ones, and his success is not measured so much by the increase in basic knowledge as by the increase in titer he achieves. Without any doubt, mutation and selection have proved to be extremely successful in this endeavor. Whereas classical Mendelian genetics played a most important role in plant breeding, yielding practical results and a good deal of fundamental knowledge, the development of antibiotic-synthesizing microbial strains, owing to the very nature of the empirical approach, yielded nothing comparable. In both cases the primary objectives were the same; the difference in the approach is obviously due to the difference in the genetic systems.

As far as industrial microorganisms are concerned, it seems highly probable that mutation and selection will continue to be the most important means for obtaining improved strains in the shortest possible time and under optimal economic conditions. For several years, attempts have been made to complement the purely empirical approach of mutation and selection with studies aiming at a comprehension of the biosynthetic mechanisms in the hope of improving and rationalizing the development of antibiotic-producing strains. The success of these activities depended greatly on the application of isotopic labeling of possible precursors coupled with improved detection techniques, especially in connection with carbon-13 (4), the isolation of blocked mutants, and the analysis of their products. Finally, the enzymatic reactions connected with the biosynthesis of the antibiotic were often analyzed. As a result of the difficulties which were frequently encountered in isolating functional enzyme systems of the secondary metabolism, protoplasts are being successfully employed with an increasing frequency (1, 10). Up to the present, formal genetics has played only a minor role (5, 11, 15), and its contribution to our understanding of the quantitative and qualitative aspects of antibiotic biosynthesis can be summarized as follows (after Elander et al. [5]):

- Yield improvement
  - Mutation and selection
  - Genetic recombination
- Biosynthetic modification
  - Mutational biosynthesis
  - Blocked mutants and product cosynthesis

Mutants for elucidation of pathways
  Blocked mutants
  Regulatory mutants

YIELD IMPROVEMENT

Whereas the improvement of the productivity of a strain has been achieved predominantly by empirical methods of mutation and selection, the biochemical analysis of high-producing strains and their ancestors has shown that the efficiency of secondary metabolism depends on the optimal availability of primary metabolites as precursors. Phenomena of overproduction of primary metabolites due to the deregulation of pathways have been observed in penicillin- and cephalosporin-producing fungi (2, 5, 11, 17). An investigation of the origin of the seven-carbon amino unit of the rifamycin chromophore (see Fig. 2) with mutants blocked in the biosynthesis of the aromatic amino acids (*aro*) showed that in the procaryote *Nocardia mediterranei* a similar regulation pattern must assure the optimal flow of the precursor (6, 7). A mutant blocked in the transketolase and another presumably deficient in its shikimate kinase (see Fig. 3) accumulate D-ribulose and shikimate, respectively. The respective amounts produced correspond stoichiometrically to the production of rifamycin B in the parent strain. These results are not only of fundamental interest but also offer possibilities for improving the productivity of strains by investigating the regulation of those parts of the primary metabolism which are directly involved in the synthesis of precursors of a given antibiotic.

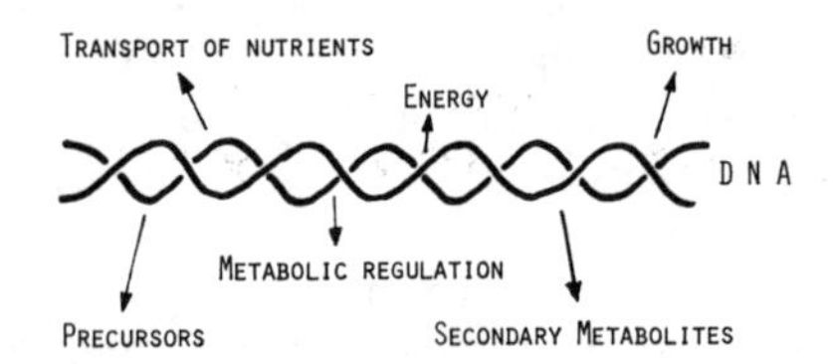

FIG. 1. *Metabolic determinants of secondary metabolites.*

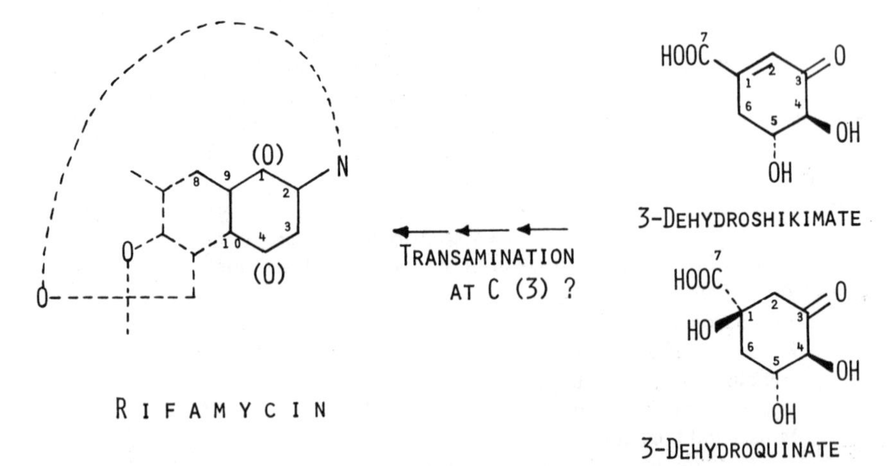

FIG. 2. *Possible precursors of the seven-carbon amino unit of rifamycin.*

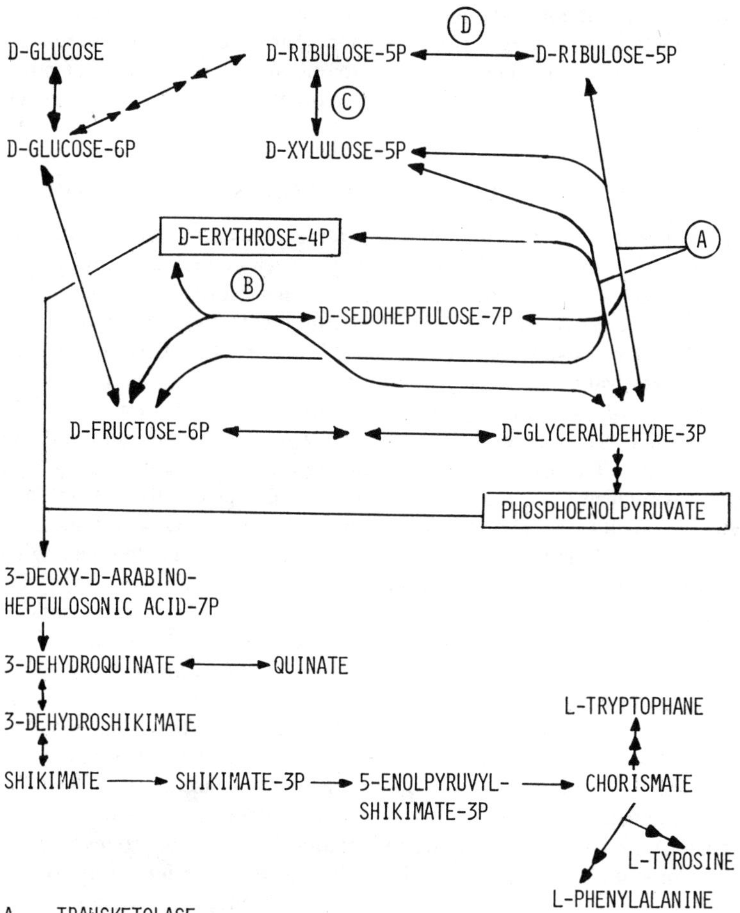

FIG. 3. *Biosynthetic pathways: pentose shunt, glycolysis, and shikimate pathway.*

BIOSYNTHETIC MODIFICATION

Mutations that interfere directly with enzymes of the secondary metabolism may lead to the accumulation of intermediates or to modified products (5, 11). The fact that many strains produce a spectrum of related compounds may be due to the existence of a spectrum of functionally slightly different enzymes of the pathway or to the relatively low specificity of the enzymes involved. Furthermore, in the aminoglycosides, as suggested by Nara (16), modifying enzymes may lead to temporary storage of the modified aminoglycosides in the cells and to excretion only after a second modification. It is obvious that in such cases mutations can lead to a variety of novel products. The formation of *N*-acetyldeacetoxycephalosporin C may be interpreted in a similar way (19). It has been shown that deacetoxycephalosporin C, unlike cephalosporin C, is accumulated in the cells and might therefore be subject to further transformations. Thanks to the relatively low specificity of enzymes of the secondary metabolism, in particular in actinomycetes, a combination of mutants and modified structural components of an antibiotic allow the synthesis of novel substances. This method is called mutational biosynthesis or mutasynthesis and has proved useful.

MUTANTS FOR ELUCIDATION OF BIOSYNTHETIC PATHWAYS

The application of blocked mutants for the elucidation of pathways for the biosynthesis of antibiotics is of great importance. Such mutants are also a prerequisite for a genetic investigation. The technique has been successfully used particularly for investigating the biosynthesis of cephalosporin and tetracycline, and most of this field has been extensively reviewed (5, 9, 11, 12). Its usefulness is further illustrated by recent investigations of rifamycin biosynthesis (6–8). Whereas the ansa chain of rifamycins and of other ansamycins such as streptovaricins is derived from acetate, propionate, and methionine, the origin of a seven-carbon amino unit including C1 to C4 and C8 to C10 of the naphthoquinone part of the rifamycins could not be established conclusively by incorporation studies. ^{14}C-labeled shikimate and aromatic amino acids failed to reveal a specific incorporation pattern. However, incorporation studies with D-[1-^{13}C]glucose were in accordance with a shikimate-type origin of the seven-carbon amino unit (14). With the aid of two mutants in the primary metabolism, one transketolase deficient and the other presumably shikimate kinase deficient, it could be clearly demonstrated that the seven-carbon amino unit derives from the basic part of the *aro*-biosynthetic pathway. The origin of the moiety must be localized between D-sedoheptulose-7-phosphate and shikimate (6, 7).

Mutants are not only useful for the investigation of the connection between primary and secondary metabolism but are also extremely valuable for the elucidation of the reaction sequences in the secondary metabolism itself. Often, mutants accumulate intermediates of an antibiotic. Their structural elucidation may help to understand a pathway and to clarify the relationship between groups of chemically related compounds. The biochemical activities responsible for certain steps in the biosynthesis are best analyzed with the aid of such mutants. Cosynthesis of strains with defined blocks in the pathways are also of particular usefulness. The interesting work of Lancini and co-workers (14) has contributed substantially to the understanding of the biosynthesis of the ansamycins. The relationship of rifamycins W, S, and B was established with incorporation studies using ^{13}C-labeled precursors and transformation of rifamycins W and S with washed mycelium from rifamycin B-producing strains into riamycin B. Rifamycin W, an early intermediate in the pathway, has fundamentally the same carbon skeleton as damavaricin, a precursor of streptovaricin (18). In this group of naphthalenic ansamycins, even earlier intermediates, the protostreptovaricins I–V, have been isolated (3). The recent discovery of protorifamycin I (8), an ansamycin structurally related to the protostreptovaricins and an immediate precursor of rifamycin W, sustains the hypothesis (Fig. 4) that all ansamycins with a naphthalenic moiety may derive from a common progenitor. On the basis of the knowledge already available, such a compound may be discovered by a thorough analysis of excretion products of appropriate mutant strains.

GENERAL OUTLOOK

If we consider the biosynthesis of antibiotics from the quantitative, industrial point of view, it seems very likely that in the future the industrial microbiologist will also

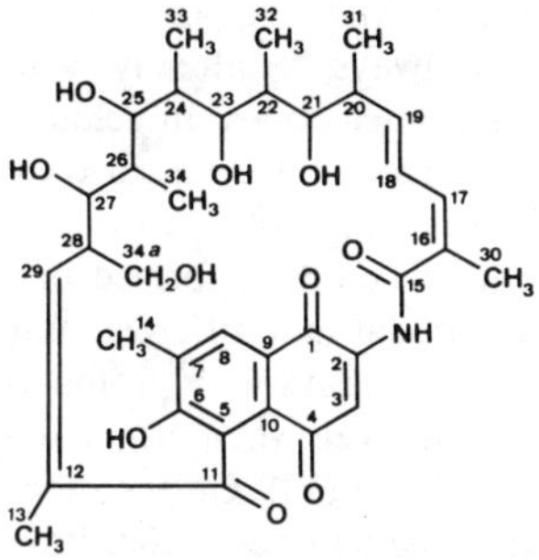

PROTORIFAMYCIN I

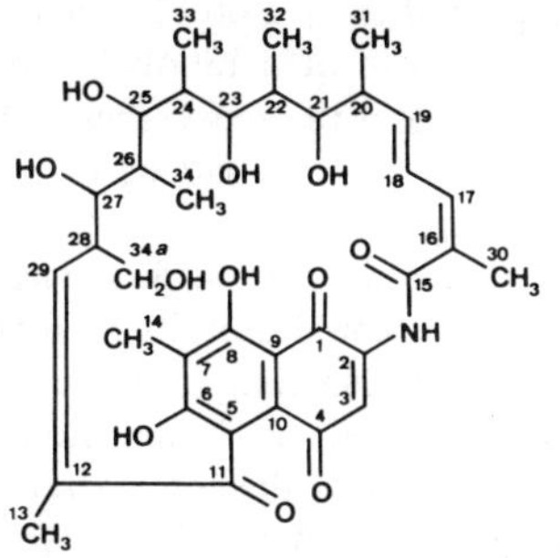

RIFAMYCIN W

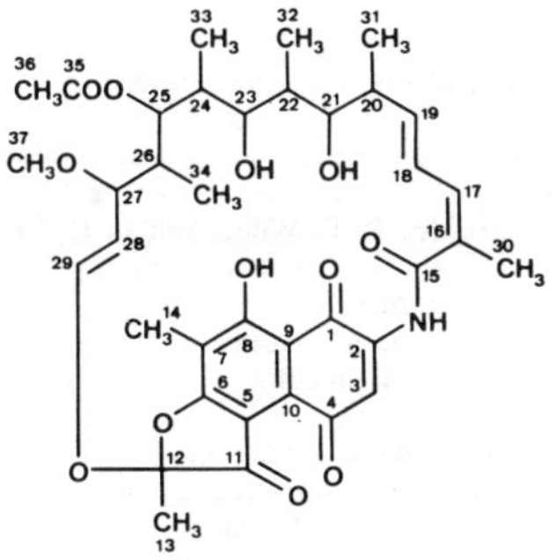

RIFAMYCIN S

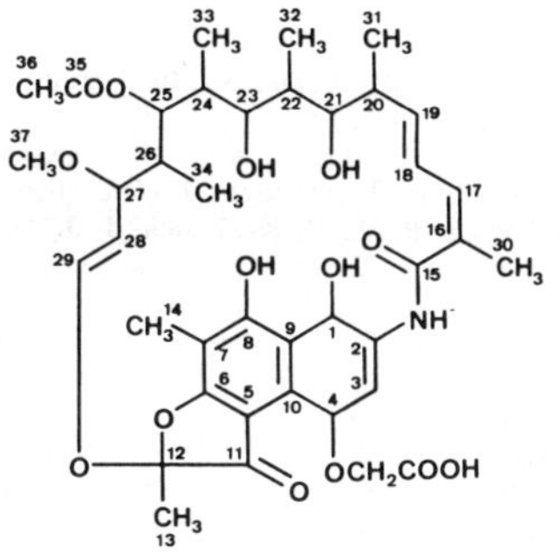

RIFAMYCIN B

FIG. 4. *Structures of protorifamycin I and rifamycins W, S, and B.*

rely on mutation and selection for strain development. Although there are no comparative trials evaluating the efficiency of breeding versus mutation and selection, the latter is very probably more successful. In particular, in the initial phase of strain development programs, a few mutagenic treatments may lead to 10- to 100-fold increases in titers. For this reason and others, such as analytical problems with low-producing strains, a genetic analysis should start with mutants that already possess reasonable productivity.

The haplontic condition of most of the antibiotic-producing microorganisms is one reason for the success of mutation in yield improvement. Very often a gene or a number of genes responsible for a certain biochemical reaction are rate-limiting for the antibiotic synthesis. The identification of the yield-determining factors is essential for the selection of specific mutants. Here, a genetic system could be of great help. On the other hand, recombination with the aim of strain improvement seems to have a limited potential. In recombining the genome of two strains, the delicate, balanced polygenic system which determines the biosynthesis of an antibiotic may be disrupted. This is probably one of the reasons why a majority of crosses between low- and high-producing strains of *Penicillium chrysogenum* led to diploid recombinants in which the positive mutations were almost completely recessive. Likewise, haploid segregants frequently had titers lower than or at the best equal to those of the parental strains (11). In the long term, it is conceivable that the biosynthesis of an antibiotic may be understood, as is already the case for various primary metabolites such as amino acids. It will then be a problem not so much of breeding, with its quantitative aspects, but much more of a rational optimalization of pathways, as already beautifully demonstrated in the biosynthesis of amino acids and nucleotides. In achieving this goal, genetic systems will be of great help. Genetic recombination, in contrast to mutation, may have an important role in the "engineering" of strains. A genome could be combined in such a way that new products could be formed or certain reaction steps could usefully be replaced by others present in a different strain with or without mating compatibility. Thanks to fusion techniques with protoplasts, the introduction of two genomes into one cytoplasmic unit, even when originating from evolutionarily very distant species, is no longer a problem (13). On the other hand, recombination on the molecular level is still far from being understood. Progress in this field will be of primary importance for a successful application of genetics in the biosynthesis of antibiotics.

REFERENCES

1. **Bost, P. E., and A. L. Demain.** 1977. Studies on the cell-free biosynthesis of β-lactam antibiotics. Biochem. J. **162:**681–687.
2. **Demain, A. L., and P. S. Masurekar.** 1974. Lysine inhibition of in vivo homocitrate synthesis in Penicillium chrysogenum. J. Gen. Microbiol. **82:**143–151.
3. **Deshmukh, P. V., K. Kakinuma, J. J. Ameel, K. L. Rinehart, Jr., P. F. Wiley, and L. H. Li.** 1976. Protostreptovaricins I-V. J. Am. Chem. Soc. **89:**870–872.
4. **Eck, C. F.** 1973. Carbon-13 the growing organic isotope. Res./Dev. **24**(8)**:**32–38.
5. **Elander, R. P., L. T. Chang, and R. W. Vaughan.** 1977. Genetics of industrial microorganisms, p. 1–40. *In* D. Perlman and G. T. Tsao (ed.), Annual reports on fermentation processes, vol. 1. Academic Press Inc., New York.
6. **Ghisalba, O., and J. Nüesch.** 1978. A genetic approach to the biosynthesis of the rifamycin chromophore in Nocardia mediterranei. I. Isolation and characterization of a pentose-excreting auxotrophic mutant of Nocardia mediterranei with drastically reduced rifamycin production. J. Antibiot. **31:**202–214.
7. **Ghisalba, O., and J. Nüesch.** 1978. A genetic approach to the biosynthesis of the rifamycin chromophore in Nocardia mediterranei. II. Isolation and characterization of a shikimate excreting auxotrophic mutant of Nocardia mediterranei with normal rifamycin production. J. Antibiot. **31:**215–225.
8. **Ghisalba, O., P. Traxler, and J. Nüesch.** 1978. Early intermediates in the biosynthesis of ansamycins. I. Isolation and identification of protorifamycin I. J. Antibiot. **31:**1124–1131.
9. **Gorman, M., and F. Huber.** 1977. β-Lactam antibiotics, p. 327–346. *In* D. Perlman and G. T. Tsao (ed.), Annual reports on fermentation processes, vol. 1. Academic Press Inc., New York.
10. **Hitchcock, M. J. M., and E. Katz. 1978.** Actinomycin biosynthesis by protoplasts derived from *Streptomyces parvulus.* Antimicrob. Agents Chemother. **13:**104–114.
11. **Hopwood, D. A., and M. J. Merrick.** 1977. Genetics of antibiotic production. Bacteriol. Rev. **41:**595–635.
12. **Hošťálek, Z., M. Blumauerová, J. Ludvík, V. Jechová, V. Běhal, J. Časlavská, and E. Čurdová.** 1976. The role of the genome in secondary biosynthesis in Streptomyces aureofaciens, p. 155–177. *In* K. D. Macdonald (ed.), Second international symposium on the genetics of industrial microorganisms. Academic Press, London.
13. **Jones, C. W., I. A. Mastrangelo, H. Smith, and H. Liu.** 1976. Interkingdom fusion between human (HeLa) cells and tobacco hybrid-protoplasts. Science **193:**401–403.
14. **Lancini, G. C., and R. J. White.** 1976. Rifamycin biosynthesis, p. 139–153. *In* K. D. Macdonald (ed.), Second international symposium on the genetics of industrial microorganisms. Academic Press, London.
15. **Macdonald, K. D., and G. Holt.** 1976. Genetics of biosynthesis and overproduction of penicillin. Sci. Prog. (London) **63:**547–573.
16. **Nara, T.** 1977. Aminoglycoside antibiotics, p. 299–326. *In* D. Perlman and G. T. Tsao (ed.), Annual reports on fermentation processes, vol. 1. Academic Press Inc., New York.
17. **Queener, S. W., J. McDermott, and A. B. Radue.** 1975. Glutamate dehydrogenase specific activity and cephalosporin C synthesis in the M8650 series of *Cephalosporium acremonium* mutants. Antimicrob. Agents Chemother. **7:**646–651.
18. **Rinehart, K. L., Jr., F. J. Antosz, P. V. Deshmukh, K. Kakinuma, P. K. Martin, B. I. Milavetz, K. Sasaki, T. R. Witty, L. H. Li, and F. Reusser.** 1976. Identification and preparation of damavaricins, biologically active precursors of streptovaricins. J. Antibiot. **29:**201–203.
19. **Traxler, P., H. J. Treichler, and J. Nüesch.** 1975. Synthesis of N-acetyldeacetoxycephalosporin C by a mutant of Cephalosporium acremonium. J. Antibiot. **28:**605–606.

Biochemical Genetics of the β-Lactam Antibiotic Biosynthesis

JUAN F. MARTÍN, JOSE M. LUENGO, GLORIA REVILLA, AND JULIO R. VILLANUEVA

Instituto de Microbiología Bioquímica and Departamento de Microbiología, Facultad de Ciencias, Universidad de Salamanca, Salamanca, Spain

BIOSYNTHESIS OF PENICILLIN AND OTHER β-LACTAM ANTIBIOTICS

In addition to the classical members of the penicillin and cephalosporin groups, several new β-lactam antibiotics have been described in recent years. These include the cephamycins, produced by members of the genus *Streptomyces,* nocardicin, formed by a species of *Nocardia,* and β-lactam-like structures such as clavulanic acid and thienamycin, which function as β-lactamase inhibitors (Fig. 1). *Penicillium chrysogenum* produces 6-aminopenicillanic acid (6-APA), isopenicillin N (with an L-α-aminoadipyl side chain), and a variety of penicillins with nonpolar side chains. *Cephalosporium acremonium* produces two different structures: penicillin N, with a D-α-aminoadipyl side chain, and cephalosporin C, with the same side chain but a different nucleus. The procaryotic *Streptomyces* species produce 7-methoxycephalosporin, also with a D-α-aminoadipyl side chain but with a methoxy group attached to C7 of the β-lactam ring. The β-lactamase inhibitor clavulanic acid and thienamycin each have a nucleus similar to that of penicillin except that an oxygen or carbon atom has been substituted for the S atom of penicillin in the thiazolidine ring.

The biosyntheses of all these members of the β-lactam group are probably related. Penicillin is synthesized by condensation of three amino acids, namely, L-α-aminoadipic acid, cysteine, and valine, to form the tripeptide δ-(L-α-aminoadipyl)-L-cysteinyl-D-valine. This tripeptide forms two rings, giving rise to the penicillin nucleus. The tripeptide theory of the origin of penicillin was established by the discovery of a tripeptide containing α-aminoadipic acid from the mycelium of *P. chrysogenum* (3–5). More recently, three peptides have been found in the mycelium of *C. acremonium* (19, 20). The main peptide (P_3) was shown to be δ-(L-α-aminoadipyl)-L-cysteinyl-D-valine. According to Fawcett et al. (11) the same peptides are also present in the mycelium of *Streptomyces clavuligerus,* which produces a cephamycin. The role of α-aminoadipic acid as an essential intermediate in penicillin synthesis has been established by the behavior of lysine auxotrophs of *P. chrysogenum* in relation to penicillin synthesis (15). Lysine auxotrophs blocked after α-aminoadipic acid were able to synthesize penicillin, whereas those blocked before α-aminoadipic acid were not. These results suggest that α-aminoadipic acid, an intermediate of the lysine biosynthetic pathway in fungi, is also a precursor of the tripeptide (1, 2, 10).

The enzymes involved in penicillin synthesis have not been characterized, but circumstantial evidence indicates that two enzymes are involved (Fig. 2). Lemke and Nash (18) reported that mutants which did not produce penicillin fell into two classes, those which were able to synthesize the tripeptide and those which were not. This suggests that the first group are blocked in "penicillin cyclase" subsequent to the formation of the tripeptide. Formation of the tripeptide may be catalyzed by the tripeptide synthetase in a form similar to that of peptide antibiotic formation in

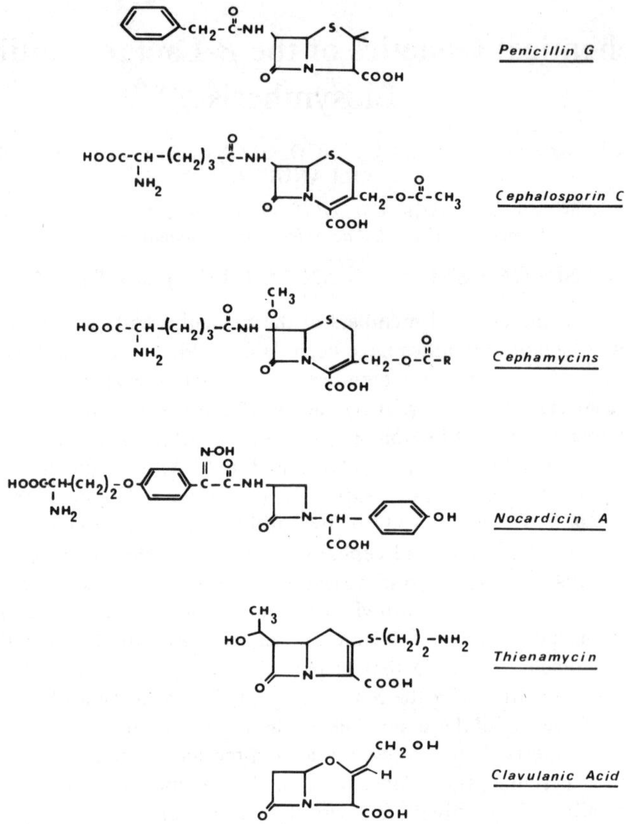

FIG. 1. *Chemical structures of penicillin G and other β-lactam antibiotics.*

Bacillus (17). The synthesis of these peptide antibiotics is different from protein synthesis, since it does not take place on ribosomes and does not require an RNA template. The proposed model for the biosynthesis of the tripeptide is similar to that of glutathione, which has been accomplished by soluble enzymes in a cell-free system, beginning from the N terminal and proceeding through the formation of enzyme-bound γ-glutamylphosphate, which later yields an enzyme-bound dipeptidyl-phosphate intermediate (25). No free dipeptide has been found, which supports the hypothesis of enzyme-bound intermediates.

The last steps in penicillin biosynthesis involve the formation of the penicillin rings. Although 6-APA is found in precursor-free fermentations, it appears to be a shunt metabolite derived by deacylation of isopenicillin N. Present evidence suggests that isopenicillin N is the direct product of the cyclization process.

The α-aminoadipic acid side chain of isopenicillin N is finally exchanged by the action of a penicillin acyl transferase which has been characterized in *P. chrysogenum* (9).

REGULATION OF PENICILLIN BIOSYNTHESIS BY LYSINE

In 1957, Demain described lysine inhibition of penicillin synthesis (8). This inhibition is reverted by α-aminoadipic acid and other intermediates of the lysine

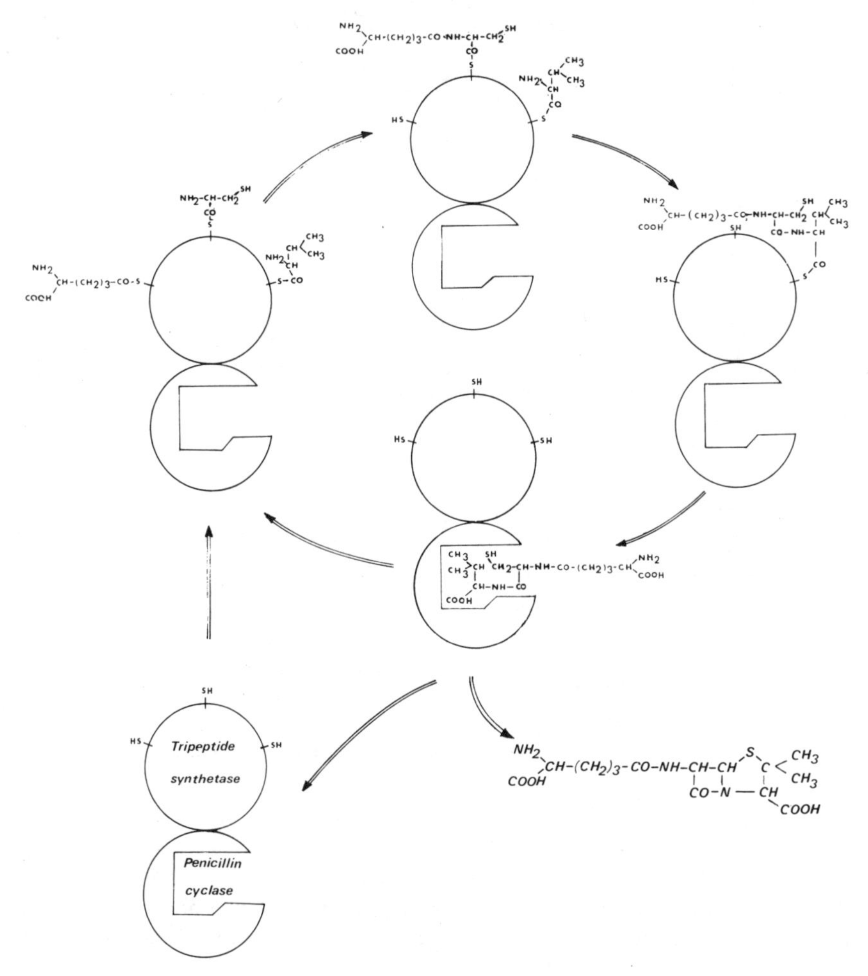

FIG. 2. *Proposed model for the biosynthesis of penicillin. The penicillin synthetase may be considered as an enzyme complex of two subunits: the tripeptide synthetase that synthesizes the tripeptide and the penicillin cyclase that forms the β-lactam and thiazolidine rings (see text).*

biosynthetic pathway (12). Both lysine and penicillin biosyntheses involve α-aminoadipic acid. It seems, therefore, that lysine and penicillin are products of a branched biosynthetic pathway (Fig. 3). Lysine feedback inhibits penicillin synthesis by suppressing the formation of α-aminoadipic acid.

Lysine biosynthesis in fungi has been studied in *Saccharomyces* and *Neurospora* (6, 13, 21, 27) and is regulated by feedback repression of homocitrate synthase, the first enzyme of the pathway (16, 28). Repression is likewise exerted at the aminoadipate-semialdehyde dehydrogenase level, the first enzyme after α-aminoadipic acid (7, 26). Lysine also produces feedback inhibition of homocitrate synthase (12, 21). Masurekar and Demain (23) suggested that lysine inhibition of penicillin biosynthesis in *P. chrysogenum* was due to the inhibition of an early enzyme in its biosynthetic pathway. They showed later an in vivo inhibition of the homocitrate synthase (10) but failed to detect any in vitro effect of lysine on the activity of this enzyme. Since

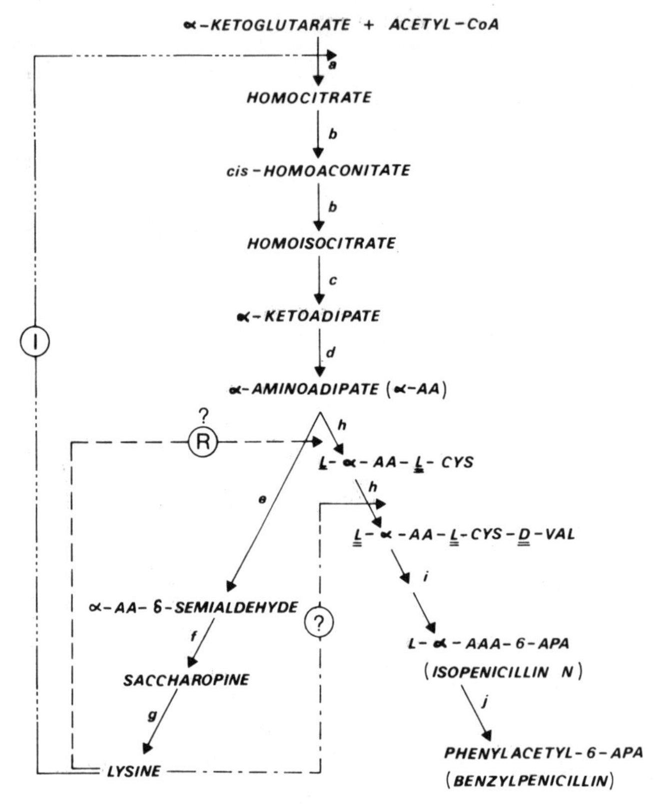

FIG. 3. *Schematic branched pathways of lysine and penicillin biosynthesis, indicating the probable regulatory mechanisms that control the pathway.*

the lysine pathway in other fungi is subject to both inhibition and repression by lysine, it was of interest to study the molecular mechanism of the lysine regulatory effect in *P. chrysogenum.*

Several mutants of *P. chrysogenum* were used, including a high-producer strain, TP-26, kindly provided by Antibioticos, S. A. (Léon, Spain). Excretion of homocitrate by lysine auxotroph *P. chrysogenum* Wis. 54–1255 L2 (Lys^-) is reduced by 23% when the culture is supplemented with 50 mM lysine at inoculation time. If supplemented with 50 mM lysine daily, homocitrate excretion is completely inhibited.

Homocitrate synthase activity has been assayed in cell-free extracts of *P. chrysogenum* Wis. 54–1255, of the Lys^- auxotroph L2, and of the high-penicillin producer TP-26. α-Ketoglutarate-dependent formation of [^{14}C]homocitrate from [^{14}C]acetyl coenzyme A was followed radiometrically, after thin-layer-chromatographic separation of homocitrate. Homocitrate synthase activities of cell-free extracts of the parental strain Wis. 54–1255 grown in complex medium supplemented with 50 mM lysine at inoculation time or with daily 50 mM lysine additions were, respectively, 10 and 42% less than the activity of control cultures (Fig. 4). Inhibition is complete, however, when strain Wis. 54–1255 is grown in a synthetic Czapek-asparagine medium, in which no penicillin synthesis takes place. Similarly, cell-free extracts of the L2 mutant grown in Czapek-asparagine medium supplemented daily with 50 mM

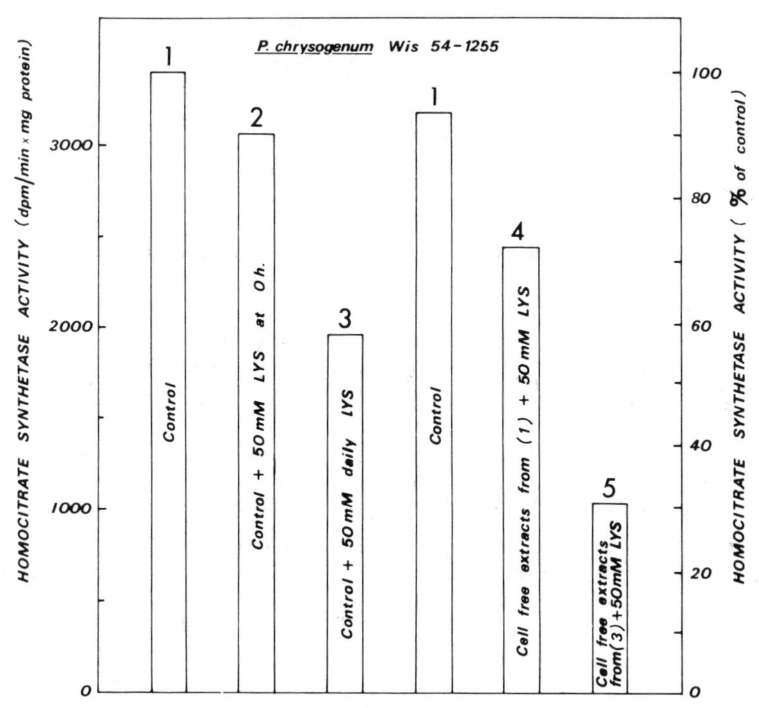

FIG. 4. *Homocitrate synthase activity in cell-free extracts of* Penicillium chrysogenum *Wis. 54–1255, grown under different conditions (see text).*

lysine show no homocitrate synthase activity. This confirms our previous in vivo results, which showed that there is a complete inhibition of penicillin synthesis when a high lysine pool is maintained throughout. The sensitivity to lysine inhibition of homocitrate synthase in cell-free extracts of the high-penicillin producer strain, TP-26, is identical. Similar results have been obtained by studying the effect of lysine on the incorporation of [^{15}C]valine into penicillin in short-term experiments, which may suggest a direct inhibition of the tripeptide synthetase by lysine or may be explained on the assumption that it is depriving the cells of the penicillin precursor α-aminoadipic acid (Fig. 3). In the cells in which protein synthesis was inhibited by cycloheximide, the lysine regulation of penicillin synthesis is still exerted, which suggests enzyme inhibition, since under these conditions no synthesis of new protein takes place.

In all cases, the percentages of inhibition of homocitrate synthase activity are lower than the percentages of inhibition of penicillin synthesis. This suggests that in addition to the inhibition of homocitrate synthase, some other regulatory effect is exerted by lysine. Moreover, when cells were grown in medium supplemented with 50 mM lysine and washed to remove lysine, the incorporation of [^{14}C]valine into penicillin in the lysine-free suspension was distinctly lower in cells grown in the presence of lysine than in those grown in the absence of the amino acid. This suggests that penicillin synthesis is also subject to repression by lysine and is in contrast with the results of Masurekar and Demain (23), who, however, used a lower lysine concentration.

FEEDBACK REGULATION OF PENICILLIN BIOSYNTHESIS BY PENICILLIN ITSELF

Feedback regulation by the end product is an important regulatory mechanism for the control of cell economy. Although end product regulation of biosynthetic pathways of primary metabolites is well known, knowledge of the feedback regulation of the biosyntheses of special metabolites has been limited by the lack of knowledge of the enzymes involved in antibiotic biosyntheses. In the last few years it has become evident that several antibiotics inhibit their own biosyntheses (chloramphenicol, aurodox, cycloheximide, staphylomycin [virginiamycin], ristomycin, puromycin, fungicidin, candihexin, and mycophenolic acid) (see ref. 22).

P. chrysogenum supplemented at the time of inoculation with high levels of exogenous penicillin does not synthesize penicillin. The concentration of exogenous penicillin required for a complete inhibition of de novo penicillin synthesis depends upon the antibiotic level usually attained by the producer. We have studied two different strains: *P. chrysogenum* Wis. 54–1255, which produces about 600 μg of penicillin per ml, and *P. chrysogenum* TP-26, which produces more than 5,000 μg of penicillin per ml in shaken flasks. Exogenous benzylpenicillin concentrations of 500 μg/ml fully inhibited penicillin synthesis by strain Wis. 54–1255, whereas 10,000 μg of exogenous penicillin per ml was required to get complete inhibition of penicillin synthesis by strain TP-26.

Using a short-term resting cell system for incorporation of [^{14}C]valine into penicillin, we studied the effects of different concentrations of benzylpenicillin and 6-APA on penicillin syntheses by both strains. In strain Wis. 54–1255, a 49% inhibition of precursor incorporation into penicillin by 1,000 μg of exogenous penicillin per ml after a 60-min incubation was found. A 44% inhibition of valine incorporation by 1,000 μg of exogenous 6-APA per ml was observed in the same strain. The high-producing strain, TP-26, was less susceptible to end product regulation, since an addition of 10,000 μg of exogenous penicillin per ml was required to obtain a 51% inhibition of [^{14}C]valine incorporation into penicillin. Inhibition of penicillin synthesis in the latter strain by 10,000 μg of exogenous 6-APA per ml was only 27%. It is concluded that penicillin formation by the high-producing strain is less sensitive to end product inhibition by the penicillin accumulated in the broth than is strain Wis. 54–1255.

Our results are similar to those reported by Gordee and Day (14). The concentration of antibiotic which is inhibitory for its own synthesis in a particular producer strain is generally similar to the production level of that strain (22).

We may conclude that a correlation exists between the production capacity of a penicillin-producing strain and the penicillin level required for complete inhibition of penicillin synthesis. Feedback regulation by the antibiotic itself may have relevance in industrial penicillin production. Removal of end product control may be a method for increasing penicillin production.

ACKNOWLEDGEMENT

We acknowledge financial support of Antibioticos, S. A. (León, Spain), and their gift of the *P. chrysogenum* TP-26 used in these studies.

REFERENCES

1. **Abraham, E. P.** 1974. Biosynthesis and enzymic hydrolysis of penicillins and cephalosporins. University of Tokyo Press, Tokyo.

2. **Abraham, E. P., and P. Fawcett.** 1974. Penicillins, cephalosporins and the biosynthesis of peptide antibiotics by fungi, p. 319–334. *In* B. Spencer (ed.), Industrial aspects of biochemistry, vol. 30. North-Holland Publishing Co., New York.
3. **Arnstein, H. R. V., N. Artman, D. Morris, and E. J. Toms.** 1960. Sulphur containing aminoacids and peptides in the mycelium of *Penicillium chrysogenum.* Biochem. J. **76:**353–357.
4. **Arnstein, H. R. V., and D. Morris.** 1960. The structure of a peptide containing α-aminoadipic acid, cystine and valine, present in the mycelium of *Penicillium chrysogenum.* Biochem. J. **76:**357–361.
5. **Arnstein, H. R. V., D. Morris, and E. J. Toms.** 1959. Isolation of a tripeptide containing L-aminoadipic acid from the mycelium of *Penicillium chrysogenum* and its possible significance in penicillin biosynthesis. Biochim. Biophys. Acta **35:**561–562.
6. **Bhattacharjee, J. K.** 1970. Leaky mutation and coordinate regulation of the accumulation of lysine precursors in *Saccharomyces.* Can. J. Genet. Cytol. **12:**785–789.
7. **Broguist, H. P., C. C. Chao, and E. C. Clevenstine.** 1965. Lack of end-product control of latter stages of lysine biosynthesis in yeasts. Fed. Proc. Fed. Am. Soc. Exp. Biol. **28:**764.
8. **Demain, A. L.** 1957. Inhibition of penicillin formation by lysine. Arch. Biochem. Biophys. **67:**244–245.
9. **Demain, A. L.** 1974. Biochemistry of penicillin and cephalosporin fermentations. Lloydia **37:**147–167.
10. **Demain, A. L., and P. S. Masurekar.** 1974. Lysine inhibition of *in vivo* homocitrate synthesis in *Penicillium chrysogenum.* J. Gen. Microbiol. **82:**143–151.
11. **Fawcett, P. A., J. J. Usher, and E. P. Abraham.** 1976. Aspects of cephalosporin and penicillin biosynthesis, p. 129–138. *In* K. D. Macdonald (ed.), Second International Symposium on the Genetics of Industrial Microorganisms. Academic Press, London.
12. **Friedrick, C. G., and A. L. Demain.** 1977. Homocitrate synthase as the crucial site of the lysine effect on penicillin biosynthesis. J. Antibiot. **30:**760–761.
13. **Glass, J., and J. K. Bhattacharjee.** 1971. Lysine biosynthesis in *Rhodotorula*: accumulation of homocitric acid, homoaconitic acid and homoisocitric acid in a leaky mutant. Genetics **67:**365–376.
14. **Gordee, E. Z., and L. E. Day.** 1972. Effect of exogenous penicillin on penicillin biosynthesis. Antimicrob. Agents Chemother. **1:**315–322.
15. **Goulden, S. A., and F. W. Chattaway.** 1968. Lysine control of α-aminoadipate and penicillin synthesis in *Penicillium chrysogenum.* Biochem. J. **110:**55p.
16. **Hogg, R. W., and H. P. Browuist.** 1968. Homocitrate formation in *Neurospora crassa.* Relation to lysine biosynthesis. J. Biol. Chem. **243:**1839–1845.
17. **Katz, E., and A. L. Demain.** 1977. The peptide antibiotics of *Bacillus:* chemistry, biogenesis, and possible functions. Bacteriol. Rev. **41:**449–474.
18. **Lemke, P. A., and C. H. Nash.** 1972. Mutations that affect antibiotic synthesis by *Cephalosporium acremonium.* Can. J. Microbiol. **18:**255–259.
19. **Loder, P. B., and E. P. Abraham.** 1971. Isolation and nature of intracellular peptides from a cephalosporin C producing *Cephalosporium* sp. Biochem. J. **123:**471–476.
20. **Loder, P. B., and E. P. Abraham.** 1971. Biosynthesis of peptides containing α-aminoadipic acid and cysteine in extracts of a *Cephalosporium sp.* Biochem. J. **123:**477–482.
21. **Maragoudakis, M. E., H. Holmes, and M. Strassman.** 1967. Control of lysine biosynthesis in yeast by a feedback mechanisms. J. Bacteriol. **93:**1677–1680.
22. **Martin, J. F.** 1978. Manipulation of gene expression in the development of antibiotic production. *In* T. Leisinger, R. Hütter, J. Nüesch, and W. Wehrli (ed.), Antibiotics and other secondary metabolites: biosynthesis and production. Academic Press, London.
23. **Masurekar, P. S., and A. L. Demain.** 1972. Lysine control of penicillin biosynthesis. Can. J. Microbiol. **18:**1045–1048.
24. **Nash, C. H., N. de la Higuera, N. Neuss, and P. A. Lemke.** 1974. Application of biochemical genetics to the biosynthesis of β-lactam antibiotics. Dev. Ind. Microbiol. **15:**114–132.
25. **Nishimura, J. S., E. A. Dodd, and A. Meister.** 1964. Intermediate formation of dipeptide phosphate anhydride in enzymatic tripeptide synthesis. J. Biol. Chem. **239:**2553–2558.
26. **Sinha, A. K., M. Kurtz, and J. K. Bhattacharjee.** 1971. Effect of hydroxylysine on the biosynthesis of lysine in *Saccharomyces.* J. Bacteriol. **108:**715–719.
27. **Trupin, J. S., and H. P. Broquist.** 1965. Saccharopine, an intermediate of the aminoadipic acid pathway of lysine biosynthesis. I. Studies on *Neurospora crassa.* J. Biol. Chem. **240:**2524–2530.
28. **Tucci, A. F.** 1969. Feedback inhibition of lysine biosynthesis in yeast. J. Bacteriol. **99:**624–625.

Genetic Approach to the Biosynthesis of Anthracyclines

M. BLUMAUEROVÁ, E. KRÁLOVCOVÁ, J. MATĚJŮ, Z. HOŠŤÁLEK, AND Z. VANĚK

Czechoslovak Academy of Sciences, Institute of Microbiology, Prague, Czechoslovakia

Research on anthracyclines has been focused chiefly on the study of their chemistry, biological effects, and pharmacological properties (for a review, see 21), whereas their biosynthesis has been studied only to a limited extent (13, 16–21). Our considerations of the reactions that transform the hypothetical decaketide precursor of anthracyclinones to appropriate end products (21) were based solely on the knowledge of the structures of these compounds. Experimental verification of these notions was carried out with *Streptomyces coeruleorubidus* (which produces metabolites of the daunomycinone series and ε-rhodomycinone) and *S. galilaeus* (which produces substances of the ε-pyrromycinone and aklavinone series) by isolating high-yielding mutants (5, 9) and mutants blocked in the biosynthesis of parent metabolites (6, 9), feeding the mutants with presumptive biosynthetic intermediates (2), and co-synthetic experiments in mixed cultures of mutant pairs (1). The structures of the substances under study are given in Fig. 1. The results (Table 1, Fig. 2), along with existing knowledge of the physiology (3, 5, 8, 11, 12) and morphology of the mutants (4–6), the data from in vitro experiments (14), and results from other laboratories (16–19), are discussed in this paper.

PATHWAYS OF ANTHRACYCLINE BIOSYNTHESIS IN *S. COERULEORUBIDUS* AND *S. GALILAEUS*

The sequence and mutual coupling of reactions participating in the biosynthesis of daunomycinone (I), 13-dihydrodaunomycinone (II), aklavinone (V), ε-rhodomycinone (VII), ε-pyrromycinone (X), and their derivatives are shown in Fig. 3. The biosynthetic scheme may be divided into three sections including the formation of the hypothetical decaketide precursor from primary building units (steps **1** and **2**), its transformation to anthracyclinones (steps **3, 4, 6, 7-9, 11-15**), and glycosidation of the resulting aglycones (steps **5, 10,** and **16**).

Steps 1 and 2. The basic skeleton of the anthracyclinone molecule is formed by a condensation of acetate (malonate) units, whereas the terminal group originates from propionate (13, 16, 17). The decaketide precursor is common to all compounds under study in the two species. The ability to form this precursor is potentially retained in nonactive mutants blocked in later biosynthetic stages (1). The reaction rate decreases in both species on addition of an exogenous propionate or propanol and, in *S. coeruleorubidus* but not in *S. galilaeus,* is negatively affected by 5,5-diethylbarbiturate and other inhibitors (3).

Step 3. Step **3** consists of three reactions whereby the decaketide precursor is transformed into II. The first of these is the decarboxylation of the initial decaketide in position 10; this reaction takes place only in *S. coeruleorubidus* and is a necessary condition for the formation of substances of the daunomycinone group. Its rate (controlled by the feedback effect of I) is one of the major factors that determines the

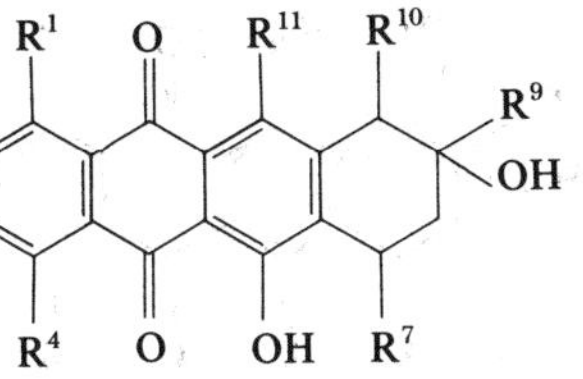

No.	Compound	R^1	R^4	R^7	R^9	R^{10}	R^{11}
I	Daunomycinone	H	OCH_3	OH	$COCH_3$	H	OH
II	13-Dihydrodaunomycinone	H	OCH_3	OH	$CHOHCH_3$	H	OH
III	7-Deoxy-13-dihydrodaunomycinone	H	OCH_3	H	$CHOHCH_3$	H	OH
IV	Carminomycinone	H	OH	OH	$COCH_3$	H	OH
V	Aklavinone	H	OH	OH	CH_2CH_3	$COOCH_3$	H
VI	7-Deoxyaklavinone	H	OH	H	CH_2CH_3	$COOCH_3$	H
VII	ε-Rhodomycinone	H	OH	OH	CH_2CH_3	$COOCH_3$	OH
VIII	ζ-Rhodomycinone	H	OH	H	CH_2CH_3	$COOCH_3$	OH
IX	ε-Isorhodomycinone	OH	OH	OH	CH_2CH_3	$COOCH_3$	OH
X	ε-Pyrromycinone	OH	OH	OH	CH_2CH_3	$COOCH_3$	H
XI	ζ-Pyrromycinone	OH	OH	H	CH_2CH_3	$COOCH_3$	H

FIG. 1. *Anthracyclinones studied in* S. coeruleorubidus *and* S. galilaeus. *Bisanhydroaklavinone (XII) and η-pyrromycinone (XIII) lack both C7 and C9 hydroxyl groups and are fully aromatized.*

ratios of these substances and VII (2). The sequence of the subsequent reactions (methylation at C4 and hydroxylation at C11) cannot be determined precisely until the blocked mutants corresponding to each of them are isolated. Methionine is the source of the methyl group for the formation of the methoxy group at C4 (17). Addition of D,L-ethionine to production strains leads to the formation of carminomycinone (IV) and its derivatives, with a concomitant lowering of the yields of methylated products (Blumauerová, unpublished data).

Steps 4 and 6. Transformation of II to I (oxidation of side chain at C9) is controlled by the feedback effect of I (2, 3, 5, 6) and proceeds at a high rate also in the opposite direction (2). The back reaction is catalyzed by a reduced nicotinamide adenine dinucleotide phosphate-dependent enzyme (14). Conversion of I and II to III (removal of hydroxyl at C7), catalyzed by a reduced nicotinamide adenine dinucleotide-dependent enzyme and inhibited by oxygen, is irreversible; the enzyme cleaves reductively also glycosides of I and II to III (14). Exogenous I is converted to II and III also in *S. galilaeus.* In both species the activity of the corresponding enzymes depends on the strain used and the age of the culture (2).

Steps 7-9. Step 7 (methylation of the decaketide precursor in position 10) is common to both species. Divergence of the pathways takes place later, in the transformation of V to VII (hydroxylation at C11 in *S. coeruleorubidus*) and to X (hydroxylation at C1 in *S. galilaeus*). Both these steps are irreversible and are not subject to feedback control (2, 9). The reaction rate of step **8** may be enhanced by genetic improvement (9) or by the addition of 5,5-diethylbarbiturate (8). Yields of VII are inversely proportional to the level of products of steps **3** and **4** (2–6).

Steps 11-15. Conversion of V, VII, and X to 7-deoxy derivatives (VI, VIII, XI) is catalyzed by substrate-specific enzymes differing in requirements for positions 1 and

TABLE 1. *Characteristics of wild and mutant strains of* S. coeruleorubidus *(W_1, P_1, A-E) and* S. galilaeus *(W_2, P_2, F, G)*[a]

Type	Morphology[b]	Activity[c] (%)	Metabolites identified[d]	Biotransformation ability[e]			Co-synthetic activity[f]		Biosynthetic step affected[g]
				Substrate	Product	Unchanged	Complementary partner	Product	
W_1	*spo-1*	100	I-S, II-S; II, III, VII	I	I-S, II-S, II, III	V, IX[h]	—	—	—
				II	I-S, II-S, III				
				X	X-S				
P_1	*whi*	>1,500[i]	I-S (il, pd), II-S (il); II, III, VII (dl)	I	I-S, II-S, II, III	V, IX[h]	—	—	**1, 2, 5**
				II	I-S, II-S, III				
				X	X-S				
A	*spo-1* *spo-3* *bld-2*	<10	I-S, II-S (tr); II, III (tr)	I	I-S, II-S, II, III (il)	IX, X[h]	B, C, D, E, F E, F	I-S, II-S (il) VII	**1, 2, 7**
				II	I-S, II-S, III (il)				
				V	VII				
B	*spo-1*	<10	I-S, II-S (tr); II, III (il); VII (tr)	V	VII	I[h] IX, X	A E, F	I-S, II-S (il) VIII (il)	**5, 7**
C	*spo-4* *bld-2*	0	No anthracycline compounds	I	II, III	VI, IX, X	A	I-S, II-S	**3, 5, 7, 9**
				II	III				
				V	VI, XII				
				VII	VIII				
D	*spo-2*	0	VII (il); VI, VIII, VII-S (tr)	I	II, III	IX, X	A	I-S, II-S	**3, 5**
E	*bld-2*	0	V, VI, XII[j]	I	II (tr), III (pd); I-S (tr)	IX, X	A B	I-S, II-S, VII VII	**1–3, 5, 6, 9**
				VII	VIII				
W_2		100	V-S, X-S; VI, XI, XII, XIII	I	II, III		—	—	—
				VII	VII-S				
				IX	IX-S				

P_2	>1,200[k]	V-S; X-S (il); VI, XI, XII, XIII	I VII IX	II, III VII-S IX-S		—	—	**1, 2, 10**
F	0	V, VI (dl); X (il); XI, XII, XIII	I	II, III	VII, IX	A B, C, D.	I-S, II-S, VII Non-anthracycline compounds[l]	**10**
G	0	No anthracycline compounds	I	II (pd), III (tr)	V, VII, IX, X	B, C, D	Non-anthracycline compounds[l]	**1, 2, 7, 8, 10–14**

[a] Symbols used: W, wild strain; P, high-yield mutants; A-G, blocked mutants; I-XIII, anthracyclinones (see Fig. 1); S, sugar derivatives (glycosides); tr, produced in trace quantities; il, increased level; dl, decreased level; pd, predominant component of the mixture. For strain origin, see references 3, 5, 6, and 9.

[b] Phenotype expression of types *spo-1* (sporulating) and *bld-2* (bald, nonsporulating) was not affected by carbon source in medium. The other types were nonsporulating on starch media and differ mutually in morphology of aerial mycelium on other carbon sources (4). In P_2, F, and G, no morphology differences from W_2 were observed.

[c] Total yield of antibiotically active glycoside complexes was determined by bioassay (in *S. coeruleorubidus*) or spectrophotometrically (in *S. galilaeus*) (3, 10).

[d] For isolation and identification of metabolites in *S. coeruleorubidus* see references 3, 6, 7, and 15.

[e] The response to feeding exogenous anthracyclinones was not followed in the types accumulating either the substrate under study or the expected products. The only exception is biotransformation of I and II in W_1 and P_1. In both types, the intensive antracycline biosynthesis started after 48 h of cultivation. Exogenous I fed to younger cultures (0 to 24 h) was converted to I-S and II-S, however, at the same time it repressed its own biosynthesis de novo in later development phases; the yields of I-S and II-S thus corresponded to the level of exogenous substrate only. Feeding I to 48-h cultures decreased the activity by 30 to 40% (in W_1) or 20% (in P_1) only. In P_1 the level of VII increased simultaneously. After feeding ^{13}C-labeled I, no radioactivity was found in VII. Exogenous II had no detectable inhibition effect on W_1 and P_1 activity (2).

[f] The level of co-synthesized I-S and II-S corresponded to about 60% (in A+B, A+C, or A+D combinations) or 25% (in A+E) of the activity of the respective parents of mutants B, C, D, or E. S-I and S-II co-synthesized in A+F combinations were qualitatively identical with metabolites of *S. coeruleorubidus.* For details, see reference 1.

[g] See Fig. 3.

[h] No glycosidation detected.

[i] In a low-glucose medium optimal for W_1 (3). Further increase in activity (by 250%) was achieved by using starch as a carbon source. Enhanced yields of I-S were accompanied by a decrease in levels of II, III, and VII (3, 5). A similar effect was observed with lactose or other disaccharides in both complex and synthetic media.

[j] For other metabolites see Fig. 2.

[k] Independent of carbon source in medium. The yields of X-S could be further increased (under a simultaneous decrease in VI, XI, XII, and XIII) by higher aeration rate (12).

[l] Typical of *S. galilaeus.*

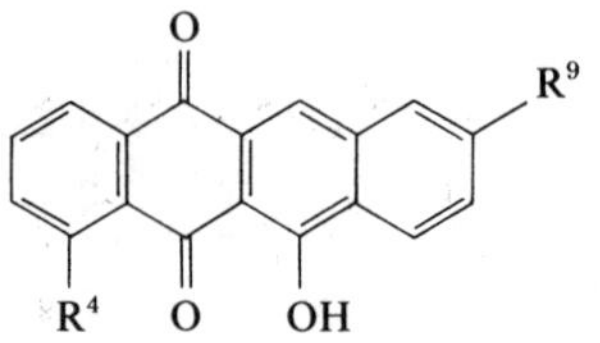

Compound	R^4	R^9
188	OCH_3	$CHOHCH_3$
190	OCH_3	CH_2CH_3
192	OH	CH_2CH_3

FIG. 2. *Metabolites isolated from* S. coeruleorubidus *mutant type E (7). For aklavinone (V), 7-deoxyaklavinone (VI), and bisanhydroaklavinone (XII), see Fig. 1. Compound 194 differs from 188 both by presence of a tertiary hydroxyl group in position 9 and by concomitant dearomatization of ring A. Compound 193 is identical with 7-deoxy-10-epiaklavinone.*

11. Anhydro derivatives (XII and XIII) are formed directly from V and X (not from VI and XI) by removing the secondary and tertiary hydroxyl in positions 7 and 9, respectively, with a concomitant aromatization of ring A (2). The rate of these reactions is increased by lowering the aeration (12) or by a mutation defect in the oxidase system (1, 7). In *S. galilaeus,* VI, XI, XII, and XIII are also formed by reductive cleavage of the glycosides (11, 12).

Steps 5, 10, and 16. The sugar moiety of glycosides is formed by a direct conversion of glucose (18) and is attached to the complete aglycone (2, 19, 20). The glycosidases of both species are substrate specific, the crucial factor being the substituents of the aglycone in positions 1, 10, and 11 (2). Mutation change of this specificity permits the formation of glycosides of VII, which is absent in the parent anthracycline producers (cf. 21). The yields of glycosides of I, II, V, and X reflect mutual competition of reactions in steps **4, 5,** and **6** (in *S. coeruleorubidus*) and steps **10–14** (in *S. galilaeus*) which are susceptible to both genetic and environmental factors (3, 5, 9, 12). In contrast to *S. galilaeus,* the glycosidation in *S. coeruleorubidus* is controlled by the carbon catabolite repression (3, 5). Carbon source in the medium also affects the morphological differentiation of cultures. Inverse correlation between the sporulating and the glycosidating activities in *S. coeruleorubidus* makes possible a rational selection of high-yielding mutants (4, 5).

CONCLUSION

Our results corroborate the original hypothesis of anthracycline biosynthesis (21), except for VII, which is not the precursor of I; methylation in position 10 prevents the decarboxylation and thus the inclusion of VII in the daunomycinone pathway. In contrast to the conventional view of the feedback effect of antibiotics on their own biosynthesis, a similar effect was found also with the antibiotically inactive aglycone moiety of anthracyclines. The two *Streptomyces* species, which produce structurally related metabolites, differ in the substrate specificity of some enzymes and in the mechanism of biosynthetic regulation. Complex changes in the activity of mutants (Table 1) suggest that the genetic loci which control separate steps of the biosynthetic sequence are probably closely clustered.

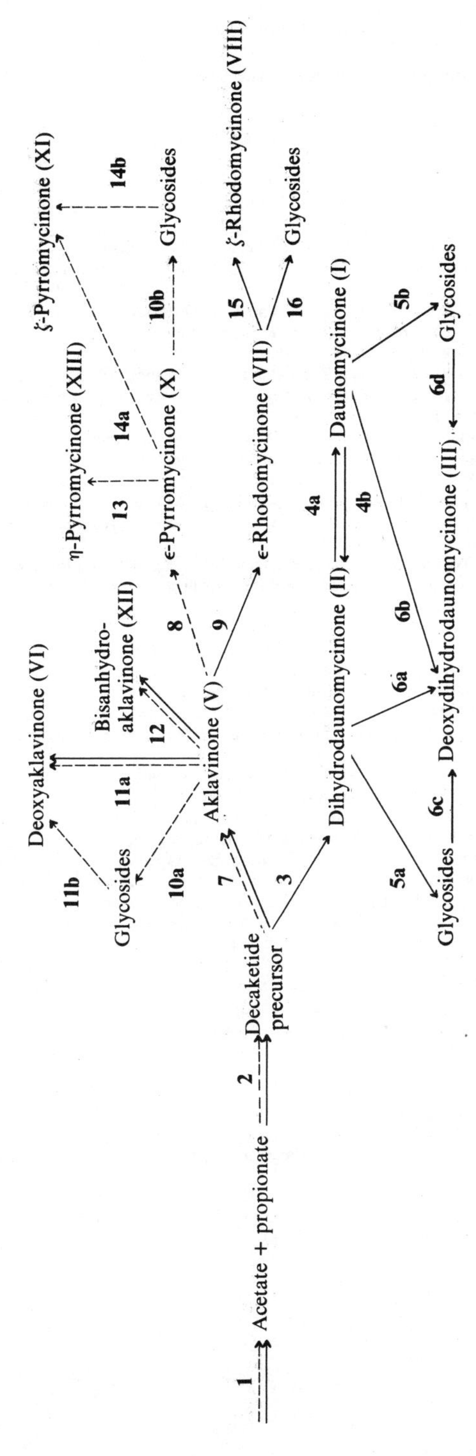

FIG. 3. *Tentative scheme of anthracycline biosynthesis in* S. coeruleorubidus *(→) and* S. galilaeus *(- - →).*

ACKNOWLEDGMENT

We thank A. L. Demain for helpful suggestions and encouragement in the course of this work.

REFERENCES

1. **Blumauerová, M., E. Královcová, Z. Hošťálek, and Z. Vaněk.** 1979. Intra- and interspecific cosynthetic activity of mutants of Streptomyces coeruleorubidus and Streptomyces galilaeus impaired in the biosynthesis of anthracyclines. Folia Microbiol. (Prague), vol. 24 (in press).
2. **Blumauerová, M., E. Královcová, J. Matějů, J. Jizba, and Z. Vaněk.** 1979. Biotransformations of anthracyclinones in Streptomyces coeruleorubidus and Streptomyces galilaeus. Folia Microbiol. (Prague), vol. 24 (in press).
3. **Blumauerová, M., J. Matějů, K. Stajner, and Z. Vaněk.** 1977. Studies on the production of daunomycinone-derived glycosides and related metabolites in Streptomyces coeruleorubidus and Streptomyces peucetius. Folia Microbiol. (Prague) **22**:275–285.
4. **Blumauerová, M., V. Pokorný, J. Šťastná, Z. Hošťálek and Z. Vaněk.** 1978. Developmental mutants of Streptomyces coeruleorubidus, a producer of anthracyclines: isolation and preliminary characterization. Folia Microbiol. (Prague) **23**:177–182.
5. **Blumauerová, M., V. Pokorný, J. Šťastná, Z. Hošťálek, and Z. Vaněk.** 1978. Improved yields of daunomycinone glycosides in developmental mutants of Streptomyces coeruleorubidus. Folia Microbiol. (Prague) **23**:249–254.
6. **Blumauerová, M., K. Stajner, V. Pokorný, Z. Hošťálek, and Z. Vaněk.** 1978. Mutants of Streptomyces coeruleorubidus impaired in the biosynthesis of daunomycinone glycosides and related metabolites. Folia Microbiol. (Prague) **23**:255–260.
7. **Jizba, J., P. Sedmera, J. Vokoun, M. Blumauerová, and Z. Vaněk.** 1979. Naphtacenequinone derivatives from a mutant strain Streptomyces coeruleorubidus. Collect. Czech. Chem. Commun., vol. 44 (in press).
8. **Královcová, E., M. Blumauerová, and Z. Vaněk.** 1977. The effect of 5,5-diethylbarbituric acid on the biosynthesis of anthracyclines in Streptomyces galilaeus. Folia Microbiol. (Prague) **22**:182–188.
9. **Královcová, E., M. Blumauerová, and Z. Vaněk.** 1977. Strain improvement in Streptomyces galilaeus, a producer of anthracycline antibiotics galirubins. Folia Microbiol. (Prague) **22**:321–328.
10. **Královcová, E., J. Tax, M. Blumauerová, and Z. Vaněk.** 1977. Bewertung der Produktion von Glykosiden des ε-Pyrromycinons (Galirubine) bei der Züchtung des Stammes Streptomyces galilaeus. Z. Allg. Mikrobiol. **17**:47–50.
11. **Královcová, E., and Z. Vaněk.** 1979. Growth and production of anthracyclines in wild type and mutant strains of Streptomyces galilaeus. Folia Microbiol. (Prague), vol. 23.
12. **Královcová, E., and Z. Vaněk.** 1979. Effect of aeration and carbon source on the production of anthracyclines in Streptomyces galilaeus. Folia Microbiol., vol. 23.
13. **Matějů, J., M. Blumauerová, and Z. Vaněk.** 1977. Biogenesis of substances of the daunomycin type by Streptomyces coeruleorubidus 39–146. 13th Annual Meeting of the Czechoslovak Society for Microbiology (Gottwaldov, 26–28 April 1977). Abstracts of Communications. Folia Microbiol. **22**:470.
14. **Matějů, J., M. Blumauerová, and Z. Vaněk.** 1979. Activity of enzymes transforming daunomycin in Streptomyces coeruleorubidus. Folia Microbiol. (Prague), vol. 24 (in press).
15. **Matějů, J., J. Vokoun, M. Blumauerová, and Z. Vaněk.** 1978. 7-Deoxy-13-dihydrodaunomycinone in cultures of Streptomyces coeruleorubidus. Folia Microbiol. (Prague) **23**:246–248.
16. **Ollis, W. D., and I. O. Sutherland.** 1960. The incorporation of propionate in the biosynthesis of ε-pyrromycinone (rutilantinone). Proc. Chem. Soc. London, p. 347–349.
17. **Paranosenkova, V. I., and V. L. Karpov.** 1976. The study of rubomycin biosynthesis. Bioorg. Khim. **1**: 1755–1759.
18. **Paranosenkova, V. I., and V. L. Karpov.** 1976. Studies on biosynthesis of carbohydrate fragment of rubomycin. Antibiotiki (Moscow) **21**:299–301.
19. **Paranosenkova, V. I., and V. L. Karpov.** 1977. Rubomycin biosynthesis: glycosylation of exogenic rubomycinone and carminomycinone by rubomycin-producing organism. Antibiotiki (Moscow) **22**: 37–41.
20. **Vaněk, Z., J. Tax, I. Komersová, and K. Eckardt.** 1973. Glycosylation of ε-pyrromycinone using the strain Streptomyces galilaeus JA 3043. Folia Microbiol. (Prague) **18**:524–526.
21. **Vaněk, Z., J. Tax, I. Komersová, P. Sedmera, and J. Vokoun.** 1977. Anthracyclines. Folia Microbiol. (Prague) **22**:139–159.

Role of Sulfur Metabolism in Cephalosporin C and Penicillin Biosynthesis

H. J. TREICHLER, M. LIERSCH, J. NÜESCH, AND H. DÖBELI

Pharmaceutical Division, Ciba-Geigy Ltd., Basel, Switzerland

This work was carried out to examine the reasons that *Penicillium chrysogenum* and *Cephalosporium acremonium* require different sulfur sources for optimal production of their respective β-lactam antibiotics. Although the direct donor of sulfur to penicillin and cephalosporin C is cysteine, addition of this amino acid to media containing inorganic sulfate does not stimulate penicillin or cephalosporin synthesis.

Penicillin-producing strains of *P. chrysogenum* can obtain sulfur for antibiotic synthesis very efficiently by the reduction pathway from inorganic sulfate. On the other hand, *C. acremonium* obtains it from methionine by reverse transsulfuration (2, 13). In the course of strain development, the inability of the *C. acremonium* wild strain to utilize sulfate efficiently for antibiotic synthesis was modified by mutation (6, 10). Mutants were obtained which still required methionine for optimal production although they could utilize sulfate to some degree (6, 10). Subsequently, mutants were prepared which could use either methionine or sulfate with equal efficiency. These mutants resembled *P. chrysogenum* in their ability to utilize inorganic sulfate for antibiotic synthesis.

In the present study specifically blocked mutants of *C. acremonium* were used to demonstrate the difference in sulfur metabolism between *P. chrysogenum* and *C. acremonium.* Antibiotic production was studied under two different conditions: in the presence of methionine and in the presence of an excess of inorganic sulfate without methionine.

Eighteen different enzymes in higher fungi have been described, which are required for an operative sulfate assimilation and the reverse transsulfuration pathway. As can be seen in Fig. 1, anabolic biosynthesis of cysteine occurs through the reaction of *O*-acetylserine with sulfide in the presence of *O*-acetylserine sulfhydrylase (cysteine synthase, step 7). Cysteine can also be synthesized by means of an alternative sulfide fixation in the presence of *O*-acetylhomoserine sulfhydrylase (methionine synthase, step 9) yielding homocysteine. Homocysteine is then converted to cysteine by the catabolic transsulfuration enzymes cystathionine-β-synthase (step 16) and cystathionine-γ-lyase (step 17). The influence of these two alternative pathways on antibiotic production was investigated in *C. acremonium.* In a comparative study, both alternative sulfide-fixation enzymes (steps 7 and 9) were also investigated in cell-free extracts of the penicillin-producing *P. chrysogenum.*

MATERIALS AND METHODS

Chemicals. *O*-acetyl-DL-homoserine was prepared by the method of Nagai and Flavin as described in a previous publication (13).

O-acetyl-L-serine was obtained from ICN Pharmaceuticals, Inc., Cleveland, Ohio.

L-Amino acid oxidase (*Crotalus atrox*, crude venom; EC 1.4.3.2) was purchased from Serva GmbH, Heidelberg, Germany.

Methane selenol (CH_3-Se-H) was prepared as follows: 30 mg of seleno-DL-methionine (Sigma Chemical Co., St. Louis, Mo.) was incubated with 10 mg of L-amino acid oxidase (EC 1.4.3.2) and 100 μg of catalase in pyrophosphate-HCl buffer (pH 8.0) in a Warburg apparatus carrying

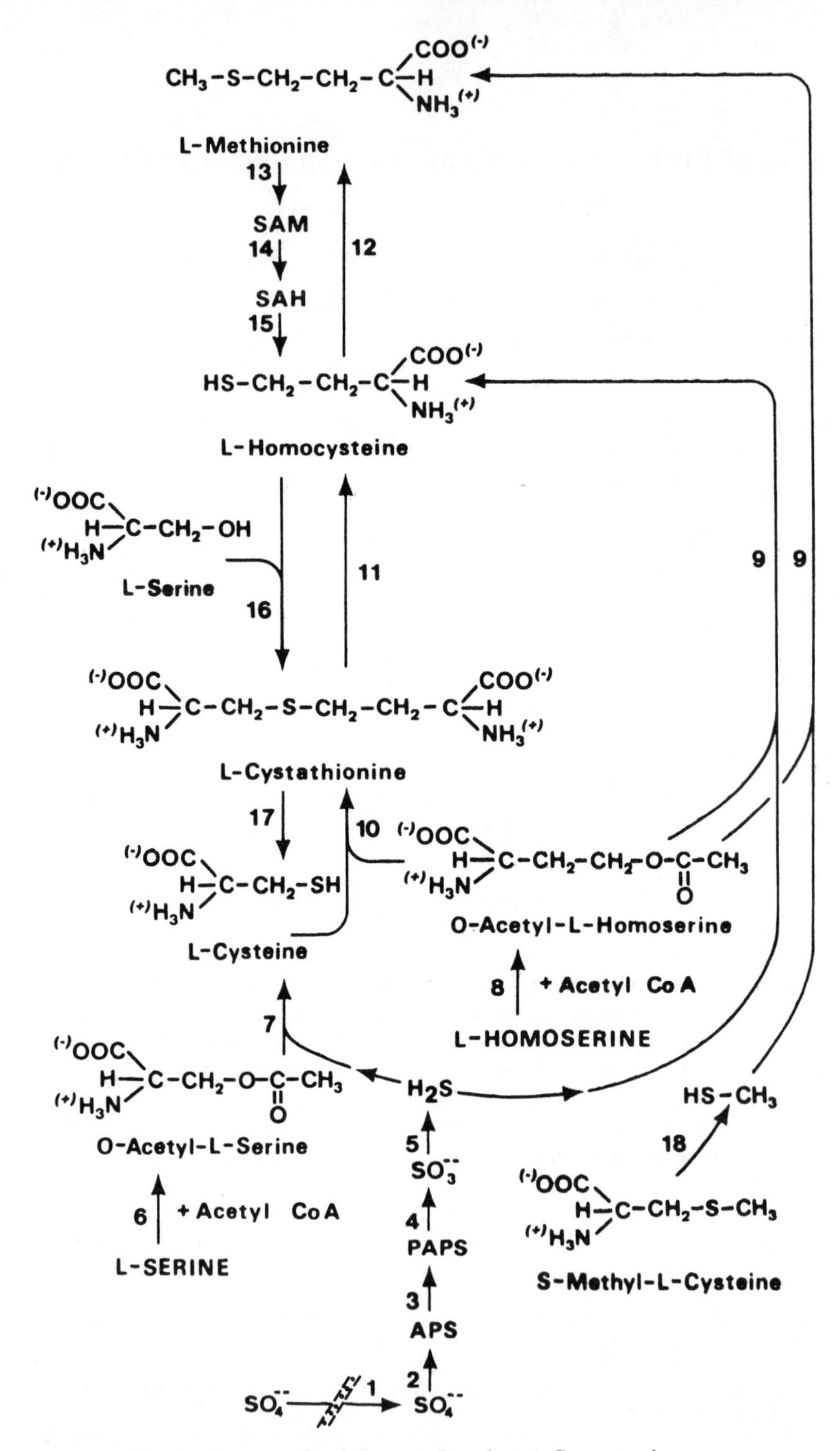

FIG. 1. *Pathways of cephalosporin C synthesis in* C. acremonium.

0.5 ml of 30% KOH in the center well. The enzymatically formed 2-keto-4-methylselenobutyric acid is unstable and readily releases methane selenol (CH_3-Se-H) which is oxidized to dimethyldiselenide (CH_3-Se-Se-CH_3) in the presence of O_2. Gas chromatographic and mass spectrometric analyses revealed CH_3-Se-Se-CH_3 in the gas phase.

Organisms and culture conditions. A superior cephalosporin C-producing strain of *C. acremonium*, designated 8650, was used as parent. Mutagenesis, enrichment procedures, and replica techniques, as well as the isolation of mutants resistant to methane selenol (CH_3-Se-H), have been described (13). The compositions of all the media and the culture conditions used were given in previous papers (5, 11–13).

Preparation of cell-free extracts. Crude cell-free extracts of mycelium grown for 72 h at 25°C were prepared as already described (1).

Analysis and separation. Determinations of cephalosporin C, packed mycelium volume, and dry cell weight, as well as the determination, fractionation, and disc gel electrophoresis of proteins, were conducted as described previously (13). The mobility (from − to +) of the enzymes assayed is given as R_m-value (migration distance divided by that of the buffer front). L-Cysteine was determined with an acidic ninhydrin reagent to which 5 μmol of Cleland's reagent had been added, by the method of Gaitonde as modified by Pieniazek (13).

Enzyme assays. Cystathionine-γ-lyase, *O*-acetyl-L-homoserine sulfhydrylase (L-methionine synthase), and *O*-acetyl-L-serine sulfhydrylase (L-cysteine synthase) activities were assayed as described previously (13). The synthesis of L-homocysteine and of L-methionine was also determined as described previously (13).

RESULTS AND DISCUSSION

A monogenic mutant 8650/*113* lacking cystathionine-γ-lyase (step 17 of Fig. 1; EC 4.4.1.1) was isolated. The mutant was unable to grow on methionine, homocysteine, and cystathionine, but showed an optimal growth response to cysteine and inorganic sulfur sources such as sulfide, thiosulfate, and sulfate. In contrast to the parent 8650, mutant 8650/*113* could not utilize sulfur from methionine, homocysteine, and cystathionine to reverse the toxicity of selenate, but in the presence of L-cysteine and sulfate it was able to reverse the toxicity. The cystathionine-γ-lyase activity of the parent strain 8650 could be precipitated out of the cell-free extract with solid $(NH_4)_2SO_4$ between 40 and 65% saturation. The enzyme (R_m = 0.18) was separated from the *S*-alkylcysteine lyase (R_m = 0.31) by disc gel electrophoresis (9, 13). L-Cystathionine acted as a substrate only for the cystathionine-γ-lyase, whereas *S*-methyl-L-cysteine (and other alkylcysteines) served as substrates for the other enzyme. No enzymatic activity could be found in mutant 8650/*113* on assaying the respective protein fractions for L-cysteine formation when L-cystathionine was used as substrate. The loss of cystathionine-γ-lyase activity has a drastic influence on antibiotic formation. To demonstrate this effect, we compared the fermentation patterns of mutant 8650/*113* and the parent 8560 (Fig. 2). A sulfate-containing synthetic minimal medium with and without DL-methionine was used. In contrast to the parent, cephalosporin production by this mutant was not stimulated by methionine. In the medium supplemented with DL-methionine, mutant 8650/*113* gave about 20% of the antibiotic potency of the parent. These results demonstrate the important function of cystathionine-γ-lyase in methionine-stimulated antibiotic synthesis. Furthermore, the antibiotic potency of the mutant was also found to be significantly lower than that of the parent strain in the sulfate-supplemented methionine-less medium. This fact that β-lactam production in the presence of sulfate as the sole sulfur source is also impaired by the loss of cystathionine-γ-lyase supports the assumption that the formation of cystathionine is a prerequisite for antibiotic synthesis. It appears that the cleavage of cystathionine constitutes an essential step in the primary metabolism, and this reaction may also act as an "inducer" of the transfer of the cysteine moiety into the pathway of secondary metabolism.

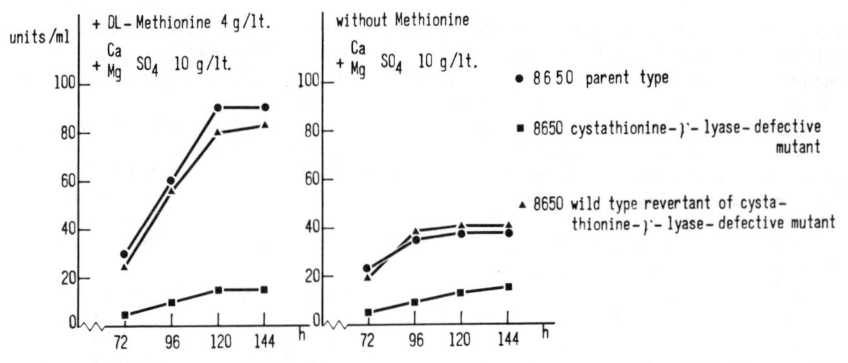

FIG. 2. *Influence of* DL*-methionine and sulfate on cephalosporin C production by* C. acremonium *parent 8650, mutant 8650/113, and revertant 8650/Rev.*[113] *in synthetic minimal medium.*

The formation of cystathionine from inorganic sulfide as the end product of the sulfur assimilation pathway can occur by two alternative routes. Both routes require two enzymatic steps, either reactions 7 and 10 or reactions 9 and 16 in Fig. 1.

A monogenic mutant was obtained, in which both routes of cystathionine formation were blocked by a single mutational event. The mutant was described in a previous publication (13) as defective in the acetylation of homoserine (step 8 in Fig. 1; homoserine-*O*-acetyltransferase, EC 2.3.1.31). The mutant was designated $OAH^-/SeMe^R$, because it has a unique requirement for *O*-acetylhomoserine (OAH) and a concomitant resistance to methane selenol (CH_3-Se-H). Conidia of *C. acremonium* are extremely sensitive to methane selenol when grown on minimal medium; its inhibitory effect can be reversed by methylmercaptan or methionine. Two distinct classes of nutritional mutants were found among the strains of stable inherited resistance to methane selenol. The latter type, $OAH^-/SeMe^R$ (mutant class I in Table 1), lacks the ability to utilize inorganic sulfur sources, cysteine, and *S*-methyl-L-cysteine when OAH is absent, but grows when OAH is present. Because of its inability to synthesize OAH as the common acceptor substrate for both metabolic routes leading to cystathionine (steps 9 and 10 in Fig. 1), this mutant cannot convert cysteine or any form of inorganic sulfur to cystathionine. Consequently, mutant $OAH^-/SeMe^R$ grows on methionine, homocysteine, and cystathionine, but not on cysteine and inorganic sulfur sources.

In synthetic minimal medium supplemented with a level of methionine low enough to support optimal biomass production but not antibiotic production, mutant $OAH^-/SeMe^R$ does not show increased cephalosporin production in the presence of an excess of cysteine or sulfate (Fig. 3). Conversely, the parent wild type 8650 or revertants of the mutant give reasonably good cephalosporin C yields on addition of excess cysteine or sulfate. However, mutant $OAH^-/SeMe^R$ could not produce cephalosporin C unless a high level of methionine was added. Since the parent was able to produce considerable amounts of cephalosporin C in cysteine or sulfate, but the mutant could not, it is evident that blocking the two alternative pathways from cysteine to cystathionine virtually eliminated cephalosporin C production from cysteine or any form of inorganic sulfur. These observations support the assumption that cystathionine is metabolically closer, and therefore more readily converted, to the β-lactam than cysteine. Experimental data obtained by Drew and Demain (3) with a double mutant with an early deficiency in sulfur assimilation and a second, as yet undetermined, block in the transsulfuration from cysteine to methionine are in

TABLE 1. *Nutritional responses of methane selenol (CH_3-SE-H)-resistant mutants of* C. acremonium *to various sulfur sources in minimal medium*

Strain	No sulfur source		Inorganic sulfur sources		Cysteine		Methionine, homocysteine, or cystathionine		S-methylcysteine		Methylmercaptan in sealed plates	
	+[a]	−	+	−	+	−	+	−	+	−	+	−
Parent wild type	0[b]	0	60	60	60	60	60	60	60	0	G[c]	G
Mutant class I	0	0	60	0	60	0	60	60	60	0	G	NG
Mutant class II	0	0	60	60	60	60	60	60	0	0	NG	NG

[a] O-acetylhomoserine added (+) or omitted (−).
[b] 0, No growth or growth inhibition; 60, maximal and extensive mycelial growth.
[c] G, Growth; NG, no growth.

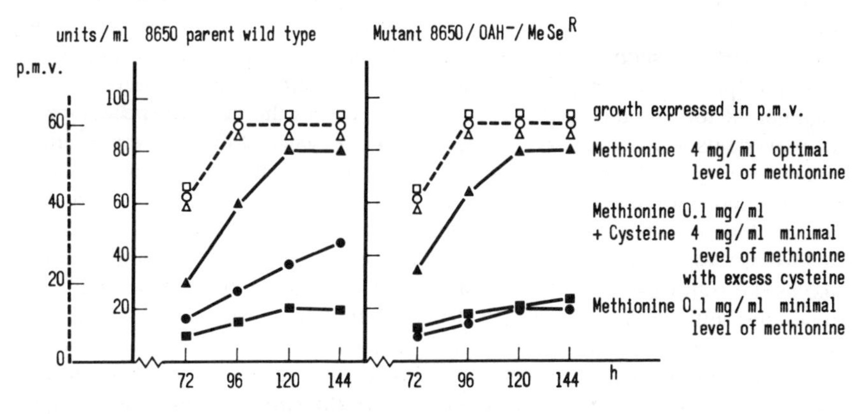

FIG. 3. *Cephalosporin C production by* C. acremonium *parent wild type 8650 and mutant 8650/OAH⁻/MeSe*[R] *in synthetic minimal medium supplemented with methionine and cysteine.*

agreement with our results. The fermentation pattern of the double mutant described (3) corresponds to the behavior of our mutant type $OAH^-/SeMe^R$.

Based on the assumption that the cleavage of cystathionine mediates the incorporation of the cysteine moiety into the β-lactam pathway, it was our aim to block one (step 9 in Fig. 1) or the other (step 7 in Fig. 1) of the two alternative sulfide fixation routes leading to cystathionine. A series of methane selenol-resistant mutants were obtained; they were designated $SMcy^-/MeSe^R$, because they are capable of utilizing the same sulfur sources as the parent except for S-methyl-L-cysteine and methylmercaptan. The mutants were described in a previous publication (13) as defective in O-acetylhomoserine sulfhydrylase (methionine synthase, step 9 in Fig. 1, EC 4.2.99.10). Methionine synthase, which is blocked in these mutants, is the only enzyme capable of condensing OAH with methylmercaptan to yield L-methionine in vivo.

No significant difference was found between the mutant and the parent strain when methionine was used as the sole sulfur source. In the sulfate-supplemented methionine-less medium, on the other hand, the antibiotic production in the $SMcy^-/MeSe^R$ mutants was significantly lower than that in the parent and the respective revertants. This impairment in productivity suggests that antibiotic production is largely obtained through cystathionine formation by means of sulfide fixation in the presence of *O*-acetylhomoserine sulfhydrylase (step 9) and cystathionine-β-synthase (step 16).

To clarify whether the simultaneous function of both routes has an additive, synergistic, or antagonistic effect on productivity, we blocked the other route of anabolic cysteine synthesis also. The only means of isolating this mutant type was the mutagenesis of a parent strain with impaired transsulfuration (step 16 or 17 in Fig. 1) or methionine synthase (step 9) and subsequent removal of the parental marker by reversion. Strain 8650/*113* (cystathionine-γ-lyase$^-$) was used as the parent. We have isolated independently several mutants which grew only on cysteine. None of the inorganic sulfur sources, methionine, homocysteine, or cystathionine, was able to replace cysteine in these double mutants. In addition, it was found that all such mutants were "methionine and homocysteine sensitive." This inhibition could only be reversed by cysteine. *O*-acetylserine sulfhydrylase activity (step 7; EC 4.2.99.8) was tested in cell-free extracts. The enzyme level was five- to sevenfold lower in the mutant than in the parent strain. It was not possible to isolate mutants which have lost all their *O*-acetylserine sulfhydrylase activity. Subsequently, revertants of the parental marker cystathionine-γ-lyase (step 17) were isolated. This reversion led to a wild-type phenotype as a result of renewed ability to utilize the alternative route of cystathionine formation (steps 9, 16, and 17). Because of the wild-type phenotype of these revertants, it was necessary to check for the presence of *O*-acetylserine sulfhydrylase activity in cell-free extracts. As expected, it corresponded to the parent double mutant in most of the revertants and was, therefore, unaffected by the reversion of the cystathionine-γ-lyase. In synthetic minimal medium no significant difference in productivity was found between the mutant and the parent when methionine was used as the sulfur source (Fig. 4A, B). On the other hand, the impairment of *O*-acetylserine sulfhydrylase activity had a positive effect on cephalosporin C production in methionine-less minimal medium supplemented with an excess of sulfate. Whereas in the parent productivity is stimulated exclusively by methionine, and remains at a low level on addition of excess sulfate (Fig. 4A), the latter mutants reached equal potency levels with an excess of sulfate or optimal amounts of methionine (Fig. 4B). Hence, impairment of this anabolic enzyme resulted in a constant tendency to give enhanced titers with excess sulfate.

This observation supports the assumption that a specific block in the final step of anabolic cysteine synthesis may facilitate the conversion of sulfide to cystathionine by the remaining alternative pathway.

On the basis of this finding, enzyme activities of cysteine biosynthesis were studied in strains of the penicillin-producing *P. chrysogenum*. No *O*-acetylserine sulfhydrylase (cysteine synthase) activity could be detected in cell-free extracts of *P. chrysogenum* during the production phase in sulfate-containing synthetic and complex media. However, strong *O*-acetylhomoserine sulfhydrylase (methionine synthase) activity was present in both media. This fact supports the previous assumption made from *C. acremonium* fermentation data, that the alternative pathway from sulfide to cystathionine via homocysteine is the main route for optimal β-lactam synthesis in *P. chrysogenum.* The ability of the wild-type strains of *P. chrysogenum* to use sulfate for antibiotic synthesis could be explained by the fact that they are capable of a preferential utilization of the alternative route. This is not the case in *C. acremonium* because cell-free extracts thereof exhibit a strong *O*-acetylserine sulfhydrylase activity. Our data from experiments with *C. acremonium* also indicate that the activity of the latter enzyme exerts an inhibitory effect on the function of the alternate pathway. This would explain why structural mutants of this enzyme show a greater ability to

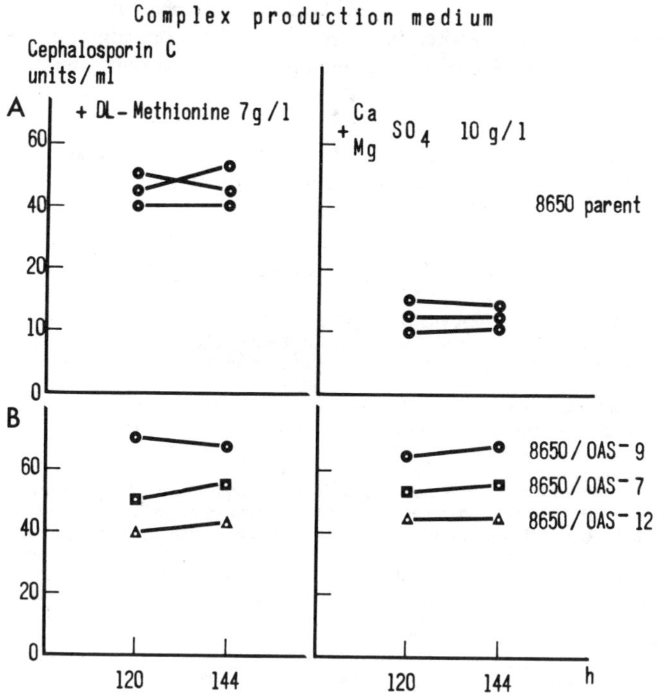

FIG. 4. *Cephalosporin C production by* C. acremonium *parent 8650 and by mutants with impaired* O-*acetylserine sulfhydrylase activity in minimal medium supplemented with methionine or with an excess of sulfate.*

utilize inorganic sulfur sources by the alternate pathway. It is still unclear whether the stimulation or inhibition of the alternate route is exerted by a single mechanism or is the result of multiple effects. A *C. acremonium* mutant with an enhanced potential to utilize sulfate for cephalosporin C production was described by Komatsu and Kodaira (6). The mutant was isolated in the course of strain development and had an enhanced cystathionine-β-synthase activity (EC 4.2.1.22, step 16 in Fig. 1). Furthermore, no significant difference in the *O*-acetylserine sulfhydrylase activity between the parent and the mutant was found. These data are in accordance with our observations that an alteration which stimulates activities of enzymes involved in the alternative route of cystathionine formation can help to bring about a higher β-lactam production, whereas direct anabolic synthesis of cysteine by *O*-acetylserine sulfhydrylase has either no effect or even a negative influence on productivity. Based on the assumption that the cleavage of cystathionine is essential for the incorporation of the cysteine moiety, any alteration which raises the cystathionine pool may facilitate antibiotic production. Moreover, a synergistic effect of sulfate and methionine on the productivity was observed in OASS$^-$ mutants (see Fig. 4B). These strains were actually capable of utilizing methionine and sulfate with equal efficiency, but methionine had to be added in a lower concentration because of the methionine sensitivity of the respective OASS$^-$ mutants. A similar observation was described by Komatsu et al. (7). A series of mutants with increased ability to produce cephalosporin C from sulfate were found to be methionine-sensitive (7).

Sulfate and methionine are eventually converted to cystathionine by two common steps (16 and 17 in Fig. 1) when the alternative route is operating, but the first step of the latter route (step 9 in Fig. 1, methionine synthase) is only operative in sulfate-mediated cystathionine formation. We have found that methionine synthase is repressed by high concentrations of exogenous methionine but is stimulated by an excess of sulfate. This would explain why mutants in which the alternative route is operating have an increasing cephalosporin C productivity from sulfate and a rather restricted tolerance of methionine.

REFERENCES

1. **Benz, F., M. Liersch, J. Nüesch, and H. J. Treichler.** 1971. Methionine metabolism and cephalosporin C synthesis in Cephalosporium acremonium. Eur. J. Biochem. **20:**81–88.
2. **Caltrider, P. G., and H. F. Niss.** 1966. Role of methionine in cephalosporin synthesis. Appl. Microbiol. **14:**746–753.
3. **Drew, S., and A. L. Demain.** 1975. Production of cephalosporin C by single and double sulfur auxotrophic mutants of *Cephalosporium acremonium.* Antimicrob. Agents Chemother. **8:**5–10.
4. **Drew, S., and A. L. Demain.** 1975. The obligatory role of methionine in the conversion of sulfate to cephalosporin C. Eur. J. Appl. Microbiol. **2:**121–128.
5. **Hinnen, A., and J. Nüesch.** 1975. Enzymatic hydrolysis of cephalosporin C by an extracellular acetylhydrolase of *Cephalosporium acremonium.* Antimicrob. Agents Chemother. **9:**824–830.
6. **Komatsu, K. -I., and R. Kodaira.** 1977. Sulfur metabolism of a mutant of Cephalosporium acremonium with enhanced potential to utilize sulfate for cephalosporin C production. J. Antibiot. **30:**226–233.
7. **Komatsu, K. -I., M. Mizuno, and R. Kodaira.** 1975. Effect of methionine on cephalosporin C and penicillin N production by a mutant of Cephalosporium acremonium. J. Antibiot. **28:**881–888.
8. **Liersch, M., J. Nüesch, and H. J. Treichler.** 1973. Methioninstoffwechsel und Cephalosporin C Synthese in Cephalosporium acremonium. Pathol. Microbiol. **39:**39.
9. **Liersch, M., J. Nüesch, and H. J. Treichler.** 1977. Enzymes of the sulfur metabolism of Cephalosporium acremonium assayed after disc electrophoresis. Experientia **33:**16, 88.
10. **Niss, H. F., and C. H. Nash.** 1973. Synthesis of cephalosporin C from sulfate by mutants of *Cephalosporium acremonium.* Antimicrob. Agents Chemother. **4:**474–478.
11. **Nüesch, J., A. Hinnen, M. Liersch, and H. J. Treichler.** 1975. A biochemical and genetical approach to the biosynthesis of cephalosporin C, p. 451–472. *In* K. D. Macdonald (ed.), Second international symposium on the genetics of industrial microorganisms. Academic Press, London.
12. **Nüesch, J., H. J. Treichler, and M. Liersch.** 1973. The biosynthesis of cephalosporin C, p. 309–334. *In* Z. Vaněk, Z. Hošťálek, and J. Cudlín (ed.), Genetics of industrial microorganisms, vol. 2. Academia, Prague.
13. **Treichler, H. J., M. Liersch, and J. Nüesch.** 1978. Genetics and biochemistry of cephalosporin biosynthesis, p. 177–199. *In* R. Hütter, T. Leisinger, J. Nüesch, and W. Wehrli (ed.), Antibiotics and other secondary metabolites. FEMS Symp., 5th, Basel, Switzerland, 1977. Academic Press, London.

Process Needs and the Scope for Genetic Methods

J. D. BU'LOCK

The University of Manchester, Manchester, England

Genetic methods, both new and old, are now part of the standard armory of the biotechnologist. The problems upon which they and other methodologies are brought to bear are conditioned, like all problems in applied science, by the double limitations of what is technically feasible and what is economically (or socially or politically) opportune, and neither the methods, nor the total technology, nor the socioeconomic bounds, are unchanging. In this survey I shall try to outline at least some of their interactions as they seem likely to occur in the immediate future. The processes I shall consider are of two kinds, typifying the extremes of the twin scales of production volume and product value on which fermentation industries are conveniently classified. I shall begin at the low-volume high-value end of the spectrum.

HORMONES AND SIMILAR PEPTIDES, ETC.

In this category I include a whole range of physiologically active substances, mostly of mammalian origin, whose appearance as targets for fermentation technology is a wholly novel phenomenon which rests almost entirely upon the promise (and, I might say, the promises) of the "new genetics." Specifically, their availability in workable amounts, far beyond what animal physiologists have found it feasible to prepare from their native sources, rests upon the possibility of "illegitimate" gene transfers. It would be entirely inappropriate, and indeed presumptuous, for me to review here the range of very refined and powerful techniques that can in principle be deployed. This has been more authoritatively done by other contributors, far better qualified to offer that account and to include in it some of the first signs of real success.

The range of targets which can be envisaged is indeed considerable: release factors, fertility control agents (inhibin), enkephalins, the "growth hormone" complex, other pituitary and hypothalamus agents, urogastrins etc., insulin, interferon, and chalones(?). As mammalian physiology advances—to be assisted not least by the availability of some of these agents themselves—others will be added. There is conjectural scope for extension into fields of nonmammalian physiology, particularly for regulators of insect and of fish development, and other conceivable targets include the synthesis of "natural" gene regulators as an approach to the treatment of some genetically determined disorders.

At present, we are only a little way advanced into this new and exciting area, and our first steps must be quite tentative. While our understanding of coding and transcription mechanisms in the higher eucaryotes remains so limited, routes based on direct transfer of mammalian genes into procaryote hosts remain very problematic.

However, it is precisely by attempting such approaches that the unknown terrain can be most effectively explored. Meanwhile, the more limited and cautious technique of rewriting the desired message in the language of the host, i.e., by synthesis of the appropriate DNA by use of the laborious but effective methods now available, is securing its first successes, notably in the example of the successful microbial synthesis of somatostatin. Other peptides of comparable size, and importance, will surely be forthcoming in the quite near future; the foreseeable success rate will surely be sufficient to confirm the ultimate correctness of the more courageous and longer-term decisions in which the more difficult hormones are a serious target. There is some urgency. Even for rather readily obtained products such as insulin the net availability of the usual animal sources is foreseeably limiting and compares unfavorably with the potentially unlimited scale of microbial production.

However, let us not overlook the fact that gene transfer will only be the first step in, for example, a successful insulin process. Both from first principles and from increasing practical experience, we can be sure that the mere insertion of a gene will not be enough to create a manufacturing process. Not all missionaries of the new faith are so frank about the difficulties as Dr. Boyer has been at the GIM-78 meeting. The gene must be expressed, and only the first part of the expression process is directly catered for by gene manipulation. Dr. Apirion has reminded us of the ribonucleases that lie in wait for the unaccustomed messenger, and beyond them are stationed the peptidases. For a working process we will also need to ensure adequate rates of substrate supply (and activation), a "rescue" system for product transport, and an effective work-up process. Given that many of the target products are rather similar, so far as these requisites are concerned, it may not be too early to begin work on the development of a really suitable multipurpose "host" fermentation, and not necessarily one using *Escherichia coli,* crippled or not. There are notorious sociopolitical issues involved here, and I do not propose to add to their discussion.

ANTIBIOTICS

For all the glamour of the mammalian hormones as targets, it is the search for new antibiotics, and for the successful realization of antibiotic production processes, which has in the past provided—and indeed still provides—the bread-and-butter for industrial genetics. This is easily confirmed by analysis of the *contributed* papers at the GIM-78 meeting; of those with an identifiable relevance to production processes, the proportions in the following major categories were as follows: antibiotics, 45%; brewing, enzymes, and foods, 20%; waste treatment, 12%; amino acids and single-cell protein (SCP), 16%.

The antibiotics industry, for reasons which are themselves not wholly devoid of sociopolitical content, has depended strongly upon innovation, always coupled with the economic pressure to maximize current productivity (in which "strain improvement" plays a vital role). Genetics, often of a rather rough-and-ready sort, but increasingly of a directed and informed nature, has always played a vital role in productivity improvement and in recent years has also been recruited into the search for innovation.

The search for new antibiotics has for several years shown a "diminishing return." This is often exaggerated, not least by the increasing reluctance of many companies to embark on new products of less-than-universal utility because of the near-prohibitive demands of the regulatory agencies. However, even allowing for this, the effect has been real enough to provoke many serious reexaminations of the primary screening processes by which new antibiotic discoveries are generated. In devising satisfactory new screening procedures, genetics takes it place alongside our knowledge of microbial and mammalian physiology as part of the foundation for more searching, more discriminating, and more relevant detector screens. The use of "supersensitive" strains, when available, allows the detection of low-level producers which one hopes can be improved at a later stage, as well as permitting the scaling-down of improvement screens to allow greater numbers of tests and higher degrees of automation. Conversely, the use of resistant strains as test screens allows the selective detection of new antibiotics that may be more useful in an increasingly acute clinical situation (largely man-made) and of agents which may well extend the usefulness of well-established antibiotics in the same circumstances. The use in suitable screening tests of organisms with defined mutational lesions, usually auxotrophs, allows a search for specific enzyme inhibitors whose possible utility has been seen in an informed view of either host or pathogen physiology. Probably all the detector screen systems in present use employ specially selected strains of microorganism as part of the test procedure, and we may confidently expect this to continue. However, the part played by an increasingly sensitive understanding of the nature of *disease*, as opposed to a simplistic view of *infection*, will be at least as great. Here, advances in mammalian, rather than microbial, physiology are in demand.

Rather more in tune with the aspirations (and speculations) of the new genetics are the uses of genetic methods to generate novelty directly. The use of blocked mutants of established antibiotic producers to accumulate novel products is, of course, well known. In the great majority of cases, the mutant products have been either direct precursors of the known antibiotic or "lateral" conversion products formed from such precursors. By such methods, much has been learned about antibiotic biosynthesis (thereby facilitating the intelligent application of other techniques), but only occasionally have the mutant products been of any intrinsic interest. Perhaps the technique has been more generally valuable in simplifying a spectrum of products, in the interest of productivity, than as a generator of novelty. Moreover, in many cases, the effective mutations have almost always been ones affecting the later stages of antibiotic biosynthesis and thus have, in general, affected mainly the arrangement of functionalities around a core structure; chemical and/or microbial modifications of the antibiotic offer at least as good a route for the generation of novelty of this kind. The underlying reason for this situation is well established; it is that in many classes of antibiotic the earliest part of the biosynthesis, from which the core structure arises, is effected in an all-or-nothing system of multiple enzymes. This common feature is found in the biosynthesis of many quite different series of antibiotics. In such cases deletion mutants affecting component enzymes of the complex nearly always result in nonproduction.

However, if we compare structures within any one such group of antibiotics—for example, the erythromycin-like macrolides, or the bacterial oligopeptides, or the polyene macrolides—we can see that the different members of the series differ in a number of independently varying respects, for example, in the sequence of precursor units, local reactions at each precursor unit site, template-directed processes over different regions of the molecule, and late-stage conversion reactions (see above). Each of these aspects must be governed by at least one, and usually several, correspondingly varied macromolecules. It is precisely the promise of "illegitimate" hybridizations that such macromolecules, coming from different species, can be re-assorted. It follows that we can look at the newer techniques, particularly those for transferring whole chromosomes or large chromosome fragments and perhaps happily for making such transfers between quite closely related species, in the reasonable expectation of generating very substantial novelties in antibiotic structure. We may indeed comment that precisely such processes seem to have occurred naturally, in generating the variety we already see, but that is another matter. The manner in which the genetic capabilities for synthesis of particular metabolites are distributed among different organisms is remarkable both for its general conformity with accepted taxonomy and for the surprisingly frequent transgressions which to some of us imply "naturally illegitimate" transfers between widely disparate species. Particularly striking examples would be the occurrence of gibberellins in a *Fusarium* species and of ergot alkaloids in morning glory, as pointed out by Leo Vining some years ago. Be that as it may, the beginnings of a new chapter in the antibiotic "saga" can already be read, and we can be rather confident that it will be an exciting episode.

Meanwhile, there is much to do in expediting the process of strain improvement for those antibiotics already found to be desirable products. Here, the old confrontation between the academic geneticist, who tells us how easy it will all be when we do it his way, and the industrial worker, who tells us that his way is neither rational nor efficient but at least it works, has eased perceptibly, but it still exists. The truth is, of course, somewhere in between. In practical terms, a rather wider range of methods than the academics allow is generally adopted. Random mutant selection is still a very useful way of combining useful improvement steps with the creation of just that degree of genetic diversity which the various recombination methods can then exploit, always provided that they can be made to work with the material at hand. Directed mutant selection is a better way of creating that diversity, but it depends upon basic understanding of both primary and secondary metabolic pathways *in relevant species*, and, surprisingly, this is still coming far too slowly. The industrial laboratories have relatively little time for such studies, while the academic laboratories still find the "traditional" organisms both more accessible and more fashionable. We know far more about the genetics of some streptomycetes than we do about their biochemistry. Only at the GIM-78 meeting, for example, have the sources of the malonyl-coenzyme A used in tetracycline biosynthesis been made clear (by Z. Hošťálek and co-workers), whereas we have seen numerous "maps" bearing accurately located genes whose biochemical expression is wholly unknown. As a recent traveller across the wide spaces of America's Midwest, I could comment that a map which only shows Howard Johnson's is good for publicity but useless for navigation. More seriously, it becomes increasingly important to identify the biochemical expression of mapped genes, and this calls for a *combination* of expertise.

Another very practical aspect to which the new genetics might well pay closer attention is the very considerable situation difference between the early steps of strain improvement for rather new antibiotics and the very late steps of strain improvement (or even productivity maintenance) that are continuously important in the production of well-established antibiotics. In the one, there is very great scope for improvement but little diversity out of which to create it; in the other, there is usually plenty of genetic diversity available but the chances of significant improvement are very low.

Possibly quite different types of recombination method are suited to these very different situations, but if so this has yet to be spelled out for us. On the other hand it has been made explicit by several contributors to the GIM-78 meeting that even the best recombinant methods must be given a fair chance; i.e., the genetically diversified strains still need to be screened in significantly larger numbers than have sometimes been examined. As to the screening itself, the contribution of genetic modifications of test organisms to the devising of faster, more sensitive, and more easily replicated screens has already been noted.

SOME MISCELLANEOUS PRODUCTS

Before considering the high-volume low-value processes at the other end of the biotechnology spectrum, a very few remarks about some intermediate cases may be useful. In the field of enzyme production, for example, the combination of both new and old genetics with other process requirements seems to be proceeding very happily, and provided that economically viable enzyme uses can be correctly identified in advance the overall prospects are bright. Not all of us, for example, are wholly convinced by the current enthusiasm (particularly in the United States) for cellulase, but the determinants here are a very subtle blend of social, political, and economic factors which vary quite sharply from one country to another. Correspondingly, the Western enthusiasm for conspicuous nonfoods, for aids to safer weight loss, and so forth, looks rather different from less "advanced" countries where the dollar cost of soymeal is a more categorical imperative. The future of the amino acid fermentations may also be in some doubt given the recent development of combined chemical and fixed-enzyme processes as alternatives, and this is an interesting situation for us, because the main process disadvantage of the established amino acid fermentations results from the same genetic considerations that have made those fermentations possible in the first place. Few if any of these fermentations can be run continuously, and this now constitutes a serious limit on their productivity. The reason is that under continuous fermentation conditions many spontaneous mutations towards lower amino acid production are likely to confer growth advantage. The greater the mutational load imposed to improve productivity, the greater this negative selection effect becomes.

SCP, ALCOHOL, ETC.

For large-volume low-cost processes, such as the production of microbial biomass or fermentation ethanol, the constraints imposed by the whole process technology take precedence over most other considerations. For example, the productivity advantage of continuous fermentation is decisive, and the corresponding need for genetic stability in the producing organism is therefore emphasized. In biomass production this particular problem will actually be less acute, since in general the chemostat conditions provide selection pressure for characteristics which in this case are desirable—the combination of high growth rate and high substrate affinity at the imposed temperature, pH, etc.—and indeed the chemostat provides a powerful selection system for just such strains. Thus, within reasonable limits other types of optimization become acceptable targets for genetic adjustment. Some of these have to do with the fermentation itself, for instance, modest adjustments of overall protein composition and RNA content (though the latter is best coped with in product recovery). Other improvement targets center on aspects not always considered as part of the fermentation, and in particular upon the product recovery process. The improvement of characteristics such as flocculation and ease of dewatering may be as important in successful process design as any other single factor, and in these respects long-term strain stability cannot be presumed. The concept of strain improvement in respect to the whole process, rather than for the fermentation step alone, is relatively new to some, though we are reminded that it has long been a governing principle in brewing.

One of the most difficult parameters in SCP processes is temperature. Under some conditions the heating and cooling costs of the process may be quite marginal, but under others, notably in tropical and semitropical zones, they can be decisive. By using the chemostat as a selector system, the temperature tolerance of most organisms can be raised, though the problem of maintaining other characteristics (notably yield factors) at higher temperatures is considerable. In general, strains with a higher temperature tolerance will also have higher temperature optima; however, observation suggests that to achieve this result, selection between wild types will be at least as effective as selection between mutants, and probably more so. Here, the tendency of genetic research to concentrate on a limited number of organisms must be guarded against insofar as it tends to close the researcher's eyes to the tremendous diversity of microorganisms "in the wild." For example, it is from studies of wild types that our increased understanding of thermotolerance mainly derives. Over the past few years, it has become increasingly clear that thermotolerance has many causes, not just one or two; i.e., it is a polygenic character.

The development of new microbiological processes for waste treatment overlaps to some extent with the development of SCP processes, and although there is no time to consider its more specialized aspects we should certainly note that this is becoming one of the most promising application fields for industrial genetics.

In contrast to SCP production, serious reexamination of the process technology of anaerobic fermentations is only just beginning. At a time when gasoline-alcohol mixture is just returning to the gas stations of Iowa, their consideration here is perhaps apt. Although major production programs for ethanol are under way in several countries, and others are committed to various stages of feasibility study and/or foreseen development, the technology of the actual production systems in current use remains largely conventional and is almost certainly, and significantly, suboptimal. For 30 years, alcohol production technology has been sheltered by the wine snob, the ale snob (a very British species), and the excise man. However, out of several radically different technologies now being advocated for industrial alcohol production no clear "winner" has yet emerged, and this in turn leaves us uncertain as to whether or not some targets for genetic improvement are relevant. For example, it is entirely a matter of process choice whether alcohol tolerance is a useful characteristic to seek to improve genetically. In a conventional batch process it is the combination of alcohol production rate and alcohol tolerance which is decisive, whereas in a continuous process with yeast recycle the sedimentation characteristics of the yeast are at least equally important (since these determine, in any given system, the biomass density which can be maintained in the reactor itself). In a plug-flow fixed-cell reactor system, both substrate tolerance (initial) and alcohol tolerance (final) are important, but neither is decisive, whereas, in a system in which alcohol is continuously removed from the fermentation, tolerance is virtually irrelevant (though other characteristics may be critical). Here, then, the successful engineering of a process takes first priority. On the other hand, other characteristics are already recognizable as targets for genetic improvement. These include the detailed optimization of sugar utilization with reference to particular substrate compositions (such as has become traditional in the potable alcohol field, where, however, the end targets are rather different) and, once again, temperature tolerance. The importance of the latter in regard to process control has already been noted; in relation to total thermal economy, however, much again depends upon other aspects of the total process. It is not generally realized, for example, that when the ultimate treatment for stillage residues is evaporation—for example, to give an animal feed concentrate—

the thermal cost of the distillation step per se is absorbed into the total evaporative heat requirement. Nevertheless, whatever the total process concept (and whatever the plant location), it will be advantageous to run the fermentation at the highest possible temperature, so that, as for SCP, a combination of wild-type selection and strain improvement for temperature tolerance is highly desirable.

In this respect, we should note that the search for procaryotic anaerobes giving a useful range of fermentation products is once again being actively pursued, and with some early signs of success. We are not necessarily "condemned" to the yeasts, attractive though some of us find them, and this is important since it opens the way not only to higher operating temperatures but also to a greater range of substrates and of products. Perhaps it also opens the way to more immediate applications for the "new genetics," in the organisms where it is still most at home.

Genetic Approaches to Nitrogen Fixation

J. E. BERINGER AND A. W. B. JOHNSTON

John Innes Institute, Norwich NR4 7UH, United Kingdom

To date only procaryotic organisms have been shown to be able to fix atmospheric nitrogen. Given the level of genetic manipulation in some procaryotes and the great economic importance of biological nitrogen fixation, it is perhaps surprising that so little has been done on the genetics of nitrogen-fixing microorganisms (see Table 1).

By far the most detailed studies have been on the nitrogen fixation (Nif) genes of *Klebsiella pneumoniae.* Nine closely linked Nif genes have been identified, although the functions of several of them remain to be determined (5). These studies have also illustrated some of the complexities in Nif gene expression; both NH_4^+ and O_2 are involved in the regulation. Clearly, there is a need for analogous work to be done with other nitrogen-fixing species; linkage maps are available only for two species of *Rhizobium* (2, 7, 8).

Several nitrogen fixers associate with eucaryotes, the most notable of these associations being the symbiosis between *Rhizobium* and legumes. In such cases a full genetic analysis of the nitrogen-fixing process must pay regard not only to the Nif genes of the type found in *K. pneumoniae* but also to the other determinants required for the establishment of the symbioses. Genetic analysis of the hosts will presumably also be required to identify the steps in the development of symbiotic relationships. Studies of two legume hosts, the pea (*Pisum* spp.) and the soybean (*Glycine max*), are already well established.

Prerequisites for biological nitrogen fixation are as follows (see 10):

1. ATP (5 to 30 mol per N_2 reduced)
2. Fe and Mo
3. Equable temperature (above 30°C unusual)
4. Exclusion of O_2
5. Exclusion of hydrogen ions (wasted energy in H_2 evolution)
6. Regulation (to avoid waste of energy)

The main technical problem of manipulating nitrogen-fixing microorganisms is the extreme sensitivity of nitrogenase to oxygen, which inactivates it irreversibly. The following methods for avoiding the problem have evolved (see 13):

1. Avoidance (anaerobes and facultative anaerobes)
2. Respiratory protection (*Azotobacteriaceae)*
3. Conformational changes (*Azotobacter*)
4. Association with other macromolecules (e.g., membranes)
5. Microaerophilic growth (e.g., slime production, etc.)
6. Compartmentalization (e.g., heterocysts)
7. Oxygen carriers (e.g., leghemoglobin in legume root nodules)

Many of these methods allow sufficient access of oxygen into the microorganism so that oxidative phosphorylation can provide the ATP to satisfy the large energy requirements for nitrogen fixation. Anaerobes and facultative anaerobes avoid the problem of oxygen denaturation by ensuring that nitrogenase is synthesized only at sufficiently low oxygen tensions. We have little idea of the complexity of oxygen-protection systems in aerobes and, in particular, of the number of genes that may be required. A genetic analysis of oxygen protection systems is an important prerequisite to establishing Nif genes in new aerobic organisms.

TABLE 1. *Nitrogen-fixing procaryotes*[a]

Family	Genus[b]
Azotobacteriaceae	*Azotobacter* Tf, Cj
	Beijerinckia
	Derxia
	Azotococcus
Enterobacteriaceae	*Klebsiella* Td, Cj
	Escherichia Tf, Td, Cj
Bacillaceae	*Bacillus* Tf, Td
	Clostridium
Rhizobiaceae	*Rhizobium* Tf, Td, Cj
Thiorhodaceae	*Chromatium*
	Ectothiorhodospira
Athiorhodaceae	*Rhodopseudomonas* Cj
	Rhodospirillum
	Rhodomicrobium
Chlorobacteriaceae	*Chlorobium*
	Chloropseudomonas
Spirillaceae	*Desulfovibrio*
	Desulfotomaculum
	Spirillum
Mycobacteriaceae	*Mycobacterium*
Frankiaceae	*Frankia*
Methane-oxidizing bacteria	
Chroococcaceae	*Gloecapsa*
Nostocaceae	*Anabaena*
	Anabaenopsis
	Aphanizomenon
	Cylrindrospermum
	Gloeotrichia
	Nodularia
	Nostoc Cj
Oscillatoriaceae	*Plectonema*
Stigonemataceae	*Mastigocladus*

[a] See Burns and Hardy (3).

[b] Cj, Conjugation; Td, transduction; Tf, transformation. *Escherichia coli* has been made into a nitrogen-fixing microorganism by transfer of *Klebsiella pneumoniae* Nif genes (6).

The supply of energy is a major limitation on nitrogen fixation because of the large amounts needed for nitrogenase activity and because of the metabolic requirements of the populations of nitrogen-fixing microorganisms that are needed to fix significant amounts of nitrogen. Photosynthetic procaryotes produce their own energy and are limited mainly by internal factors. Other nitrogen-fixing microorganisms are dependent upon supplies of energy-rich substances produced by plants. Competition for these substances with other soil organisms and their relatively low level in the soil are significant limitations on fixation by free-living procaryotes. The efficiency with which available energy can be utilized for nitrogen fixation is greatly increased when there is a specific interaction between the plant and microorganism. At its most efficient this can be within a specific structure, such as *Rhizobium* spp. within a root nodule. All the available energy can be utilized directly by the microorganism, and the plant does not need to compete for the fixed nitrogen.

Regulation of the activity of nitrogenase is fundamental to an efficient metabolism in nitrogen-fixing microorganisms, and "wild-type" strains do not usually excrete fixed nitrogen into the environment. The exploitation of nitrogen fixation requires that microorganisms fix more nitrogen than they require and that they excrete this excess. For *Rhizobium* (in root nodules) and *Anabaena* (in the aquatic fern *Azolla*), it has been shown that the host not only provides the energy but also regulates the assimilation of fixed nitrogen, and a significant proportion of that which is fixed is taken up by the plant (9, 11). Mutants of *K. pneumoniae* that excrete large amounts of fixed nitrogen have been produced (11), which suggests that this may be a potentially valuable method for increasing the yield of fixed nitrogen from other procaryotes.

METHODS FOR INCREASING BIOLOGICAL NITROGEN FIXATION

Isolation of nitrogenase and its industrial use to catalyze ammonia production. The use of nitrogenase to catalyze ammonia production is primarily an industrial problem and is very much influenced by the stability and efficiency of the enzyme. Mutants producing an enzyme that is significantly more oxygen resistant, wastes less energy in hydrogen evolution, or is intrinsically more efficient in its energy requirements have not been reported. This would be the role of genetics in this process and may not be feasible since oxygen sensitivity and hydrogen evolution may be fundamental to the structure of an efficient enzyme. Indeed, there must have been a strong selection for oxygen resistance and increased efficiency in nature.

Use of free-living nitrogen-fixing procaryotes to produce ammonia. The use of photosynthetic nitrogen fixers to produce ammonia is particularly appealing, especially if mutants that excrete excess nitrogen are available. A major limitation is to provide a sufficient surface area for photosynthesis to occur. Used in an extensive manner, such as growth on irrigation reservoirs, as suggested by Shanmugan et al. (11), this process may produce significant amounts of nitrogen, but only in agricultural systems where extensive irrigation is practiced. Genetic manipulation of photosynthetic genes and of genes involved in nitrogenase function and in the regulation of fixed nitrogen and its assimilation could have a major impact on this process.

Increased yield of fixed nitrogen from the *Azolla-Anabaena* symbiosis. The *Azolla-Anabaena* symbiosis already contributes significant amounts of nitrogen to rice crops in Asia (9). There is a potential for increasing the efficiency of nitrogen fixation since this system has only recently been subjected to a simple rigorous screening and selection process to determine the optimal species for particular cultural systems (12). The genetic manipulation of both the host and symbiont has yet to be carried out, and little is known about the genetics of either partner.

Increased efficiency of intercropping and rotations. Our present knowledge of the value of intercropping nitrogen-fixing plants with other crops is very limited. Selection and the breeding of legumes and other crops that respond particularly well to intercropping remains to be done.

Increased efficiency of loose associations. The stimulation of nitrogen-fixing microorganisms in the soil by grasses and other plants has only recently been studied in detail (4), and the value of the resulting fixed nitrogen to the plants is still a subject for debate. The potential for exploiting this system is enormous, particularly in terms of making the association closer, so that the nitrogen-fixing bacteria are more closely associated with the plant roots. Essentially no conscious genetic work has been done to improve the response of these associations; thus, they are still at a stage where the selection and breeding of improved strains should be beneficial.

Increased efficiency of existing symbiotic associations. The most important symbiosis for crop plants is that between *Rhizobium* and legumes. Despite many years of research, the screening of plants for their response to different *Rhizobium* isolates has not been done for all legume crops, though specifically selected *Rhizobium* strains for the inoculation of some legumes are available. Genetic variability in the nitrogen-fixing ability of different host-*Rhizobium* combinations has been demonstrated, but, as yet, crosses between bacteria or plants to select improved strains have apparently not been used for commercial purposes. The selection and breeding of legumes for increased efficiency in nitrogen fixation and the breeding of *Rhizobium* strains that carry hydrogenases to utilize the hydrogen evolved from nitrogenase are examples of feasible approaches to improving the efficiency of this symbiosis.

New associations. The efficiency of intimate symbiotic associations has already been stated. Any extension in the host range of *Rhizobium* species or the actinomycetes that form nitrogen-fixing nodules on other dicotyledonous plants would be extremely valuable. Genetic studies of host-range determinants in the *Rhizobium*-legume symbiosis have only just started. As yet, there have been no reports of genetic work with any of the non-*Rhizobium* nodule-forming symbioses; indeed, there have been enormous difficulties in even isolating and culturing the microorganism. The potential for increasing the range of nodulated plants is uncertain, but, for the *Rhizobium*-legume symbiosis, genetic and biochemical studies are at last shedding some light on host-range determinants.

Introduction of nitrogen fixation genes into plants. Recent reports that *Agrobacterium tumefaciens* plasmid DNA can be stably maintained and expressed in plants has stimulated the idea of introducing Nif genes into plants. Nif genes have been cloned onto small plasmids (1), which is a useful preliminary step to the introduction into the plant. Oxygen protection for nitrogenase and the requirement for a complex regulation are two important problems to consider if the genes can be introduced.

CONCLUSIONS

Very little has been done genetically to improve biological nitrogen fixation, and the potential for increasing yields by genetic means is significant. In most cases selection and breeding of improved strains and varieties have yet to be tried. In the long term it seems likely that fundamental changes, such as alterations in host range and the transfer of nitrogen fixation genes to plants, will be feasible.

REFERENCES

1. **Ausubel, F., G. Riedel, F. Cannon, A. Peskin, and R. Margolskee.** 1977. Cloning nitrogen fixing genes from *Klebsiella pneumoniae in vitro* and the isolation of *nif* promoter mutants affecting glutamine synthetase regulation, p. 111–128. *In* A. Hollaender (ed.), Genetic engineeiing for nitrogen fixation. Basic life sciences, vol. 9. Plenum Press, New York.
2. **Beringer, J. E., S. A. Hoggan, and A. W. B. Johnston.** 1978. Linkage mapping in *Rhizobium leguminosarum* by means of R plasmid-mediated recombination. J. Gen. Microbiol. **104**:201–107.

3. **Burns, R. C., and R. W. F. Hardy.** 1975. Nitrogen fixation in bacteria and higher plants. Molecular biology, biochemistry, and biophysics, vol. 21. Springer-Verlag, New York.
4. **Burris, R. H.** 1977. A synthesis paper on nonleguminous N_2-fixing systems, p. 487–511 *In* W. Newton, J. R. Postgate, and C. Rodriguez-Barrueco (ed.), Recent developments in nitrogen fixation. Academic Press Inc., New York.
5. **Dixon, R., C. Kennedy, A. Kondorosi, V. Krishnapillái, and M. Merrick.** 1977. Complementation analysis of *Klebsiella pneumoniae* mutants defective in nitrogen fixation. Mol. Gen. Genet. **157:**189–198.
6. **Dixon, R. A., and J. R. Postgate.** 1972. Genetic transfer of nitrogen fixation from *Klebsiella pneumoniae* to *Escherichia coli.* Nature (London) **237:**102–103.
7. **Kondorosi, A., G. B. Kiss, T. Forrai, E. Vincze, and Z. Banfalvi.** 1977. Circular linkage map of *Rhizobium meliloti* chromosome. Nature (London) **268:**525–527.
8. **Meade, H. M., and E. R. Signer.** 1977. Genetic mapping of *Rhizobium meliloti.* Proc. Natl. Acad. Sci. U.S.A. **74:**2076–2078.
9. **Peters, G. A.** 1977. The *Azolla-Anabaena azolae* symbiosis, p. 231–258. *In* A. Hollaender (ed.), Genetic engineering for nitrogen fixation. Basic life sciences, vol. 9. Plenum Press, New York.
10. **Postgate, J. R.** 1974. Prerequisites for biological nitrogen fixation in free-living heterotrophic bacteria, p. 663–686. *In* A. Quispel (ed.), The biology of nitrogen fixation. North Holland Publishing Co., Amsterdam.
11. **Shanmugan, K. T., F. O'Gara, K. Andersen, C. Morandi, and R. C. Valentine.** 1978. Control of biological nitrogen fixation, p. 393–416. *In* D. R. Nielson and J. G. MacDonald (ed.), Nitrogen in the environment, vol. 2. Academic Press Inc., New York.
12. **Talley, S. N., B. J. Talley, and D. W. Rains.** 1977. Nitrogen fixation by *Azolla* in rice fields, p. 259–281. *In* A. Hollaender (ed.), Genetic engineering for nitrogen fixation. Basic life sciences, vol. 9. Plenum Press, New York.
13. **Yates, M. G.** 1977. Physiological aspects of nitrogen fixation, p. 219–270. *In* W. Newton, J. R. Postgate, and C. Rodriguez-Barrueco (ed.), Recent developments in nitrogen fixation. Academic Press Inc., New York.

Genetic Approaches to New Streptomycete Products

W. F. FLECK

Forschungszentrum für Molekularbiologie und Medizin, Akademie der Wissenschaften der Deutschen Demokratischen Republik, Zentralinstitut für Mikrobiologie und Experimentelle Therapie, 69 Jena, German Democratic Republic

The search for new chemotherapeutic agents, especially antibiotics, is an important prerequisite for the health of man in the future. From more than 3,000 antibiotics described in the literature, about 50 are produced commercially. In spite of these impressive numbers, there are no antibiotics available for combatting some of the infectious diseases, and a serious lack of effective antiviral and antitumor agents still exists. The problem of multiple drug resistance further stresses the need for new antibiotics.

In order to detect new potential antitumor and antiviral antibiotics from microbial sources, a complex screening program was developed in my laboratory during 1966 through 1972. This program consisted of models of molecular biology and proved to be of value in the search for new antitumor antiobiotics which are able to disturb selectively the mechanisms operating in both microbial and mammalian cells, e.g., replication, transcription, recombination, excision, and DNA integration. The results and efficiency of this program have been described (20). The new antibiotics that have resulted from our continued search for biodynamic metabolites of different species of streptomycetes are the anthracycline antibiotics trypanomycin (26), leukaemomycins (21), violamycins (22, 23), the chromoglycoside antibiotic lambdamycin (25), and the nitrogen-free pigment antibiotic maduramycin (24).

In the second phase of this work, we dealt with strains producing low levels of potentially new antibiotics and with the ways of improving them.

The third phase of our work was concerned with the preparation of new derivatives of anthracycline antibiotics with improved biological activities and reduced toxicity (including cardiotoxicity).

There are at least three effective ways to prepare new derivatives of antibiotics. One way is the total synthesis or the semisynthetic complementation of new derivatives (30, 32, 34, 37, 41, 51). A second approach is the modification of the original structure of the antibiotic by means of cells or their enzymes; the bioconversion or biotransformation of anthracycline antibiotics has been reported (28, 31, 36, 49, 50). The third method is genetic manipulation, which implies changes in the genome of the antibiotic-producing strains (see 29). The techniques that have been used to bring about genome variability in *Streptomyces* include transfer or elimination of plasmids, genetic recombination, and random mutagenesis. The last has been made particularly efficient by the adoption of new mutagens (48), e.g., nitrosoguanidine (19), and of new procedures to induce mutations in specific regions of the map, e.g., comutation (14, 40) or synchronized cultures (27).

My objective is to focus attention on cosynthetic mutants useful in obtaining new anthracycline antibiotics by mutasynthesis and, furthermore, on interspecific hybridization as a tool for detection of new antibiotics.

MUTANTS OF ANTIBIOTIC-PRODUCING CULTURES

Mutation of antibiotic-producing microorganisms has resulted in the isolation of mutants which elaborate new antibiotics. These new materials usually possess structural features of the parent antibiotics but lack some of their functional groups. Well-known examples of antibiotic derivatives obtained by mutation are those of tetracyclines, penicillins, cephalosporins, erythromycins, formycins, rifamycins, celesticetins, novobiocins, and others (for review, see 42). One of the most striking examples of the production of new and therapeutically valuable antibiotics by mutation is the formation of adriamycin (14-hydroxydaunomycin) (6, 7). This antibiotic is synthesized by *S. peucetius* var. *caesius*, a mutant isolated from a population of the daunomycin-producing organism which had been exposed to *N*-nitro-*N*-methylurethane. Adriamycin has striking antineoplastic properties, being active in animals and humans against leukemia and different kinds of solid tumors.

MUTATIONAL BIOSYNTHESIS (MUTASYNTHESIS)

Recently, the usefulness of a technique called mutational biosynthesis (A. L. Demain and K. Nagaoka, U.S. Patent 3,956,275, 11 May 1976, and U.S. Patent 3,993,544, 23 November 1976) or mutasynthesis (43, 45) for the production of new derivatives of aminocyclitol antibiotics, novobiocins, actinomycins, and other antibiotics has been demonstrated (Table 1). This technique is illustrated by the following example.

A culture ordinarily makes an anthracycline antibiotic, e.g., violamycin BI composed of four moieties ABCD (A, aglycone; B, rhodosamine; C, 2-deoxy-L-fucose; D, rhodinose) from simple carbon and nitrogen sources. A mutant is obtained that does not produce the antibiotic because it cannot synthesize the aglycone A. When analog A′ is added to the medium instead of A, the mutant will produce a new anthracycline A′BCD, if A′ can penetrate into the cell, if the incorporating enzyme is not too specific, and if the new anthracycline A′BCD has antibiotic activity.

TABLE 1. *Mutational biosynthesis*

Idiotroph	Blocked in biosynthesis	Mutasynthon	Mutasynthetic	Reference
S. fradiae	Neomycin	Streptamine	Hybrimycin A_1 and A_2	Shier et al. (44)
S. fradiae	Neomycin	Epistreptamine	Hybrimycin B_1 and B_2	Shier et al. (44)
S. rimosus subsp. *paromomycinus*	Paromomycin	Streptamine	Hybrimycin C_1 and C_2	Shier et al. (43, 45)
S. kanamyceticus	Kanamycin	Streptamine	Kanamycins	Kojima and Satoh (33)
B. circulans	Butirosin	2-Deoxystreptamine	Butirosins	Claridge et al. (15)
M. inyoensis	Sisomycin	Streptamine	Mutamicins	Testa et al. (46, 47)
S. griseus	Streptomycin	2-Deoxystreptidine	Streptomutin A	Nagaoka and Demain (38)
S. griseus	Streptomycin	2-Deoxystreptidine	Deoxystreptomycin A and B	Demain and Nagaoka[a]
S. fradiae	Neomycin	2,6-Dideoxystreptamine	6-Deoxyneomycin B and C	Cleophax et al. (16)
S. rimosus subsp. *paromomyceticus*	Paromomycin	2,6-Dideoxystreptamine	6-Deoxyparomomycin I and II	Cleophax et al. (16)
S. niveus	Novobiocin	8-Demethyl-8-chlorocoumarin derivative	Chlornovobiocin	Sebek (42)
M. purpurea	Gentamicin C complex	Streptamine	2-Hydroxygentamicin C complex	Daum et al. (17, 18)

[a] U.S. Patent 3,956,275, 11 May 1976; U.S. Patent 3,993,544, 23 November 1976.

Similar to the observations on aminocyclitol antibiotic-producing streptomycetes (Table 1), blocked mutants of the leukaemomycin-producing *S. griseus* were selected (C. Stengel and W. F. Fleck, unpublished data) which could not synthesize the aglycone daunomycinone but were able to produce the amino sugar daunosamine. When the aglycone ε-rhodomycinone or other aglycones of the rhodomycinone type were fed into the fermentation medium instead of daunomycinone, the mutant produced a red pigment which differed from the parent antibiotics in antibacterial activity and chromatographic behavior. Studies are in progress to elucidate whether the new anthracyclinone glycosides are new daunosaminyl-rhodomycinones.

In our other preliminary experiments, mutants of the violamycin-producing *S. violaceus* were isolated which were blocked in the biosynthesis of rhodomycinones. When daunomycinone was fed into the medium of blocked mutants instead of rhodomycinone aglycones, new anthracyclines with biological activity were produced. Studies are in progress to determine whether these mutants are able to glycosylate daunomycinone fed with amino sugars and sugars synthesized by the parent *S. violaceus* (e.g., rhodosamine, 2-deoxy-L-fucose, rhodinose). Rhodosaminyl-2-deoxy-L-fucosyl- and/or rhodinosyl-daunomycinones have not yet been described as natural members of the anthracycline family. We are confident that new anthracycline antibiotics will be generated by these techniques.

INTERSPECIFIC RECOMBINANTS

Interspecific recombination in *Actinomycetales* is not easy to demonstrate. Some authors failed to detect prototroph formation in heterologous mixtures of auxotrophs of different species. On the other hand, there is good evidence for interspecific recombination in certain *Streptomyces* strains (Table 2). In our continued search for blocked mutants of the turimycin-producing *S. hygroscopicus* and the violamycin-producing *S. violaceus,* interspecific cosynthesis was detected since cosynthetic activity was generated by combinations of an inactive mutant of *S. hygroscopicus* and an inactive mutant of *S. violaceus.* In all combinations tested, a halo of inhibition was seen on the side of the *S. violaceus* mutants, which thus presumably acted as convertors (B. Schlegel and W. F. Fleck, unpublished data). Because of this demonstration on solid medium, we tried to obtain the biologically active principle in

TABLE 2. *Interspecific recombinants of antibiotic-producing streptomycetes*

Crossed species	Secondary metabolite produced	Reference
S. rimosus and *S. aureofaciens*	Oxytetracycline, chlortetracycline	Alačević (1, 2)
S. rimosus and *S. coelicolor*	Oxytetracycline	Alačević (1, 2)
S. aureofaciens and *S. coelicolor*	Chlortetracycline	Alačević (1, 2)
S. aureofaciens and *S. violaceoruber*	Chlortetracycline	Bradley (10)
S. coelicolor and *S. rimosus*	Red pigments, oxytetracycline	Alačević (3, 4)
S. rimosus and *S. aureofaciens*	Oxytetracycline, chlortetracycline	Alačević et al. (5)
S. rimosus and *S. aureofaciens*	Oxytetracycline, chlortetracycline	Polsinelli and Beretta (39)
S. rimosus and *S. aureofaciens*	Aureovocin	Blumauerová et al. (9)
S. coelicolor and *S. griseus*	Grisin	Lomovskaya et al. (35)

larger amounts under submerged conditions, but without success. We have, therefore, tried to cross both strains in order to form stable interspecific antibiotic-producing recombinants.

Indeed, we have obtained colonies from mixed cultures of blocked mutants of *S. hygroscopicus* and *S. violaceus* which were able to grow on selective media. In our crosses the blocked mutant of the turimycin-producing species was auxotrophic and resistant to streptomycin (*met tur str*). The blocked mutant of the violamycin-producing species was prototrophic but susceptible to streptomycin (*vio* str^+). Depending on the mutants crossed, 4 to 90 colonies per 2×10^6 total spores plated were isolated from selective medium (minimal medium plus streptomycin). Some of the recombinant phenotypes produced a red pigment antibiotic which has chromatographic behavior markedly different from that of the parent antibiotic violamycin BI, and three isolates showed higher antibiotic activity than the original strain of *S. violaceus.* Free aglycones have never been found when crude extracts from recombinant phenotypes were chromatographed in parallel with violamycin extracts from high-yielding strains. Further studies will be necessary to evidence a true recombination between the two different species, but it appears that the new antibiotic differs in both the sugar and aglycone parts of the violamycin BI molecule.

The interspecific recombinant phenotypes are not able to produce the main constituents of the violamycin BI complex (ϵ-rhodomycinone and ϵ-isorhodomycinone), and the typical sugars of the violamycin BI molecule (2-deoxy-L-fucose, rhodinose) were not detected in hydrolysates of the purified glycoside. One of the constituents isolated and purified thus far has been identified as a γ-rhodomycinone glycoside (Schlegel and Fleck, unpublished data). γ-Rhodomycins have been described (8, 11–13), but no information is available on their biological activity. In contrast to the described γ-rhodomycins, our γ-rhodomycinone glycoside which contains an as yet unidentified amino sugar is formed by an *in*terspecific *re*combinant, *S. violaceus* subsp. *iremyceticus* IMET 43 615, and possesses antimicrobial activity, especially against gram-positive bacteria and mycobacteria. Furthermore, the γ-rhodomycinone glycoside which was named iremycin inhibited the multiplication of λ phages in cells of *Escherichia coli* in a manner similar to that known for other anthracyclines when estimated by the BIP test method (20). Inactivation of iremycin was observed after in vitro complex formation between iremycin and calf thymus DNA. Studies on the therapeutic and toxic activity of iremycin in tumor-bearing animals are in progress.

The identification of the anthracyclines formed by the other interspecific phenotypes has not yet been completed, but the compounds will most likely represent new structures. We believe that further genetic work of this kind (generation of blocked and cosynthetic mutants, interspecific recombination) will yield new modified antibiotic analogs of industrial interest and of chemotherapeutic value.

REFERENCES

1. **Alačević, M.** 1963. Interspecific recombination in Streptomyces. Nature (London) **197:**4874, 1323.
2. **Alačević, M.** 1965. Differences in morphological properties of interspecific recombinants in Streptomycetes in relation to parent strains. Mikrobiologija **2:**143–151.
3. **Alačević, M.** 1965. Pigment variations of some interspecific recombinants in Streptomycetes. Mikrobiologija **2:**159–162.
4. **Alačević, M.** 1969. Interspecific recombination in Streptomycetes in relation to their classification. Mikrobiologija **6:**9–16.
5. **Alačević, M., D. Vlasic, and I. Spada-Sermonti.** 1966. Interspecific recombination among antibiotic-producing streptomycetes, p. 720–722. *In* Antibiotics—advances in research, production and clinical use. Czechoslovak Medical Press, Prague.

6. **Arcamone, F.** 1977. New antitumor anthracyclines. Lloydia **40:**45–66.
7. **Arcamone, F., G. Cassinelli, G. Fantini, A. Grein, P. Orezzi, C. Pol, and C. Spalla.** 1969. Adriamycin, 14-hydroxydaunomycin, a new antitumor antibiotic from S. peucetius var. caesius. Biotechnol. Bioeng. **11:**1101–1110.
8. **Biedermann, E., and H. Bräuniger.** 1972. Untersuchungen zur Isolierung und Konstitutionsaufklärung der Rhodomycin-Antibiotica des Streptomyces-Stammes JA 8467. Pharmazie **27:**782–789.
9. **Blumauerová, M., A. A. Ismail, Z. Hošťálek, and Z. Vaněk.** 1971. Mutation studies in Streptomyces aureofaciens. Radiation Radioisotopes Industrial Microorganisms, p. 157–166. International Atomic Energy Agency, Vienna.
10. **Bradley, S. G.** 1964. Genetic analysis of an unstable mutant of Streptomyces violaceoruber. Dev. Ind. Microbiol. **6:**296.
11. **Brockmann, H., P. Boldt, and J. Niemeyer.** 1963. Beta-Rhodomycinon und Gamma-Rhodomycinon. Chem. Ber. **96:**1356–1372.
12. **Brockmann, H., and J. Niemeyer.** 1961. Zur Konstitution des beta und gamma-Rhodomycinons. Naturwissenschaften **48:**570–571.
13. **Brockmann, H., and T. Waehnelt.** 1961. Eine neue Gruppe von Rhodomycinen. Naturwissenschaften **48:** 717.
14. **Carere, A., and R. Randazzo.** 1976. Co-mutation in Streptomyces, p. 573–581. *In* K. D. Macdonald (ed.), Second international symposium on the genetics of industrial microorganisms. Academic Press, London.
15. **Claridge, C. A., J. A. Busch, M. D. Defuria, and K. E. Price.** 1974. Fermentation and mutation studies with a butirosin-producing strain of Bacillus circulans. Dev. Ind. Microbiol. **15:**101–113.
16. **Cleophax, J., St. D. Gero, et al.** 1976. A chiral synthesis of D-(+)-2,6-dideoxystreptamine and its microbial incorporation into novel antibiotics. J. Am. Chem. Soc. **98:**7110–7112.
17. **Daum, S. J., D. Rosi, and W. A. Goss.** 1977. Production of antibiotics by biotransformation of 2,4,6/3,5-pentahydroxycyclohexanone and 2,4/3,5-tetrahydroxycyclohexanone by a deoxystreptamine-negative mutant of Micromonospora purpurea. J. Am. Chem. Soc. **99:**283–284.
18. **Daum, S. J., D. Rosi, and W. A. Goss.** 1977. Mutational biosynthesis of idiotrophs by Micromonospora purpurea. II. Conversion of non-amino containing cyclitols to aminoglycoside antibiotics. J. Antibiot. **30:**98–105.
19. **Delić, V., D. A. Hopwood, and E. J. Friend.** 1970. Mutagenesis by N-methyl-N′-nitro-N-nitrosoguanidine (NTG) in Streptomyces coelicolor. Mutat. Res. **9:**167–182.
20. **Fleck, W.** 1974. Development of microbiological screening methods for detection of new antibiotics. Postepy Hig. Med. Dosw. **28:**479–498.
21. **Fleck, W., and D. Strauss.** 1975. Leukaemomycin, an antibiotic with antitumor activity. I. Screening, fermentation, and biological activity. Z. Allg. Mikrobiol. **15:**495–503.
22. **Fleck, W., D. Strauss, W. Koch, and H. Prauser.** 1974. Violamycin, a new red-pigment antibiotic. Z. Allg. Mikrobiol. **14:**551–558.
23. **Fleck, W., D. Strauss, W. Koch, and H. Prauser.** 1975. Fermentation and isolation of new anthracycline antibiotics: violamycin A, B II, B I and their aglycones. Antibiotiki (Moscow) **11:**966.
24. **Fleck, W. F., D. G. Strauss, J. Meyer, and G. Porstendorfer.** 1978. Fermentation, isolation, and biological activity of maduramycin: a new antibiotic from Actinomadura rubra. Z. Allg. Mikrobiol. **18:**389–398.
25. **Fleck, W., D. Strauss, H. Prauser, W. Jungstand, H. Heinecke, W. Gutsche, and K. Wohlrabe.** 1976. Lambdamycin, ein Antibioticum aus dem Stamm IMET 31 118 von Streptomyces glaucoachromogenes PRAUSER. Z. Allg. Mikrobiol. **16:**521–528.
26. **Fleck, W., D. Strauss, C. Schönfeld, W. Jungstand, C. Seeber, and H. Prauser.** 1972. Screening, fermentation, isolation, and characterization of trypanomycin, a new antibiotic. Antimicrob. Agents Chemother. **1:**385–391.
27. **Godfrey, O. W.** 1974. Directed mutation in Streptomyces lipmanii. Can. J. Microbiol. **28:**1479–1485.
28. **Hamilton, B. K., M. S. Sutphin, M. C. Thomas, D. A. Wareheim, and A. A. Aszalos.** 1977. Microbial N-acetylation of daunorubicin and daunorubicinol. J. Antibiot. **30:**425–426.
29. **Hopwood, D. A., and M. J. Merrick.** 1977. Genetics of antibiotic production. Bacteriol. Rev. **41:**595–635.
30. **Jung, M. E., and J. A. Lowe.** 1978. Synthetic approaches to aclacinomycin and pyrromacin antitumor antibiotics via Diels-Alder reactions of 6-alkoxy-2-pyrones. Total synthesis of chrysophanol, helminthosporin, and pachybasin. J. Chem. Soc. D **3:**95.
31. **Karnetová, J., J. Matějů, P. Sedmera, J. Vokoun, and Z. Vaněk.** 1976. Microbial transformation of daunomycinone by Streptomyces aureofaciens B-96. J. Antibiot. **29:**1199–1202.
32. **Kende, A. S., Y. G. Tsay, and J. E. Mills.** 1976. Total synthesis of (+)-daunomycinone and (+)-carminomycinone. J. Am. Chem. Soc. **98:**1967–1969.
33. **Kojima, M., and A. Satoh.** 1973. Microbial semisynthesis of aminoglycosidic antibiotics by mutants of S. ribosidificus and S. kanamyceticus. J. Antibiot. **26:**784–786.
34. **Krohn, K., and A. Rösner.** 1978. Synthese des 4-desoxy-gamma-rhodomycinons. Tetrahedron Lett., no. 4, p. 353.
35. **Lomovskaya, N. D., T. A. Voeykova, and N. M. Mkrtumian.** 1977. Construction and properties of hybrids obtained in interspecific crosses between Streptomyces coelicolor A3(2) and Streptomyces griseus Kr.15. J. Gen. Microbiol. **98:**187–198.
36. **Marshall, V. P., E. A. Reisender, L. M. Reineke, J. H. Johnson, and P. F. Wiley.** 1976. Reductive microbial conversion of anthracycline antibiotics. Biochemistry **15:**4139–4145.

37. **Miller, D. G.** 1977. Approaches to the synthesis of anthracycline antibiotics. Diss. Abstr. Int. B **37:** 3931–3932.
38. **Nagaoka, K., and A. L. Demain.** 1975. Mutational biosynthesis of a new antibiotic, streptomutin A, by an idiotroph of Streptomyces griseus. J. Antibiot. **28:**627–635.
39. **Polsinelli, M., and M. Beretta.** 1966. Genetic recombination in crosses between *Streptomyces aureofaciens* and *Streptomyces rimosus.* J. Bacteriol. **91:**63–68.
40. **Randazzo, R., G. Sermonti, A. Carere, and M. Bignami.** 1973. Comutation in *Streptomyces.* J. Bacteriol. **113:**500–501.
41. **Raynolds, P. W., M. J. Manning, and J. S. Sweton.** 1977. A regiospecific synthesis of anthracyclinones. Tetrahedron Lett. **28:**2383.
42. **Sebek, O. K.** 1976. Use of mutants for the synthesis of new antibiotics, p. 522–525. *In* D. Schlessinger (ed.), Microbiology—1976. American Society for Microbiology, Washington, D.C.
43. **Shier, W. T., S. Ogawa, M. Hickens, and K. L. Rinehart, Jr.** 1973. Chemistry and biochemistry of the neomycins. XVII. Bioconversion of aminocyclitols to aminocyclitol antibiotics. J. Antibiot. **25:**551–561.
44. **Shier, W. T., K. L. Rinehart, Jr., and D. Gottlieb.** 1969. Preparation of four new antibiotics from a mutant of Streptomyces fradiae. Proc. Natl. Acad. Sci. U.S.A. **63:**198–204.
45. **Shier, W. T., P. C. Schaefer, D. Gottlieb, and K. L. Rinehart, Jr.** 1974. Use of mutants in the study of aminocyclitol antibiotics. Biosynthesis and preparation of the hybrimycin complex. Biochemistry **13:** 5073–5078.
46. **Testa, R. T., and B. C. Tilley.** 1975. Biotransformation, a new approach to aminoglycoside biosynthesis. I. Sisomycin. J. Antibiot. **28:**573–579.
47. **Testa, R. T., G. H. Wagman, P. J. L. Daniels, and M. J. Weinstein.** 1974. Mutamicins: biosynthetically created new sisomycin analogues. J. Antibiot. **27:**917–921.
48. **Townsend, M. E., H. M. Wright, and D. A. Hopwood.** 1971. Efficient mutagenesis by near ultraviolet light in the presence of 8-methoxypsoralen in Streptomyces. J. Appl. Bacteriol. **34:**799–801.
49. **Vaněk, Z., J. Tax, J. Cudlín, M. Blumauerová, N. Steinerová, J. Matějů, I. Komersová, and K. Stajner.** 1976. Biogenesis of linear tri- and tetracyclic oligoketides and their glycosides, p. 473–495. *In* K. D. Macdonald (ed.), Second international symposium on the genetics of industrial microorganisms. Academic Press, London.
50. **Wiley, P. F., and V. P. Marshall.** 1975. Microbial conversion of anthracycline antibiotics. J. Antibiot. **28:** 838–840.
51. **Yamaguchi, T., and M. Kojima.** 1977. Synthesis of daunosamine. Carbohydr. Res. **59:**343.

Some Recent Developments in *Streptomyces* Genetics

K. F. CHATER

John Innes Institute, Norwich NR4 7UH, England

This review will be concerned with "nonchromosomal" genetics. The omission of normal chromosomal genetics does not reflect a decline in its importance, but rather the absence of major advances since previous reviews (29, 30), with the important exception of the development of protoplast fusion (31), which is dealt with by D. A. Hopwood elsewhere in this volume. My approach is to consider what is known about the occurrence and genetic determination of "nonessential" (and therefore possibly plasmid-specified) functions in streptomycetes, and then to discuss aspects of plasmids and temperate phages relevant to the development of potential "recombinant DNA" systems in streptomycetes.

ANTIBIOTIC RESISTANCE

General. A number of streptomycetes have been examined for idiosyncratic drug resistance patterns which might indicate the possession of specific resistance determinants (24, 25; Table 1). Some of these drug resistances were lost at rather high frequencies which were increased by UV irradiation, and in the best-studied case (chloramphenicol resistance in *Streptomyces coelicolor*) resistance versus sensitivity did not show linkage with chromosomal markers. Some mutants were able to revert to the original unstable resistance phenotype, albeit at a lower frequency than forward mutations to sensitivity. However, some streptomycin-sensitive derivatives of *S. glaucescens* were not revertible (25, 34), and the resistance determinant was unlinked with chromosomal markers (34). In such cases, loss of, or deletion from, a plasmid might indeed have occurred, and the reversible mutations might have been due to DNA rearrangements (as with phase determination in *Salmonella*; 67) or insertional inactivation, affecting expression of relevant plasmid-borne genes. As will be mentioned, deletions and internal duplications of *Streptomyces* plasmids are probably common.

Resistance of antibiotic producers to their own antibiotics. Antibiotic resistance has an additional special significance for streptomycetes since they need to resist self-inhibition by their own antibiotics. The first example of plasmid-specified resistance to an endogenous antibiotic was the resistance to methylenomycin encoded by the genetically well-defined plasmid SCP1 of *S. coelicolor* A3(2) (37). For oxytetracycline resistance in a strain of *S. rimosus* (8; A. M. Boronin, unpublished data quoted in 28) and for leucomycin resistance in *S. reticuli* (57), the correlation between sensitivity and the loss of extrachromosomal DNA nicely complements evidence from curing and infectious transfer that plasmids are involved.

In none of the examples discussed so far is the mechanism of resistance understood. Among biochemically well-defined systems, some (e.g., resistance of *S. azureus* to thiostrepton through specific methylation of 23S rRNA; 21) are still genetically unexplored; however, high-frequency loss of neomycin resistance by the producer *S. fradiae* was correlated with loss of aminoglycoside phosphotransferase (66).

TABLE 1. *Naturally occurring antibiotic resistance in streptomycetes*[a]

Organism	Resistance to	% Sensitive		% Reversion to resistance (spontaneous)	Loss of antibiotic production	Infectious transfer	Correlation with plasmid DNA	References and comments
		Spontaneous	Curing					
S. coelicolor A3(2)	Chloramphenicol	0.5	5 (UV)	0.001	NA	—	—	CAT not involved (24)
S. acrimycini IPV1610	Chloramphenicol	1	4 (UV)	NT	NA	NT	NT	CAT not involved (25, 63)
	Chloramphenicol	?	(0.01; NTG)	(0.0001; NTG)	NA	—	NA	CAT⁻ chromosomal mutant (63)
	Erythromycin	<0.1–0.6	0.7–5.2 (UV)	NT	NA	NT	NT	(25)
S. lividans 66	Chloramphenicol	<0.6	5.2 (UV)	0.0001	NA	NT	—	CAT not involved (25)
S. griseus CUB609	Streptomycin	0.2	1.8–10 (UV)	0.0001	NT	NT	NT	(25)
S. glaucescens ETH22794	Streptomycin	0.3	2.8 (UV)	(a) 0.001	NA	NT	NT	(25)
				(b) <0.0001	NA	NT	NT	
S. glaucescens ETH22794	Streptomycin	?	?	<0.0001	NA	—	—	Unlinked to chromosome (34)
S. coelicolor A3(2)	Methylenomycin	0.1	2 (UV)	Undetectable	Lost	+	—	SCP1 carries genes for methylenomycin synthesis and resistance (36, 37, 64)
S. rimosus	Oxytetracycline	0.2–0.3	2.4–11.2 (AF, AO, PF)	Undetectable	Lost	+	+	(8, 10)
S. bikiniensis	Streptomycin	<0.3	2–16 (EB, AF)	?	Lost	—	NT	(58)
S. reticuli	Leucomycin	?	1% (EB)	?	Lost	+	+	Plasmid loss also affects sporulation, melanin production, and fertility (57)
S. fradiae	Neomycin	?	+ (AO)	?	Lost	NT	?	Plasmid DNA detected, relation to resistance unspecified; resistance due to phosphotransferase (66)

[a] NTG, *N*-methyl-*N'*-nitro-*N*-nitrosoguanidine; AF, acriflavine; AO, acridine orange; PF, proflavine; EB, ethidium bromide; NA, not applicable; NT, not tested; ?, not clear; CAT, chloramphenicol acetyltransferase.

ANTIBIOTIC PRODUCTION

In the examples discussed in the previous section, the drug-sensitive derivatives are also nonproducers, and involvement of plasmids in antibiotic synthesis is therefore indicated. Table 2 lists nine more examples of putative plasmid involvement in antibiotic synthesis. However, the nature of this involvement is not usually clear. Only for methylenomycin synthesis is there good evidence that structural genes are on a plasmid: transfer of SCP1 to SCP1$^-$ *S. coelicolor, S. parvulus* or *S. lividans* recipients is always accompanied by the production of and resistance to methylenomycin as nonselected characters, and co-synthesis of methylenomycin occurs with certain pairs of nonproducing strains carrying SCP1-linked mutations (36). In the few other systems for which data are available, pathway genes are chromosomally located, the plasmid presumably encoding some regulatory or export function (M. Okanishi, this volume). In a single case [actinorhodin in *S. coelicolor* A3(2)] it appears that plasmids are not involved at all since all mutations to nonproduction map in a short region of the chromosome (54, 65).

DIFFERENTIATION

Poorly sporulating derivatives of several streptomycetes arise at significant frequencies after UV or chemical curing agent treatment (48, 52, 53, 57, 58). Only for *S. reticuli* (57; H. Schrempf and W. Goebel, personal communication) are there both physical evidence that plasmid DNA is lost and good genetic data for reinfection. In *S. alboniger, S. scabies, S. violaceus-ruber,* and *S. coelicolor* A3(2) (52, 53; M. J. Merrick, unpublished data), the morphological abnormality is accompanied in a variable proportion of isolates by a requirement for arginine, which correlates with loss of argininosuccinate synthetase activity (B. M. Pogell, personal communication). Exogenous arginine does not suppress the morphological abnormality. Reversion to wild-type morphology or (where relevant) to arginine independence has never been observed (B. M. Pogell, personal communication). It should be noted that at least four genes involved in aerial mycelium formation and at least eight involved in its metamorphosis into spores have chromosomal locations in *S. coelicolor* A3(2) (16, 44).

EXTRACELLULAR ENZYMES

The first indication of plasmids in streptomycetes was a high-frequency loss of melanin production in *S. scabies* and its transfer in heterokaryons (27). Melanin production involves the action of extracellular tyrosinase. In *S. glaucescens* the structural gene for tyrosinase is chromosomally located, but export of tyrosinase appears to involve a nonchromsomal gene(s) since export-defective derivatives occur at high frequency after curing treatment, and the defective phenotype does not segregate with chromosomal markers in crosses (3). The latter mutants also lack an extracellular β-glucanase, have become constitutive for an extracellular chitinase, and are more sensitive to streptomycin (see above) and to certain mutagens (60). Plasmid DNA could not be detected in *S. glaucescens,* but the plasmid of *S. reticuli* was altered in derivatives defective in melanin production (57).

Surveys of the ability of different wild isolates to utliize specific insoluble macromolecular substrates might reveal more evidence of nonchromosomal determinants.

High-frequency curing by acriflavine or sodium dodecyl sulfate of production of another kind of extracellular enzyme, β-lactamase, has also been observed (46).

TABLE 2. *Plasmid involvement in antibiotic synthesis in streptomycetes*[a]

Organism	Antibiotic	Curing agent[a]	Infectious transfer	Plasmid DNA	References[b]
S. coelicolor A3(2)	Methylenomycin	UV	Yes	Not detected	36, 37, 64
S. rimosus[c]	Oxytetracycline	AO, AF, PF	Yes	Yes	8, 9, 10, A
S. bikiniensis	Streptomycin	AF, EB	No	?	58
S. reticuli	Leucomycin	EB	Yes	Yes	57
S. fradiae	Neomycin	AO	?	?	66
S. rimosus subsp. *paromomycinus*	Paromomycin	AO	?	?	62
S. hygroscopicus[c]	Turimycin	AO	Yes	?	33
S. clavuligerus[c]	Holomycin	UV	Yes	Not detected	35, B
S. venezuelae[c]	Chloramphenicol	AF, T	No[d]	"Flower shape"	1, 50, C
S. venezuelae 3022a	Chloramphenicol	AF, EB	?	?	11
S. kasugaensis	Kasugamycin	AF, T	?	15-μm "flower shape"	50, C
S. kasugaensis	Aureothricin	AF, T	?	3.35 μm	50, C
S. griseus	Streptomycin	AF	?	?	50
S. kanamyceticus	Kanamycin	AF, AO, EB	?	?	12

[a] AO, Acridine orange; AF, acriflavine; PF, proflavine; EB, ethidium bromide; T, high temperature.
[b] A = A. M. Boronin, personal communication in reference 28; B = R. Kirby, personal communication; C = M. Okanishi, this volume.
[c] At least some probable pathway genes chromosomal.
[d] Nonlinkage with chromosome.

SEX FACTORS

At least four plasmids have sex factor activity in streptomycetes. Two of these, SCP1 (61) and SCP2 (5), have certain properties in common. They are highly self-transmissible to recipients lacking the plasmid within a given species, and they can also be transferred between species. Their transfer is accompanied by what appears to be "lethal zygosis," manifested on agar plates as a narrow zone of recipient growth inhibition around areas where plasmid transfer has occurred (5). They promote chromosome transfer to an extent that depends both upon which plasmid is used and upon its particular state in both parent strains (5, 29).

The *S. reticuli* plasmid involved in leucomycin production and resistance, and other characters, is also highly self-transmissible within *S. reticuli* and promotes the transfer of chromosomal genes (H. Schrempf and W. Goebel, personal communication). A fourth example of a *Streptomyces* sex factor is the *S. rimosus* plasmid SRP1 (26), for which no properties other than self-transmission and sex factor activity have been described. Spontaneous loss of SCP1, SCP2, SRP1, and the *S. reticuli* plasmid occurs at appreciable frequencies (at least 0.1%).

STREPTOMYCES PLASMID DNA

All methods attempted so far have failed to reveal DNA specifying SCP1 properties or chloramphenicol resistance in *S. coelicolor* A3(2) (24; J. Westpheling, unpublished data), holomycin production in *S. clavuligerus* (R. Kirby, personal communication), or streptomycin resistance or melanin production in *S. glaucescens* (34). A method not depending on dye-buoyant density gradient centrifugation was needed before plasmid DNA involved in chloramphenicol synthesis by *S. venezuelae* and in aureothricin synthesis by *S. kasugaensis* was found. These DNA molecules had a flower-like appearance by electron microscopy (50). However, conventional dye-buoyant density gradients have allowed the isolation of covalently closed circular (CCC) DNA molecules from a number of streptomycetes (Table 3).

Several of these plasmids tend to lose or amplify segments of their DNA. Thus, reduced derivatives of the *S. reticuli* 48-megadalton plasmid (57), the *S. fradiae* 55-megadalton plasmid (20), and the *S. kasugaensis* 6.7-megadalton plasmid (50), and an internal tandem repeat of SCP2 (6), have been recognized; dimers and trimers can often be detected in SCP2 and SLP1 preparations (5; M. J. Bibb, unpublished data).

TRANSFORMATION WITH PLASMID DNA

In the presence of polyethylene glycol, CCC DNA of a high-fertility variant of SCP2, SCP2*, has been efficiently introduced into *Streptomyces* protoplasts (7). Transformants were detected by the rather plaquelike manifestation of lethal zygosis (see above). Thus, it should soon be possible to introduce "foreign" DNA, inserted in vitro into SCP2* or perhaps other plasmid vectors, into *Streptomyces* cultures. It should be borne in mind that, for the purposes outlined by D. A. Hopwood elsewhere in this volume, either wide host range cloning vectors or a number of vectors with different host ranges will be needed: as yet, wide host range has not been demonstrated for any *Streptomyces* plasmid.

Since the very high frequency of transformation obtained (up to 20%) by this method may well apply to any species of CCC DNA capable of stable maintenance in a given recipient, transformation of "cryptic" plasmids into new host strains

TABLE 3. *Closed circular DNA isolated from streptomycetes*

Organism	Plasmid designation	Functions	Mol wt	Restriction enzyme analysis	Aberrations observed	Reference[a]
S. coelicolor A3(2)	SCP2 (pSH1)	Transmissible sex factor	$18–20 \times 10^6$	Map	Dimers, trimers, internal duplication	5, 6, 56
S. coelicolor sp.	pHS2	Unknown	55×10^6	—	—	55
	pHS3	Unknown	8×10^6	—	—	55
S. lividans 66	SLP1	Causes "pocks" ("lethal zygosis"?)	7.9×10^6	Map	Dimers, trimers	A
S. reticuli	—	Transmissible sex factor; sporulation; leucomycin production and resistance; melanin production	48×10^6	Cleavage patterns	Deletions, fragmentation	57, B
S. venezuelae 3022a	—	Unknown	18×10^6	—	—	11, 43
S. kasugaensis	—	Aureothricin production	$\sim 6.7 \times 10^6$	—	Deletion?	50
	—	Kasugamycin production	$\sim 30 \times 10^6$	—	—	50
	—	Unknown	$\sim 1.2 \times 10^6$	—	—	50
S. fradiae sp.	—	Unknown	22×10^6	—	—	66
		Unknown	15×10^6	—	—	66
S. fradiae sp.	—	Unknown	55×10^6	Map	Deletion	20
S. virginiae	—	Unknown	Unknown	—	—	39

[a] A = M. J. Bibb and J. M. Ward, unpublished data; B = H. Schrempf and W. Goebel, personal communication.

without selection may not be difficult; the acquisition of plasmid DNA might then be correlated with novel phenotypic characters.

TEMPERATE BACTERIOPHAGES AS POTENTIAL CLONING VECTORS

For some purposes (for example, gene amplification and high levels of transcription of introduced DNA), bacteriophage DNA may be preferred as a vector of introduced DNA. For this and other reasons, the understanding of *Streptomyces* phages, especially temperate phages, is needed. The best studied is φC31, for which there is now a linear 20-gene linkage map of the vegetative genome (19) and analysis of the DNA by both electron-microscopic denaturation (59) and restriction enzyme studies (Table 4; N. D. Lomovskaya, personal communication). In Norwich, we have so far recognized and studied DNA from temperate phages of four immunity groups, all attacking *S. coelicolor* A3(2) (Table 4). Of these, R4 is notable for its relatively wide host range: it formed plaques on more than half the wild-type streptomycetes tested from the collection of D. A. Hopwood, and lysogeny was demonstrable in most cases (K. F. Chater, A. T. Carter, and L. C. Wilde, unpublished data). For φC31 (45) and VP5 (17) heat-inducible mutants are available, which would facilitate control of copy number and expression of inserted DNA should either of these phages emerge as a useful vector.

The generation of genetic and restriction enzyme maps, the identification of dispensable regions of phage DNA, the removal of unwanted restriction enzyme target sites, and the development of phages with reduced genomes allowing larger pieces of "foreign" DNA to be inserted would all be helped by the availability of viable phage deletion mutants. It is likely that these will be found among chelating agent-resistant mutants, as for φ105 of *Bacillus subtilis* (23) and coliphage λ (51). VP5, φC31, and R4 (but not S14 or φ448) are very sensitive to certain chelating agents, and resistant mutants have been isolated. A reduction in size of the largest *Eco*RI fragment has been detected with DNA from one of the VP5 mutants (E. A. Page and K. F. Chater, unpublished data).

Previous reports of transfection of streptomycetes with phage DNA used simple DNA/protoplast mixing (47) or addition of DNA to late-exponential-phase, competent *S. virginiae* cultures (40), followed in each case by further incubation in liquid medium. For cloning purposes, direct detection of transfection events by plaque formation would be more useful. We have recently found that VP5 DNA can be introduced into *S. coelicolor* A3(2) protoplasts by the polyethylene glycol procedure

TABLE 4. *Temperate phages of* S. coelicolor *A3(2) and properties of their DNA*

Temperate phage	First described (reference)[a]	Host range	Mol wt of DNA	G+C[b] in DNA (%)	Cleavage sites in DNA			
					*Eco*RI	*Hin*dIII	*Sal*GI	*Sal*PI
φC31	41	Narrow	27×10^6	62	6 or 7	~12	~31	0
VP5	22	Narrow	27×10^6	59	5	?	?	?
R4	A	Wide	30×10^6	67	1	0	~27	1
Homoimmune { φ448	B	Narrow	34×10^6	59	8	~10	5	1 or 2
Homoimmune { S14	C	Narrow	35×10^6	59	4	?	7	0

[a] A = K. F. Chater and A. T. Carter, submitted for publication; B = N. D. Lomovskaya, personal communication; C = C. Stuttard, unpublished data.

[b] Guanine plus cytosine.

(7), the events being detected as plaques formed in soft agar overlays containing the transfection mixture, on R2 agar (49) base plates (K. F. Chater, unpublished data).

RESTRICTION AND MODIFICATION

Restriction and modification of phages have been detected in *S. coelicolor* A3(2), *S. griseus* 20, *S. griseus* Kr.15, and *S. griseofovillus* (42; N. D. Lomovskaya, personal communication); *S. olivaceus* ATCC 31126, *S. cattleya* NRRL 8057, and *S. lipmanii* NRRL 3584 (F. Flett, S. F. Wootton, and R. Kirby, personal communication); *S. rimosus* NRRL 2234 (K. F. Chater, unpublished data); *S. albus* G (18); and *S. albus* PCMI52766 (15a). The enzymes responsible for restriction have been identified in the last two cases as *Sal*GI (2) and *Sal*PI (14), respectively. A second Mg^{2+}-dependent site-specific endodeoxyribonuclease is present in *S. albus* G (*Sal*GII; 2). Such enzymes are also present in *S. achromogenes* ATCC 12767 (*Sac*I, *Sac*II, *Sac*III) and *S. griseus* sp. (J. R. Arrand, P. Myers, R. J. Roberts, and K. F. Chater, unpublished data); *S. stanford* (*Sst*I, *Sst*II, *Sst*III, isoschizomers of the *Sac* enzymes; S. Goff and A. Rambach, unpublished data); and *S. varsoviensis* CUB608 (K. F. Chater, unpublished data). Genetic studies on *S. albus* G (15; K. F. Chater and L. C. Wilde, unpublished data) show that the restriction-modification determinants are not highly transmissible, though it is still possible that they may be plasmid-borne (as for *Eco*RI; 4).

Since the *Eco*RI restriction enzyme can function in vivo with the *Escherichia coli* DNA ligase to accomplish *Eco*RI site-specific recombination both within and between plasmid molecules (13), restriction-modification systems may become useful tools for bacterial genetics. The stimulation of such genetic reassortment by heat treatment of mutants temperature sensitive for modification has been envisaged (32). Such a mutant has been obtained for *Sal*GI-specific modification (15).

SOME PRESSING QUESTIONS

It should be clear from this review that the preoccupations and techniques of *Streptomyces* genetics are rapidly moving closer to those of modern *E. coli* genetics. Some of the more obvious questions now posed are as follows.

What other functions involve plasmids?

Why are some plasmids difficult to isolate?

How are plasmids transmitted in normal cultures, heterokaryons, and fusing protoplasts?

What are the biochemical mechanisms of plasmid involvement in antibiotic synthesis, exo-enzyme production, and differentiation?

How do plasmids mobilize chromosomal genes?

Can given plasmids survive and their genes be normally expressed in a wide range both of *Streptomyces* and eubacterial hosts?

Will restriction systems be a significant barrier in cloning experiments?

Can eubacterial plasmids survive and their genes be normally expressed in *Streptomyces*?

Can transposons be detected, and, if so, can they be used as felicitously as in gram-negative bacteria (38)?

ACKNOWLEDGMENTS

I thank Mervyn Bibb, David Hopwood, and Janet Westpheling for their helpful comments on the manuscript.

REFERENCES

1. **Akagawa, H., M. Okanishi, and H. Umezawa.** 1975. A plasmid involved in chloramphenicol production in *Streptomyces venezuelae*: evidence from genetic mapping. J. Gen. Microbiol. **90:**336–346.
2. **Arrand, J. R., P. A. Myers, and R. J. Roberts.** 1978. A new restriction endonuclease from *Streptomyces albus* G. J. Mol. Biol. **118:**127–135.
3. **Baumann, R., and H. P. Kocher.** 1976. Genetics of *Streptomyces glaucescens* and regulation of melanin production, p. 535–551. *In* K. D. Macdonald (ed.), Second international symposium on the genetics of industrial microorganisms. Academic Press, London.
4. **Betlach, M., V. Hershfield, L. Chow, W. Brown, H. M. Goodman, and H. W. Boyer.** 1976. A restriction endonuclease analysis of the bacterial plasmid controlling the *Eco*RI restriction and modification of DNA. Fed. Proc. Fed. Am. Soc. Exp. Biol. **35:**2037–2043.
5. **Bibb, M. J., R. F. Freeman, and D. A. Hopwood.** 1977. Physical and genetical characterisation of a second sex factor, SCP2, for *Streptomyces coelicolor* A3(2). Mol. Gen. Genet. **154:**155–166.
6. **Bibb, M. J., and D. A. Hopwood.** 1978. Genetic and physical studies of a *Streptomyces* plasmid. Abstr. Int. Symp. Genet. Ind. Microorganisms, 3rd, Madison, Wis., Abstr. no. 74, p. 37.
7. **Bibb, M. J., J. M. Ward, and D. A. Hopwood.** 1978. Transformation of plasmid DNA into Streptomyces at high frequency. Nature (London), **274:**398-400.
8. **Boronin, A. M., A. N. Borisoglebskaya, and L. G. Sadovnikova.** 1974. Oxytetracycline-sensitive mutants of *Streptomyces rimosus*, the producer of oxytetracycline. Abstr. Int. Symp. Genet. Ind. Microorganisms, 2nd, p. 103.
9. **Boronin, A. M., and S. Z. Mindlin.** 1971. Genetic analysis of *Actinomyces rimosus* mutants with impaired synthesis of antibiotic. Genetika **7:**125–131.
10. **Boronin, A. M., and L. G. Sadovnikova.** 1972. Elimination by acridine dyes of oxytetracycline resistance in *Actinomyces rimosus*. Genetika **8:**174–176.
11. **Canham, P. L., A. M. Michelson, and L. C. Vining.** 1978. Plasmids and chloramphenicol production by *Streptomyces* species 3022a. Abstr. Int. Symp. Genet. Ind. Microorganisms, 3rd, Madison, Wis., Abstr. no. 72, p. 36.
12. **Chang, L. T., D. A. Behr, and R. P. Elander.** 1978. Effects of plasmid-curing agents on cultural characteristics and kanamycin formation in a production strain of *Streptomyces kanamyceticus*. Abstr. Int. Symp. Genet. Ind. Microorganisms, 3rd, Madison, Wis., Abstr. no. 73, p. 36.
13. **Chang, S., and S. N. Cohen.** 1977. *In vivo* site-specific genetic recombination promoted by the *Eco*RI restriction endonuclease. Proc. Natl. Acad. Sci. U.S.A. **74:**4811–4815.
14. **Chater, K. F.** 1977. A site-specific endodeoxyribonuclease from *Streptomyces albus* CMI52766 sharing site-specificity with *Providencia stuartii* endonuclease *Pst*I. Nucleic Acids Res. **4:**1989–1998.
15. **Chater, K. F.** 1978. The genetic determination of restriction in *Streptomyces albus* G. Abstr. Int. Symp. Genet. Ind. Microorganisms, 3rd, Madison, Wis., Abstr. no. 4, p. 2.

15a. **Chater, K. F., and A. T. Carter.** 1978. Restriction of a bacteriophage in *Streptomyces albus* P (CMI52766) by endonuclease *Sal*PI. J. Gen. Microbiol. **109:**181-185.

16. **Chater, K. F., and M. J. Merrick.** 1976. Approaches to the study of differentiation in *Streptomyces coelicolor* A3(2), p. 583–593. *In* K. D. Macdonald (ed), Second international symposium on the genetics of industrial microorganisms. Academic Press, London.
17. **Chater, K. F., and K. I. Sykes.** 1976. Induction of prophage during germination and sporulation in *Streptomyces coelicolor,* p. 131–142. *In* A. N. Barker, J. Wolf, D. J. Ellar, G. J. Dring, and G. W. Gould (ed.), Spore research 1976, vol. 1. Academic Press, London.
18. **Chater, K. F., and L. C. Wilde.** 1976. Restriction of bacteriophage of *Streptomyces albus* G involving endonuclease *Sal*I. J. Bacteriol. **128:**644–650.
19. **Chinenova, T. A., and N. D. Lomovskaya.** 1975. Temperature-sensitive mutants of ϕC31 actinophage of *Streptomyces coelicolor* A3(2). Genetika **11:**132–141.
20. **Chung, S. T., and R. L. Morris.** 1978. Isolation and characterization of plasmid deoxyribonucleic acid from *Streptomyces fradiae.* Abstr. Int. Symp. Genet. Ind. Microorganisms, 3rd, Madison, Wis., Abstr. no. 78, p. 39.
21. **Cundliffe, E.** 1978. Mechanism of resistance to thiostrepton in the producing organism *Streptomyces azureus.* Nature (London) **272:**792–795.
22. **Dowding, J. E., and D. A. Hopwood.** 1973. Temperate bacteriophages for *Streptomyces coelicolor* A3(2) isolated from soil. J. Gen. Microbiol. **78:**349–359.
23. **Flock, J.-I.** 1977. Deletion mutants of temperate *Bacillus subtilis* bacteriophage ϕ105. Mol. Gen. Genet. **155:**241–247.
24. **Freeman, R. F., M. J. Bibb, and D. A. Hopwood.** 1977. Chloramphenicol acetyltransferase-independent chloramphenicol resistance in *Streptomyces coelicolor* A3(2). J. Gen. Microbiol. **98:**453–465.
25. **Freeman, R. F., and D. A. Hopwood.** 1978. Unstable naturally occurring resistance to antibiotics in *Streptomyces.* J. Gen. Microbiol. **106:**377–381.
26. **Friend, E. J., M. Warren, and D. A. Hopwood.** 1978. Genetic evidence for a plasmid controlling fertility in an industrial strain of *Streptomyces rimosus.* J. Gen. Microbiol. **106:**201–206.
27. **Gregory, K. F., and J. C. C. Huang.** 1964. Tyrosinase inheritance in *Streptomyces scabies.* I. Genetic recombination. J. Bacteriol. **87:**1281–1286.
28. **Hopwood, D. A.** 1978. Extrachromosomally determined antibiotic production. Annu. Rev. Microbiol. **32:** 373–392.
29. **Hopwood, D. A., K. F. Chater, J. E. Dowding, and A. Vivian.** 1973. Recent advances in *Streptomyces*

coelicolor genetics. Bacteriol. Rev. **37:**371–405.

30. **Hopwood, D. A., and M. J. Merrick.** 1977. Genetics of antibiotic production. Bacteriol. Rev. **41:**595–635.
31. **Hopwood, D. A., H. M. Wright, M. J. Bibb, and S. N. Cohen.** 1977. Genetic recombination through protoplast fusion in *Streptomyces.* Nature (London) **268:**171–173.
32. **Humphreys, G. O., G. A. Wilshaw, H. R. Smith, and E. S. Anderson.** 1976. Mutagenesis of plasmid DNA with hydroxylamine: isolation of mutants of multicopy plasmids. Mol. Gen. Genet. **145:**101–108.
33. **Kähler, R., and D. Noack.** 1974. Action of acridine orange and ethidium bromide on growth and antibiotic activity of *Streptomyces hygroscopicus* JA 6599. Z. Allg. Mikrobiol. **14:**529–533.
34. **Kieser, T., R. Hütter, and M. Suter.** 1978. Genetic characterization of albino mutants of *Streptomyces glaucescens.* Abstr. Int. Symp. Genet. Ind. Microorganisms, 3rd, Madison, Wis., Abstr. no. 71, p. 35.
35. **Kirby, R.** 1978. Genetic studies on *Streptomyces clavuligerus.* Abstr. Int. Symp. Genet. Ind. Micoorganisms, 3rd, Madison, Wis., Abstr. no. 76, p. 38.
36. **Kirby, R., and D. A. Hopwood.** 1977. Genetic determination of methylenomycin synthesis by the SCP1 plasmid of *Streptomyces coelicolor* A3(2). J. Gen. Microbiol. **98:**239–252.
37. **Kirby, R., L. F. Wright, and D. A. Hopwood.** 1975. Plasmid-determined antibiotic synthesis and resistance in *Streptomyces coelicolor.* Nature (London) **254:**265–267.
38. **Kleckner, N., J. Roth, and D. Botstein.** 1977. Genetic engineering *in vivo* using translocatable drug-resistance elements. J. Mol. Biol. **116:**125–159.
39. **Konvalinkova, V.** 1977. Transfection et transformation chez un actinomycete: cas de *Streptomyces virginiae.* Ph.D. Thesis, University of Liège, Liège, Belgium.
40. **Konvalinkova, V., P. Roelants, and M. Mergeay.** 1977. Transfection in *Streptomyces virginiae.* Biochem. Soc. Trans. **5:**941–943.
41. **Lomoskaya, N. D., N. M. Mkrtumian, N. L. Gostimskaya, and V. N. Danilenko.** 1972. Characterization of temperate actinophage ϕC31 isolated from *Streptomyces coelicolor* A3(2). J. Virol. **9:**258–262.
42. **Lomovskaya, N. D., T. A. Voeykoya, and N. M. Mkrtumian.** 1977. Construction and properties of hybrids obtained in interspecific crosses between *Streptomyces coelicolor* A3(2) and *Stretomyces griseus* Kr.15. J. Gen. Microbiol. **98:**187–198.
43. **Malik, V. S.** 1977. Preparative method for the isolation of supercoiled plasmid DNA from a chloramphenicol-producing streptomycete. J. Antibiot. **30:**897–899.
44. **Merrick, M. J.** 1976. A morphological and genetic mapping study of bald colony mutants of *Streptomyces coelicolor.* J. Gen. Microbiol. **96:**299–315.
45. **Novikova, N. L., O. N. Kapitanova, and N. D. Lomovskaya.** 1973. Thermal prophage induction in germinating spores of *Streptomyces coelicolor* A3(2). Mikrobiologiya **42:**713–718.
46. **Ogawara, H., and S. Nozaki.** 1977. Effect of acriflavine on the production of β-lactamase in Streptomyces. J. Antibiot. **30:**337–339.
47. **Okanishi, M., K. Hamama, and H. Umezawa.** 1968. Factors affecting infection of protoplasts with deoxyribonucleic acid of actinophage PK-66. J. Virol. **2:**686–691.
48. **Okanishi, M., T. Ohto, and H. Umezawa.** 1970. Possible control of formation of aerial mycelium and antibiotic production in *Streptomyces* by episomic factors. J. Antibiot. **23:**45–47.
49. **Okanishi, M., K. Suzuki, and H. Umezawa.** 1974. Formation and reversion of Streptomycete protoplasts: cultural condition and morphological study. J. Gen. Microbiol. **80:**389–400.
50. **Okanishi, M., and H. Umezawa.** 1978. Plasmids involved in antibiotic production in streptomycetes, p. 19–36. *In* E. Freerksen, I. Tárnok, and J. H. Thumin (ed.), Genetics of the Actinomycetales. Proceedings of the International Colloquium at the Forschungsinstitut Borstel. G. Fischer Verlag, Stuttgart.
51. **Parkinson, J. S., and R. J. Husky.** 1971. Deletion mutants of bacteriophage lambda. I. Isolation and initial characterisation. J. Mol. Biol. **56:**369–384.
52. **Redshaw, P. A., P. A. McCann, and B. M. Pogell.** 1977. Loss of multiple differentiated functions accompanies appearance of aerial mycelia-negative (am$^-$) trait in streptomycetes. Abstr. Annu. Meet. Am. Soc. Microbiol., 77th, Abstr. no. K251, p. 228.
53. **Redshaw, P. A., P. A. McCann, L. Sankaran, and B. M. Pogell.** 1976. Control of differentiation in streptomycetes: involvement of extrachromosomal deoxyribonucleic acid and glucose repression in aerial mycelia development. J Bacteriol. **125:**698–705.
54. **Rudd, B. A. M., and D. A. Hopwood.** 1978. Genetics of actinorhodin biosynthesis in *Streptomyces coelicolor* A3(2). Abstr. Int. Symp. Genet. Ind. Microorganisms, 3rd, Madison, Wis., Abstr. no. 19, p. 10.
55. **Schrempf, H.** 1978. Isolation and characterisation of plasmids in *Streptomyces,* p. 13. *In* E. Freerksen, I. Tarnok, and J. H. Thumin (ed.), Genetics of the Actinomycetales. Proceedings of the International Colloquium at the Forschungsinstitut Borstel. G. Fischer Verlag, Stuttgart.
56. **Schrempf, H., H. Bujard, D. A. Hopwood, and W. Goebel.** 1975. Isolation of covalently closed circular deoxyribonucleic acid from *Streptomyces coelicolor* A3(2). J. Bacteriol. **121:**416–421.
57. **Schrempf, H., and W. Goebel.** 1978. Plasmids in *Streptomyces.* Abstr. Int. Symp. Genet. Ind. Microorganisms, 3rd, Madison, Wis., Abstr. 80, p. 40.
58. **Shaw, P. D., and J. Piwowarski.** 1977. Effects of ethidium bromide and acriflavine on streptomycin production by *Streptomyces bikiniensis.* J. Antibiot. **30:**404–408.
59. **Sladkova, I. A., N. D. Lomovskaya, and T. A. Chinenova.** 1977. The structure and size of the genome of actinophage ϕC31 of *Streptomyces coelicolor* A3(2). Genetika **13:**342–344.

60. **Suter, M., R. Hütter, and T. Leisinger.** 1978. Mutants of *Streptomyces glaucescens* affected in the production of extracellular enzymes, p. 61–64. *In* E. Freerksen, I. Tarnok and J. H. Thumin (ed.), Genetics of the Actinomycetales. Proceedings of the International Colloquium at the Forschungsinstitut Borstel. G. Fischer Verlag, Stuttgart.
61. **Vivian, A.** 1971. Genetic control of fertility in *Streptomyces coelicolor* A3(2): plasmid involvement in the interconversion of UF and IF strains. J. Gen. Microbiol. **69:**353–364.
62. **White, T. J., and J. Davies.** 1978. Possible involvement of plasmids in the biosynthesis of paromomycin. Abstr. Int. Symp. Genet. Ind. Microorganisms, 3rd, Madison. Wis., Abstr. no. 79, p. 39.
63. **Wright, H. M., and D. A. Hopwood.** 1977. A chromosomal gene for chloramphenicol acetyltransferase in *Streptomyces acrimycini.* J. Gen. Microbiol. **102:**417–421.
64. **Wright, L. F., and D. A. Hopwood.** 1976. Identification of the antibiotic determined by the SCP1 plasmid of *Streptomyces coelicolor* A3(2). J. Gen. Microbiol. **95:**96–106.
65. **Wright, L. F., and D. A. Hopwood.** 1976b. Actinorhodin is a chromosomally-determined antibiotic in *Streptomyces coelicolor* A3(2). J. Gen. Microbiol. **96:**289–297.
66. **Yagisawa, M., T-S. R. Huang, and J. Davies.** 1978. The possible role of plasmids in neomycin biosynthesis and modification. Abstr. Int. Symp. Genet. Ind. Microorganisms, 3rd, Madison, Wis., Abstr. no. 75, p. 37.
67. **Zeig, J., M. Silverman, M. Hilmen, and M. Simon.** 1977. Recombinational switch for gene expression. Science **196:**170–172.

Plasmids and Antibiotic Synthesis in Streptomycetes

M. OKANISHI

Department of Antibiotics, National Institute of Health, Kamiohsaki, Shinagawa-ku, Tokyo 141, Japan

In 1970, my colleagues and I reported a preliminary experiment suggesting the possibility of plasmid involvement in antibiotic production (10). Thereafter, many researchers became interested in this phenomenon, and to date plasmid involvement has been studied in the formation of aureothricin (1), cephamycin (5), chloramphenicol (1, 9), holomycin (5), kanamycin (6), kasugamycin (9), methylenomycin A (15, 17), oxytetracycline (2, 5), puromycin (13), streptomycin (14), and turimycin (7). In some of these reports, however, the proof of the plasmid involvement is rather indirect and not completely satisfactory. The only strong evidence that any structural genes for antibiotic synthesis are plasmid-borne is provided for methylenomycin A production by *Streptomyces coelicolor* A3(2) (5, 15, 17). The involvement of other plasmids may be in the regulation of antibiotic production or resistance to the antibiotic produced rather than directly in the structural genes for the biosynthetic steps. This paper concerns the experiments we have performed to clarify this situation.

INVOLVEMENT OF A PLASMID IN CHLORAMPHENICOL PRODUCTION

The ability of *S. venezuelae* ISP5230 to produce chloramphenicol was lost with high frequency after treatment with acriflavine (1). To obtain genetic evidence for plasmid involvement in chloramphenicol production, we crossed the nonproducing variant SVM3, which has auxotrophic markers, with the chloramphenicol-producing strain SVM1, which carries the complemental auxotrophic markers, and observed that the chloramphenicol-nonproducing mutation, *cpp*, did not show a linkage with a set of chromosomal markers (1). Recently, it was found that most of the *cpp* variants arising after acriflavine treatment still produce a small amount of chloramphenicol in the normal medium. We then attempted to clarify the role of plasmids in chloramphenicol production.

Strains SVM1 and SVM2 carrying auxotrophic markers produced chloramphenicol in the normal medium and in synthetic medium. Strains SVM3, SVM4, and SVM2-T8, which are representatives of the nonproducing variants obtained with high frequency after acriflavine treatment or incubation at high temperature and which are presumed to be plasmid deficient by mating experiments, produced a very small amount of chloramphenicol (Table 1). In this situation, it was suspected that the so-called cured strains still have a few plasmids in their mycelium. Therefore, the acriflavine treatment, the high-temperature incubation, or the induction of mutation was reapplied to the producing strain SVM2, and nonproducing mutants were selected. All of the mutant loci (*cpp*) were mapped on the chromosome between *met* and *ilv*. Among the nonproducing mutants selected, a novel mutant, SVM2-2A7, was isolated. It produced 1-deoxychloramphenicol instead of chloramphenicol, and the mutant locus was also located on the chromosome between *met* and *ilv* (Table 1).

To examine whether or not these Cpp$^-$ chromosomal mutants carry the plasmid, we mated them with the plasmid-deficient variant SVM3 or SVM4. Among the nutritional recombinants selected, the chloramphenicol-producing recombinants appeared with high frequency in all the crosses. The degree of their chloramphenicol production was almost the same as that of the original strain SVM2. However, the nutritional recombinants from crosses between the plasmid-deficient variants SVM3 and SVM4 or SVM2-T8 did not produce chloramphenicol (Table 1). Hence, these Cpp$^-$ chromosomal mutants still have the proper plasmid.

When chloramphenicol was added to a growing culture or a cell-free system of mutant SVM2-2A7, we could not detect the change to 1-deoxychloramphenicol in the medium. Therefore, the hydroxylase which oxidizes the hydrogen at the C1 position of *p*-aminophenylalanine to the hydroxyl in *p*-aminophenylserine may be missing in this mutant (16). It can also be concluded that all or most of the structural genes for the biosynthetic steps, including the oxidation of *p*-aminophenylalanine, exist between *met* and *ilv* on the chromosome and that the plasmid plays a regulatory role in the chloramphenicol production.

PLASMID INVOLVEMENT IN AUREOTHRICIN BIOSYNTHESIS

Aureothricin is chemically known as *N*-propionylpyrrothine, and isobutyropyrrothine, thiolutin, and holomycin are known as the pyrrothine-containing antibiotics (3). *S. kasugaensis* strains M338 and MB273 were used in our experiments, and both produced kasugamycin, aureothricin, and thiolutin. At first, the substance stimulating the aureothricin production was investigated in a synthetic medium, and a large increase in product was found in the medium containing inorganic sulfate, L-cysteine, L-cystine, and L-methionine. A similar experiment was performed with washed mycelia suspended in a mineral salt solution, but L-cystine was the only substance effective for aureothricin production. From these results, we proposed the pathway for aureothricin biosynthesis shown in Fig. 1. Further support for this pathway was obtained by use of radioactive L-cystine and pyrrothine (9).

All of the aureothricin- and thiolutin-nonproducing variants used in our experiments were obtained by acriflavine treatment. When they were incubated in an antibiotic-producing medium, neither aureothricin nor pyrrothine was detected in the beers and in mycelium. When the nonproducing variants were incubated in an antibiotic-producing medium supplemented with pyrrothine and propionate, however, more than 50% of the pyrrothine was converted into aureothricin. The same ability was demonstrated for *S. venezuelae* and *S. erythreus.* The nonproducing variants were equally as resistant to aureothricin as the original aureothricin-producing strain. These results suggested that a plasmid plays a role in the biosynthesis of the pyrrothine moiety and that a chromosome is responsible for its acylation. The contour length of the plasmid DNA isolated from strain M338 was 3.35 ± 0.05 μm.

Whereas all of the experiments mentioned above were carried out with strain M338 and its variants, the following experiments were carried out with strain MB273 because it forms a greater amount of aerial mycelium and the isolation of plasmid DNA is easier. As shown in Table 2, the aureothricin-nonproducing variants selected from a population treated with curing agents were classified into three groups depending on the characters of aerial mycelium and productivity of kasugamycin. Since the closed circular DNA was isolated from the aureothricin nonproducers belonging to classes I and III but not from those in class II, the plasmid was considered to have been cured in each of the three variants of class II. The contour length of the plasmid isolated from this strain was 3.2 ± 0.05 μm.

To study the role of the plasmid in aureothricin biosynthesis, we attempted to

TABLE 1. *Characterization of Cpp mutants in* S. venezuelae

Typical mutant obtained	Original strain	Mutagen used[a]	Selection medium for Cpp^-	CAP[b] produced (μg/ml)		*cpp* locus mapped	cpp^+ in $his^+ \cdot lys^+$
				SP[c]	GNa[c]		
SVM1[d]				20–25	100–120		
SVM2[e]				20–25	100–120		
SVM3	SVM2	AF	SP	0.2	5–10	Plasmid loss	0/601
SVM4	SVM1	AF	SP	0.2	5–10	Plasmid loss	0/601
SVM2-T8	SVM2	HT	SP	0.2	5–10	Plasmid loss	0/47
SVM2-1A1	SVM2	AF	GNa	0	0.5	ND[f]	ND
SVM2-HT3	SVM2	HT	GNa	0	0.5	*met-ilv*	3/42
SVM2-2A7	SVM2	AF	SP	0[g]	0[g]	*met-ilv*	33/422
SVM2-U26	SVM2	UV	SP	0	0	*met-ilv*	55/216
SVM2-N7	SVM2	NTG	SP	0	0	*met-ilv*	++/182

[a] AF, Acriflavine; HT, high temperature; NTG, nitrosoguanidine.
[b] CAP, Chloramphenicol.
[c] SP, Normal medium; GNa, synthetic medium.
[d] SVM1: *lys ilv pro met* str^r cpp^+.
[e] SVM2: *his leu ade* cpp^+.
[f] ND, Not done.
[g] 1-Deoxychloramphenicol is produced instead of chloramphenicol.

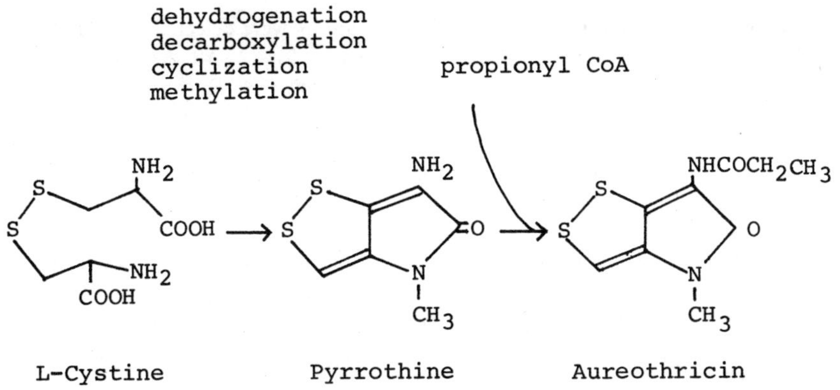

dehydrogenation
decarboxylation
cyclization
methylation

propionyl CoA

L-Cystine → Pyrrothine → Aureothricin

FIG. 1. *Pathway of aureothricin biosynthesis.*

TABLE 2. *Classification of aureothricin-nonproducing variants*

Class	Organism	Mutagen used[a]	Aerial mycelium	Antibiotic produced[b]		CCC DNA[c]
				KSM	AT·TL	
	MB273		+++ (greenish gray)	++	+++	+
I	5-1	AF	−	−	−	+
II	5-3	EB	−	−	−	+
II	5-4	EB	−	+++	−	−
II	5-13 (*arg*⁻)	Spont	−	+++	−	−
II	5-18 (*arg*⁻)	Spont	−	+++	−	−
III	5-189	AO	+++ (white)	+++	−	+
III	5-266	AO	+++ (white)	+++	−	ND[d]
	5-2	EB	+++ (grayish pink)	−	+++	ND
	5-7	EB	+++ (grayish pink)	−	+++	+

[a] AF, Acriflavine; EB, ethidium bromide; Spont, spontaneous; AO, acridine orange.
[b] KSM, Kasugamycin; AT, aureothricin; TL, thiolutin.
[c] Covalently closed circular DNA having contour length of 3.1 μm.
[d] ND, Not done.

produce aureothricin in a cell-free system. As shown in Table 3, two kinds of reaction mixtures were developed: one was used for the conversion of pyrrothine into aureothricin and the other was intended for the production of aureothricin from L-cysteine. Each reaction mixture was incubated at 30°C for 2 h and extracted with ethyl acetate (pH 9.0). The extract was concentrated and analyzed by thin-layer chromatography with $CHCl_3$-methanol (19:1) as the solvent system. Aureothricin-related substances were detected with the fluorescein-mercuric acetate reagent, which is specific for S-S bonds, and by bioautography with *Escherichia coli* as the test organism.

Depending on the acyl coenzyme A supplied, strain MB273 converted pyrrothine into aureothricin or thiolutin. The extract from the other reaction mixture gave products 1, 1′, and 3 (Fig. 2). When the complete reaction mixture was incubated in the presence of mercaptoethanol, products 5, 6, and 6′ were detected instead of 1, 1′, and 3. The control experiments are also shown in Fig. 2. All the products were ninhydrin negative. Products 1′ and 3 were active against *E. coli, Staphylococcus aureus,* and *Trichophyton rubrum,* but product 1 was not. Among the products formed in the presence of mercaptoethanol, product 6 showed the same activity. Products 1′, 3, and 6 were all positive with the fluorescein-mercuric acetate reagent. These substances are believed to be structurally related to aureothricin, which, like thiolutin, was not formed.

An identical experiment was performed with the cell-free system in nonproducing

TABLE 3. *Reaction mixtures and solution for the preparation of cell-free extracts*

Prepn	Component[a]	Amt or concn
Reaction mixture for conversion of pyrrothine (total volume, 2.5 ml; pH 6.5)	Cell-free fraction	2.0 ml
	Pyrrothine HCl	2.5 μmol
	Acetyl CoA or propionyl CoA	2.5 μmol
Reaction mixture for production of aureothricin group of antibiotics (total volume, 2.5 ml; pH 6.5)	Cell-free fraction	2.0 ml
	L-Cysteine	5.0 μmol
	PALP	2.5 μmol
	FAD	1.0 μmol
	Acetyl CoA	1.0 μmol
	S-adenosylmethionine	2.5 μmol
Solution for preparation of cell-free extracts	PIPES buffer	0.05 M
	NaCl	0.5%
	$MgCl_2 \cdot 6H_2O$	0.05%
	$CuCl_2 \cdot 2H_2O$	0.002%
	(2-Mercaptoethanol)	(0.1%)

[a] Abbreviations: CoA, coenzyme A; FAD, flavine adenine dinucleotide; PIPES, piperazine-*N,N′*-bis(2-ethanesulfonic acid); PALP, pyridoxal phosphate.

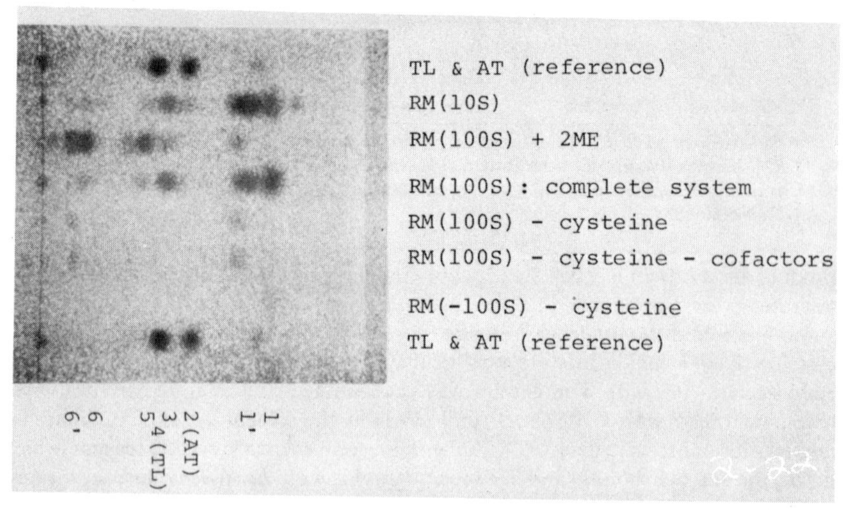

FIG. 2. *Thin-layer chromatography of reaction mixtures (RM) for the production of the aureothricin group of antibiotics. Incubation was at 30°C for 2 h. Mixtures were extracted with ethyl acetate (pH 9.0). Products were separated by thin-layer chromatography ($CHCl_3$-methanol, 19:1) and were visualized by use of the fluorescein-mercuric acetate reagent and by antibacterial activity. TL, Thiolutin; AT, aureothricin.*

variants of *S. kasugaensis* and in other antibiotic-producing species such as *S. venezuelae, S. erythreus,* and *S. griseus.* Unexpectedly, they showed the same thin-layer chromatographic pattern as the aureothricin producer.

DISCUSSION

Cell-free systems of other antibiotic-producing species (*S. venezuelae, S. erythreus,* and *S. griseus*) as well as the nonproducing variants of *S. kasugaensis* formed substances related to aureothricin. Assuming that the products, especially product 3, are related to aureothricin, the nonproducers should have biosynthetic enzymes encoded on the chromosome. All of the nonproducing variants (and other streptomycetes), however, failed to produce aureothricin or related substances in growing cultures. Therefore, it may be concluded that in some way the plasmid plays a role in the regulation of aureothricin formation. This conclusion should also apply to the plasmid involved in chloramphenicol production. Holomycin is a demethythiolutin with a structure similar to that of aureothricin. Variants that do not produce holomycin are formed spontaneously or are induced by UV irradiation. The producing ability transfers with high frequency when Hol^+ and Hol^- are mated. However, the nonproducing variants still produce the antibiotic in liquid culture. Furthermore, the nonproducing mutants isolated after nitrosoguanidine treatment did not produce the antibiotic under any cultural conditions, and their mutational loci were located on a chromosome (5).

On the other hand, most antibiotic-nonproducing variants obtained by treatment with curing agents show a pleiotropic effect on their phenotypes, such as loss of aerial mycelium, appearance of rough or wrinkled colonies, and loss of pigment. In the case of *S. coelicolor,* however, the loss of aerial mycelium was caused by the mutation of the chromosome, and most of those mutations also led to the loss of methylenomycin and actinorhodin production (4, 8). Therefore, the loss of aerial mycelium may be caused by either chromosomal mutation or loss of plasmid and often has a pleiotropic effect.

A plausible interpretation of these results is that pleiotropic variation may be caused by chromosome- or plasmid-directed alteration of the cell membrane. This interpretation is supported by results reported by Okazaki et al. (11, 12) in which the alteration of fatty acid components in the cell membrane caused the excretion of an intracellular inducer or precursor of neomycin or streptothricin, with a concomitant loss of antibiotic production.

REFERENCES

1. **Akagawa, H., M. Okanishi, and H. Umezawa.** 1975. A plasmid involved in chloramphenicol production in *Streptomyces venezuelae:* evidence from genetic mapping. J. Gen. Microbiol. **90:**336–346.
2. **Boronin, A. M., and L. G. Sadovnikova.** 1972. Elimination by acridine dyes of oxytetracycline resistance in *Actinomyces rimosus* (in Russian). Genetika **8:**174–176.
3. **Büchi, G., and G. Lukas.** 1963. A synthesis of holomycin. J. Am. Chem. Soc. **85:**647–648.
4. **Chater, K. F.** 1972. A morphological and genetic mapping study of white colony mutants of *Streptomyces coelicolor.* J. Gen. Microbiol. **72:**9–28.
5. **Hopwood, D. A.** 1978. Annu. Rev. Microbiol., in press.
6. **Hotta, K., Y. Okami, and H. Umezawa.** 1977. Elimination of the ability of a kanamycin-producing strain to biosynthesize deoxystreptamine moiety by acriflavine. J. Antibiot. **30:**1146–1149.
7. **Kahler, R., and D. Noack.** 1974. Action of acridine orange and ethidium bromide on growth and antibiotic activity of *Streptomyces hygroscopicus* JA6599. Z. Allg. Mikrobiol. **14:**529–533.
8. **Merrick, M. J.** 1976. A morphological and genetic mapping study of bald colony mutants of *Streptomyces coelicolor.* J. Gen. Microbiol. **96:**299–315.
9. **Okanishi, M.** 1977. Plasmid involvement in the production of secondary metabolites with plasmids (in

Japanese). Amino Acid and Nucleic Acid, no. 35, p. 15–30.
10. **Okanishi, M., T. Ohta, and H. Umezawa.** 1970. Possible control of formation of aerial mycelium and antibiotic production in *Streptomyces* by episomic factors. J. Antibiot. **23:**45–47.
11. **Okazaki, H., T. Beppu, and K. Arima.** 1974. Induction of antibiotic formation in *Streptomyces* sp. No. 362 by the change of cellular fatty acid spectrum. Agric. Biol. Chem. **38:**1455–1461.
12. **Okazaki, H., H. Ono, K. Yamada, T. Beppu, and K. Arima.** 1973. Relationship among cellular fatty acid composition, amino acid uptake and neomycin formation in a mutant strain of *Streptomyces fradiae.* Agric. Biol. Chem. **37:**2319–2325.
13. **Pogell, B. M., L. Sankaran, P. A. Redshaw, and P. A. McCann.** 1976. Regulation of antibiotic biosynthesis and differentiation in streptomycetes, p. 543–547. *In* D. Schlessinger (ed.), Microbiology—1976. American Society for Microbiology, Washington, D.C.
14. **Shaw, P. D., and J. Piwowarski.** 1977. Effect of ethidium bromide and acriflavine on streptomycin production by *Streptomyces bikiniensis.* J. Antibiot. **30:**404–408.
15. **Vivian, A.** 1971. Genetic control of fertility in *Streptomyces coelicolor* A3(2): plasmid involvement in the interconversion of UF and IF strains. J. Gen. Microbiol. **69:**353–364.
16. **Westlake, D. W. S., and L. C. Vinning.** 1969. Biosynthesis of chloramphenicol. Biotechnol. Bioeng. **11:** 1125–1134.
17. **Wright, L. F., and D. A. Hopwood.** 1976. Identification of the antibiotic determined by the SCP1 plasmid of *S. coelicolor* A3(2). J. Gen. Microbiol. **95:**96–106.

Genetic Relationship Between Actinomycetes and Actinophages[1]

N. D. LOMOVSKAYA, T. A. VOEYKOVA, I. A. SLADKOVA, T. A. CHINENOVA, N. M. MKRTUMIAN, AND E. V. SLAVINSKAYA

Institute of Genetics and Selection of Industrial Microorganisms, Moscow, USSR

The isolation of a temperate actinophage, øC31, which infects genetically marked *Streptomyces coelicolor* A3(2), enabled us to investigate the relationship between actinomycetes of the genus *Streptomyces* and actinophages (3). In line with these studies, we have carried out further work in the following areas: (i) genetics of øC31 actinophage and genetic control of lysogeny; (ii) physical studies of actinophage DNAs; (iii) the application of actinophages for genetic analysis of actinomycetes; (iv) interaction between actinomycetes and actinophages, using analysis of phage action on interspecific hybrids; (v) detection of restriction-modification (RM) systems in actinomycetes.

One of the main purposes of these studies was transfer of genetic material between various actinomycetes by means of actinophages. To achieve this, it is evident that information was needed to identify modes of gene exchange in *Streptomyces,* such as transformation and transduction.

Isolation of a series of øC31 thermosensitive (*ts*) mutants and those defective for ability to lysogenize (*c* mutants) made it possible to localize 20 genes controlling phage intracellular growth and maintenance of the lysogenic state. From three-factor crosses of type $ts_x c \times ts_y c^+$ and $ts_x h \times ts_y h^+$, the relative map order of markers was established.

High frequency of recombination between terminal phage markers suggested that the genetic map of øC31 might be linear rather than circular. To determine if this high frequency resulted from the lack of markers on the map or if the phage chromosome is nonpermuted, physical studies of øC31 phage DNA were carried out. When heated, phage particles were seen to disrupt and eject DNA in linear form. After incubation in 0.2 M NaCl at 60°C for 30 min, linear molecules converted into a circular form as the temperature was gradually lowered. This implies that linear molecules bear two readily associating cohesive ends. Both linear and circular molecules were of similar length and molecular weight, i.e., 37.7 ± 0.5 kilobases (kb) and $25.0 \pm 0.3 \times 10^6$, respectively. These findings support genetic evidence for linear structure of the øC31 genetic map.

We undertook the study of genetic control of lysogeny in this system since we had a set of *c* mutants and virulent mutants of actinophage øC31. Previously the prophage øC31 was shown to be a genetic determinant of lysogeny located on the *S. coelicolor* A3(2) chromosome, and the effect of zygotic induction was observed, suggesting the cytoplasmic nature of the øC31 repressor (2). A portion of *c* mutants had transdominant properties both upon mixed infection with wild-type phage and in a lysogen infected at a multiplicity of infection of >1. This effect of negative complementation showed the repressor synthesized under the control of a *c* gene to be a protein and to consist of identical subunits. In a population of *c* mutants, virulent mutants of two types (*vir* and *vd*) arose which were capable of growing in lysogens on single infection. The two types of virulent mutants shared one important characteristic, namely, the

[1] An unexpected illness prevented the senior author (N.D.L.) from delivering this paper at the symposium.

ability to cause prophage induction upon infection of a lysogen. The *vd* mutants were easily distinguished from *vir* mutants by their inability to grow in nonlysogenic strains without a helper. Resident prophage was shown to serve as a helper in lysogens infected by a *vd* mutant. Infection of lysogens containing prophage *ts* mutations in various genes with a *vd* mutant under nonpermissive conditions did not lead to phage production, which implies that all prophage genes that could be tested by *ts* mutants are required for the growth of *vd* mutants in lysogens. Neither prophage induction was observed under these conditions, suggesting that the *vd* mutant failed to provide products of prophage genes impaired by *ts* mutations.

The most plausible explanation to account for inability of the *vd* mutants to grow in nonlysogenic strains is that a region of the chromosome is deleted in these mutants. The heteroduplex technique has been used for mapping the position and length of this putative deletion mutation in øC31 DNA. The mutant DNA is 36.2 ± 0.5 kb in length and 24.0 ± 0.3 × 10^6 daltons in mass. Double-stranded *vd*/wild-type heteroduplexes were seen to contain a single-stranded loop at the position 9.4 ± 0.4 kb from the terminus of the molecule, which confirmed that approximately 3% of øC31 DNA (1.2 ± 0.1 kb) is deleted in øC31 *vd.* Wild-type phage and the *vir* mutant were found to have a complete homology.

To test the specificity of antirepression caused by superinfection with *vir* and *vd,* lysogenic strains containing heteroimmune (VP5, ø448) and homoimmune (øC43, øC62) prophages were superinfected with these mutants. Prophage induction and virulent phage growth took place only in lysogens for homoimmune phages, *S. lividans* 803 (øC43) and *S. lividans* 66 (øC62), upon infection with *vir* mutants (Table 1) and the *vd* mutant. It was inferred from the data with the *vd* mutant that prophages øC43 and øC62 provide gene products essential for mutant growth. Phages øC43 and øC62 show phenotypical resemblances to øC31. However, in contrast to øC31 and øC62, phage øC43 is not capable of lysogenizing *S. lividans* 66 and *S. coelicolor* A3(2). Moreover, øC62 is able to overcome immunity of strains 66 and A3(2) lysogenic for øC31 whereas phages øC43 and øC31 are not.

Comparative studies of phage DNAs revealed that DNAs of øC43 and øC62, like that of øC31, have cohesive ends. The DNA lengths and molecular weights of øC62 are 39.4 ± 0.6 kb and 26.1 ± 0.4 × 10^6, respectively, whereas those of øC43 are 39.6 ± 0.8 kb and 26.2 ± 0.5 × 10^6. Despite the fact that these three phages have been isolated independently, there is a high degree of homology among their DNAs. Molecules of øC31 DNA differ from those of øC62 in containing a small deletion seen in a heteroduplex micrograph (Fig. 1a) as a single-stranded loop of 1.3 ± 0.2 kb. The regions of nonhomology in the heteroduplex øC43/øC62 were visualized as two loops closely situated at the end of the molecule (Fig. 1b). Interestingly, one of the

TABLE 1. *Prophage induction and superinfecting* vir *mutant production in lysogens*[a]

Lysogen	No. of plaques: After *vir* mutant superinfection: *vir* mutant	No. of plaques: After *vir* mutant superinfection: Prophage	No. of plaques: Spontaneous prophage induction
S. lividans 66 (øC31)	98	122	7
S. lividans 66 (øC62)	56	82	6
S. lividans 803 (øC43)	110	49	8
S. lividans 66 (VP5)	165	3	3
S. lividans 66 (ø448)	92	5	4

[a] The type experiment is described.

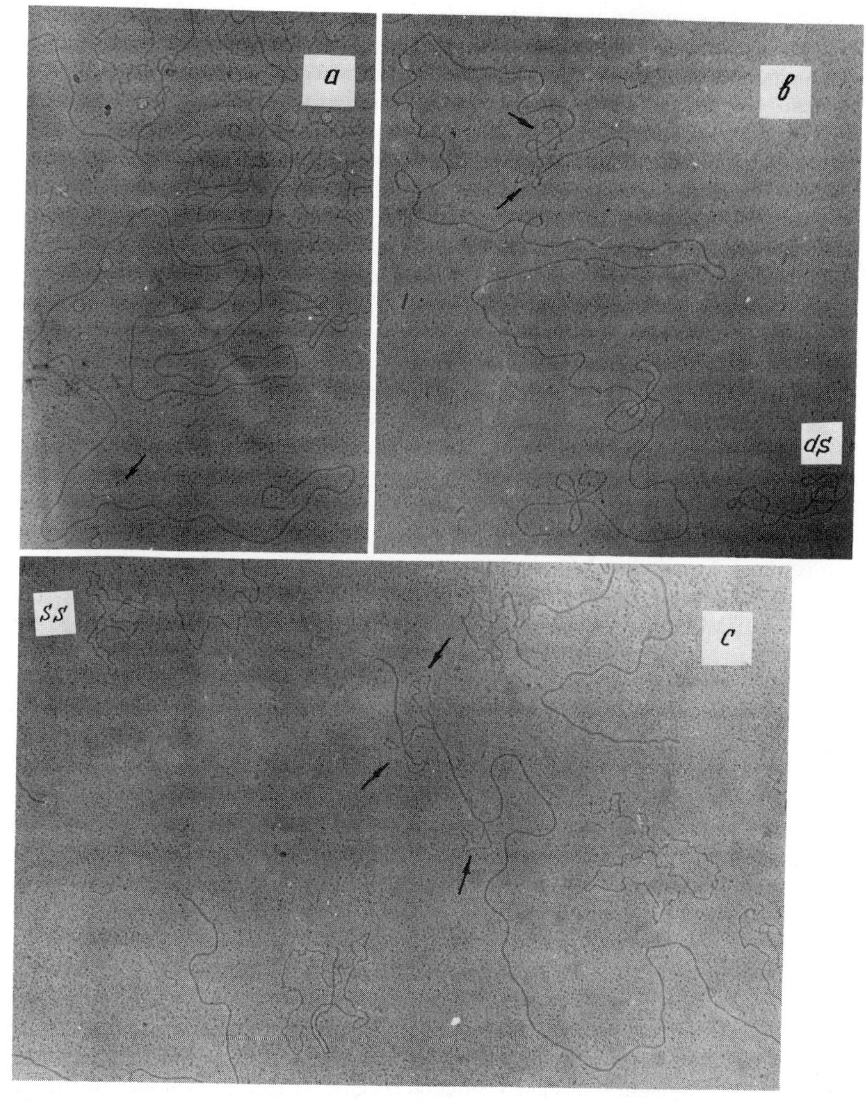

FIG. 1. *Electron micrographs of (a) øC31/øC62, (b) øC43/øC62, and (c) øC31/øC43 heteroduplex DNA molecules. Regions of nonhomology are indicated by arrows. Small double-stranded circular DNA is colE1, and single-stranded DNA is f1 phage DNA. Heteroduplexes were mounted from 75% formamide solution (cation concentration, 0.060 M) and layered onto double-distilled water. The grids were stained with uranyl acetate and shadowed with platinum.*

nonhomology regions in heteroduplexes øC43/øC62 and øC43/øC31 (Fig. 1b and c) is of unique appearance, which suggests that two sequences of different lengths are located next to one another and are flanked by short (~100 base pairs) inverted repetitions. Further research may elucidate whether the structural modifications of DNA are related to phenotypic properties of these phages.

Recent advances in genetics of actinomycetes and the availability of novel techniques in experimental work have provided a basis for exploitation of the methods designed to construct recombinants within the genus *Streptomyces.* Genetic studies of such recombinants along with the development of cloning systems within actinophages and actinomycete plasmids may improve the breeding of producers of biologically active substances.

We have obtained interspecific recombinants between *S. coelicolor* A3(2) and *S. lividans* 66 (1) as well as between A3(2) and *S. griseus* Kr.15, a producer of the antibiotic grisin (4). The use of recombinants made it possible to map on the actinomycete genome the locations of genes having different phenotypic expression in the parental strains, such as those controlling antibiotic properties, phage adsorption, or intracellular growth. Thus, we were able to position markers without having to obtain mutations in the above genes.

Furthermore, we succeeded in obtaining recombinants of *S. griseus* Kr.15 which produce grisin and are resistant to phages specific for *S. griseus* and *S. coelicolor.* Being an evolutionally stable characteristic, the phage resistance thus acquired can be more advantageous than that conferred by mutation.

Analysis of recombinants *S. coelicolor* × *S. griseus* for the presence of parental contributions by the use of phages specific for either of the two parental strains was also helpful in identifying genes controlling phage DNA RM in *Streptomyces.* We first tested *S. coelicolor* A3(2) and 60 strains representative of 22 species of the genus *Streptomyces* with actinophages øC31 and VP5. Phages grown in the A3(2) host, øC31·A3(2), and VP5·A3(2), were restricted by a number of strains. An RM system was found in strain *S. lividans* 67 tested with VP5. This phage was restricted by strain 67 when grown in A3(2), whereas VP5·67 was not restricted by either strain 67 or A3(2). This could mean that strain 67 had at least two RM systems, one of which was specific for A3(2), or that A3(2) had no RM system at all. Another possible explanation is that VP5 may possess no sites recognized by the RM system of A3(2). It should be noted that phages øC31 and VP5 were plated on some 26 strains with normal efficiency. If both of the phages possessed specific recognition sites for the A3(2) RM system, the latter would have been detected unless at least one strain of those examined had the same system. However, this was not the case. The most likely explanation is the absence in phages VP5 and øC31 of specific sites recognized by RM systems in A3(2) and other strains.

Virulent phage Pg81 of *S. griseus* was assayed for plaque-forming efficiency on a number of indicator strains, including hybrid Rcg2 constructed in a cross between *S. coelicolor* A3(2) and *S. griseus* Kr.15. Rcg2 had been deduced to contain a fragment of A3(2) chromosome carrying the *strA1* marker (4). It is not unlikely that Rcg2 retained the mechanism of Pg81 restriction specific for A3(2). As judged by the data in Fig. 2 and Table 2, the phage modified by growth in strain 15 is restricted by Rcg2 and, vice versa, the phage modified by growth in strain Rcg2 fails to plate on strain 15. It can be concluded, therefore, that two different RM systems (RM-15 and RM-2) exist in these strains.

Among Pg81·Rcg2 variants, mutants designated Pg81rs have been found which, after propagation in strain 15, give increased plating efficiency ($\sim 10^{-2}$) on Rcg2 (Fig. 2a, Table 2). The mutants may occur because they have lost a portion of the sites recognized by the Rcg2 RM system. Further studies with Pg81rs mutants should demonstrate whether one more RM-2 system can be detected or whether there are only a few sites on the phage DNA recognized by RM-2, so that the loss of one of them would markedly increase the rate of the modified variants.

TABLE 2. *Relative plaque-forming efficiency of phages Pg81 and Pg81rs*

Indicator strain	Plating efficiency of actinophage: Pg81				Pg81rs			
	15[a]	Rcg2	15-13	20	15	Rcg2	15-13	20
S. griseus Kr.15	1.0	2.2×10^{-6}	1.0	1.5×10^{-6}	1.0	2.0×10^{-6}	1.0	1.0×10^{-6}
Rcg2	2.5×10^{-6}	1.0	2.0×10^{-6}	1.0	8.5×10^{-3}	1.0	6.4×10^{-3}	1.0
S. griseus Kr.15-13	1.0	5.0×10^{-1}	1.0	2.0×10^{-1}	1.0	5.0×10^{-1}	1.0	2.4×10^{-1}
S. griseus Kr.20	5.0×10^{-5}	1.0	2.4×10^{-5}	1.0	8.0×10^{-3}	1.0	1.0×10^{-2}	1.0

[a] Strain in which phage was propagated.

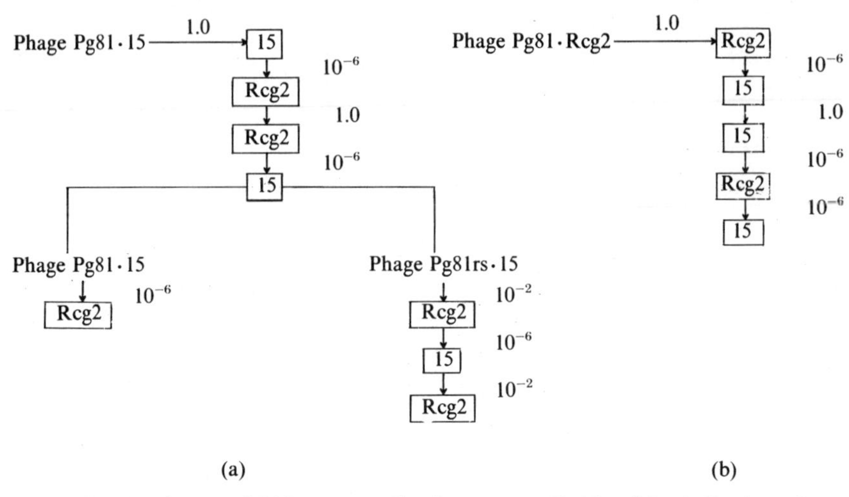

FIG. 2. *Identification of RM systems in* S. griseus *strains Kr.15 and Rcg2. Numbers close to arrows refer to approximate phage efficiency of plating on the strain indicated.*

The RM system in *S. griseus* Kr.20 was detected by means of phage Pg81 (5). The behavior of Pg81·15, Pg81·Rcg2, and Pg81·20 indicated that *S. griseus* Kr.15 and *S. griseus* Kr.20 had different RM systems and that those of strains 20 and Rcg2 are of the same type (Table 2). Mutants of *S. griseus* 15 were obtained after treatment with nitrosoguanidine, which proved to be r^-m^+ (Table 2, strain 15-13).

The results of these studies indicate that phage Pg81 of *S. griseus* is useful in testing RM systems in a number of *Streptomyces* strains. Further studies of the genetic control of RM might enhance the possibility of genetic manipulation in *Streptomyces.*

REFERENCES

1. **Alikhanian, S. I., N. D. Lomovskaya, and V. N. Danilenko.** 1976. *In* K. D. Macdonald (ed.), Second International Symposium on the Genetics of Industrial Microorganisms, p. 595–606. Academic Press, London.
2. **Lomovskaya, N. D., L. K. Emeljanova, N. M. Mkrtumian, and S. I. Alikhanian.** 1973. J. Gen. Microbiol. **77:**455.
3. **Lomovskaya, N. D., N. M. Mkrtumian, N. L. Gostimskaya, and V. N. Danilenko.** 1972. Characterization of temperate actinophage øC31 isolated from *Streptomyces coelicolor* A3(2). J. Virol. **9:**258–262.
4. **Lomovskaya, N. D., T. A. Voeykova, and N. M. Mkrtumian.** 1977. J. Gen. Microbiol. **98:**187.
5. **Zvenigorodsky, V. I., T. A. Voeykova, L. S. Yustratova, E. I. Smirnova, and N. D. Lomovskaya.** 1975. Antibiotiki **5:**409.

Plasmid-Determined Alkane Oxidation in *Pseudomonas*

JAMES SHAPIRO, MICHAEL FENNEWALD, AND SPENCER BENSON

Department of Microbiology, University of Chicago, Chicago, Illinois 60637

The discovery of metabolic plasmids in soil bacteria is now almost a decade old. Recent studies have shown that plasmid genes can endow bacterial species with the capacity to degrade a wide variety of substrates, particularly hydrocarbons and their derivatives. To make effective use of these extrachromosomal metabolic capabilities, it will be necessary to carry out more detailed genetic studies of the particular plasmids. Most practical goals require an understanding of the specific metabolic pathways, the cellular localization of biochemical activities, the regulation of the enzymes involved, and means for in vivo or in vitro manipulation of the genes encoding these enzymes. With such knowledge, we can envisage the construction of bacterial strains with the following useful characteristics: (i) multiple catabolic pathways providing the capacity to degrade a broad range of substrates, (ii) oversynthesis of specific enzymes, (iii) improved ability to metabolize new substrates, and (iv) accumulation of specific biochemical intermediates.

THE *alk* SYSTEM

For the past 4 years, we have been examining the plasmid genes involved in alkane oxidation by *Pseudomonas putida.* These genes were first discovered on the nonconjugative OCT plasmid (9), but further studies have shown that the alk^+ genes which allow growth on alkane substrates will recombine into conjugative plasmids related to OCT (8, 10, 16). Fortunately, these plasmids carry easily selected markers (e.g., camphor utilization, antibiotic resistance) so that mutant *alk* genes can be transferred from strain to strain without requiring alk^+ expression.

One of the difficulties encountered in analyzing the alk^+ loci is that they are found only on plasmids of the IncP-2 incompatibility group (16). All plasmids of this group are very large (>200 megadaltons), and only recently have methods become available for isolating IncP-2 plasmid DNA (11, 14). Thus, physical isolation and in vitro mutagenesis have been slow to develop. In addition, the replication of alk^+ loci only on IncP-2 plasmids presents another difficulty in genetic analysis—namely, the absence of a stable diploid system for complementation studies of *alk* mutations.

Despite some technical problems, we have isolated a number of *alk* mutant plasmids, and, by transferring them to various hosts, we have worked out a preliminary picture of the plasmid and chromosomal loci involved in alkane oxidation (Fig. 1). Basically, Fig. 1 shows that chromosomally encoded activities will oxidize primary alcohols to acetyl-coenzyme A and propionyl-coenzyme A, while plasmid-determined activities oxidize alkanes to primary alcohols and primary alcohols to aldehydes. There is an overlap at the second step of the pathway (alcohol to aldehyde). The chromosomal activities appear to be expressed constitutively, and the plasmid activities are inducible (13). The plasmid alkane hydroxylase resolved into two fractions by either genetic or physical methods (3, 4). For convenience, we refer to the plasmid-determined oxidizing activities by the names of the corresponding genes: $AlkA^+$ (soluble alkane hydroxylase component), $AlkB^+$ (membrane hydroxylase

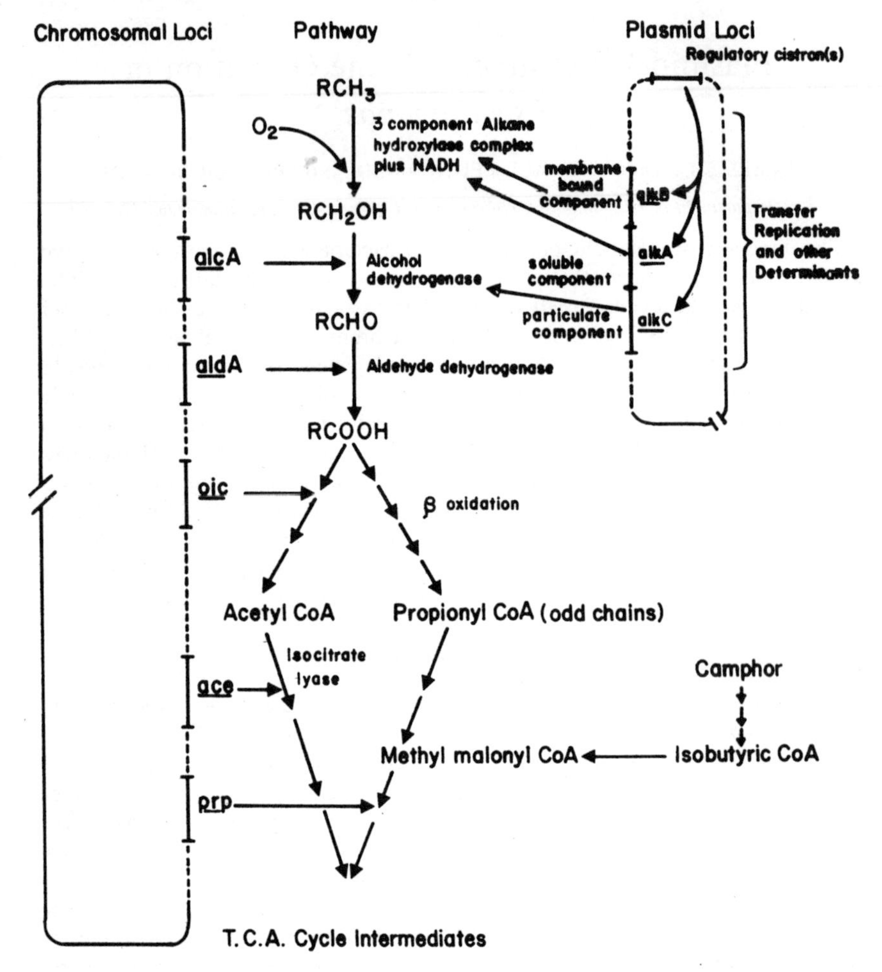

FIG. 1. *Plasmid and chromosomal loci directly involved in alkane oxidation by* P. putida. *Corrected from reference 10 for the order of* alkBAC. *NADH, Reduced nicotinamide adenine dinucleotide; CoA, coenzyme A; T.C.A., tricarboxylic acid cycle.*

component), and $AlkC^+$ (membrane alcohol dehydrogenase). In addition to the chromosomal loci encoding enzymatic activities for the metabolism of alkane oxidation products, there are chromosomal genes whose proteins are involved in determining membrane structure and expression of alkane-oxidizing activities (10). The ability to induce the $AlkA^+B^+C^+$ activities and the specificity of inducer recognition are governed by one or more plasmid genes (12). There are also plasmid mutations not shown here (*alkD*) which affect expression of the $AlkB^+$ and $AlkC^+$ membrane activities but not expression of the soluble $AlkA^+$ activity (M. Fennewald, S. Benson, and J. Shapiro, unpublished data).

Knowledge of the various loci involved in determining the oxidation pathway activities made it possible to construct a strain which specifically accumulated primary alcohol as an alkane oxidation product under laboratory conditions (17).

INSERTION ELEMENT ANALYSIS OF THE *alk* REGION

To analyze genetic polarity and have known null phenotype mutants, we chose transposable antibiotic resistance determinants as mutagenic agents (7). Insertion of a transposon into a gene disrupts the continuity of the coding sequence and so almost invariably results in a complete absence of gene function. Insertion mutations are also generally polar on cistrons promoter-distal to the mutated site and thus provide a means of analyzing transcription units. Because the transposon determines a readily selected phenotype, insertion *near* a genetic locus (but not within it) provides an outside marker useful for mapping.

We knew that a β-lactamase transposon, Tn*401,* would insert into degradative plasmids (thus converting them into resistance plasmids). But in CAM-OCT, Tn*401* would induce only *cam* mutations, not *alk* mutations (2). We then found that the Tn*7* element, conferring trimethoprim (Tp) and streptomycin (Sm) resistances (1), would insert into and near the *alk* genes of the CAM-OCT plasmid (Fennewald, Benson, and Shapiro, in preparation). The Tn*7* Tp^rSm^r phenotypes are poorly expressed in *P. putida* because of high intrinsic resistance; hence, we selected transposition of trimethoprim resistance into CAM-OCT in a *P. aeruginosa* background.

Analysis of the Tn*7*-induced *alk* mutations after transfer to *P. putida* revealed three main classes (Table 1) as well as some leaky mutants we have not yet explained. Class I has normally inducible $AlkB^+$ activity, has no $AlkA^+$ or $AlkC^+$ activity, and will revert to $AlkB^+A^+C^+$, $AlkB^+A^-C^+$, and $AlkB^-A^-C^+$. All revertants have lost Tp^r. The latter two kinds of revertants have inducible $AlkC^+$ activity and appear to result from imprecise excision of the Tn*7* element inserted in *alkA* to relieve polarity on $AlkC^+$ expression. We have shown that the $AlkB^-A^-C^+$ revertant has a deletion of *alkA* and *alkB* (Fig. 2). Class II has no $AlkA^+$, $AlkB^+$, or $AlkC^+$ activity and will revert to $AlkB^+A^+C^+$ and $AlkB^-A^+C^+$ phenotypes (both Tp^s). The second kind of revertant appears to result from imprecise excision of an *alkB*::Tn*7* insertion. Class III also has no $AlkB^+$, $AlkA^+$, or $AlkC^+$ activity but will only revert to the wild-type phenotype. This indicates that the Tn*7* element has caused a pleiotropic defect by insertion into a regulatory gene rather than by a polar effect. All these results are consistent with the existence of an *alkBAC* operon under the positive control of one or more regulatory cistrons.

By using the CAM-OCT::Tn*7* plasmids with insertions linked to the *alk* region, we have been able to begin construction of a transductional genetic map of the *alk* regulon in *P. aeruginosa* (Fig. 2). We had to use Tn*7* as a linked marker because we could not find another plasmid gene linked to *alk* mutations by F116L transduction. We originally thought that this regulon occupied a very small piece of DNA (12), but that conclusion was mistakenly based on the assumption that *alkD*208 ($B^-A^+C^-$) was a polar mutation of the *alkB* cistron. Instead, we found that the *alk* mutations fall into at least two clusters. One contains regulatory mutations and *alkD*208; this cluster is tightly linked to Tn*7* insertion number 320 ($Tn7_{320}$). Another cluster is far from $Tn7_{320}$ but moderately linked to $Tn7_{322}$ and contains *alkA* and *alkB* mutations. The very low but reproducible cotransduction of $Tn7_{320}$ with *alkB* indicates that the entire region is less than 30 megadaltons in size. (Tn*7* measures 8.5 megadaltons and the transducing phage, F116L, will encapsidate 38 megadaltons of DNA [1, 15].) We estimate the distance between the regulatory genes and the *alkBAC* operon at 10 to 20 megadaltons (based on the assumption that phage F116L encapsidates plasmid

TABLE 1. *Characterization of* alk::*Tn7 mutants*

Strain	Enzymatic profile			Revertant types		Enzymatic profile of revertants			Class
	alkB$^+$	*alkA*$^+$	*alkC*$^+$	WT	(A/B)$^-$C$^+$	*alkB*$^+$	*alkA*$^+$	*alkC*$^+$	
1117	+	−	−	Yes	Yes	−	−	+	I
						+	−	+	
						+	+	+	
1118	−	−	−	Yes	No	+	+	+	III
1119	−	−	−	Yes	No	+	+	+	III
1120	−	−	−	Yes	Yes	−	+	+	II
						+	+	+	
1121	−	−	−	Yes	Yes	−	+	+	II
						+	+	+	
1122	+	−	−	Yes	Yes	+	−	+	I
						+	+	+	
1123	−	−	−	Yes	No	+	+	+	III
1124	−	−	−	Yes	No	+	+	+	III
1125	−	−	−	Yes	No	+	+	+	III
1126	−	−	−	Yes	No	+	+	+	III
1130	−	−	−	Yes	Yes	−	+	+	II
						+	+	+	
1131	+	−	−	Yes	No	+	+	+	I(?)

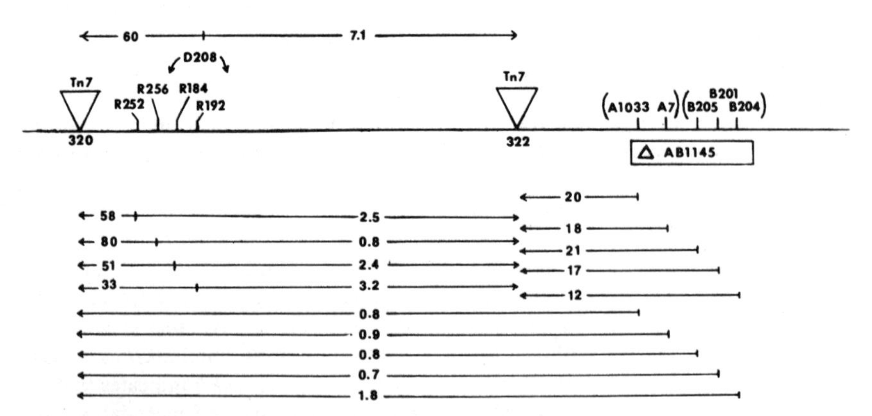

FIG. 2. *Preliminary genetic map of the* alk *regulon. The numbers in the arrows indicate cotransduction frequencies of the Tn7 insertions with various* alk *mutations (percent Tp*^{r} *among* alk$^+$ *transductants with phage grown on* alk$^+$ *Tn7 donors). We used* P. aeruginosa *phage F116L (15). ΔAB1145 was isolated as an* alkB$^-$A$^-$C$^+$ *polarity revertant of an* alkA::*Tn7 mutant.*

DNA randomly). Three-factor crosses have given us the order of markers shown. The ambiguities indicated in Fig. 2 are the position of *alkD*208 and the order of mutations within *alkA* and *alkB*.

These surprising results tell us that in vitro and in vivo manipulations of the *alk* genes will be more complex than we first envisaged. To achieve complete *alk* expression, a cell must contain the regulatory (*alkR, alkD*) and the structural (*alkBAC*) clusters. Putting them on a single plasmid would entail cloning or transposing 10 to 20 megadaltons of DNA. However, it should be feasible to isolate each cluster independently by complementation of known mutants. Moreover, the possibility of separating regulatory from structural genes may make it possible to alter expression of the *alk* regulon in useful ways.

MEMBRANES AND ALKANE OXIDATION

One of our main interests in the alkane system has been to work out the role of membranes in the oxidation pathway and its regulation. Hydrophobic substrates, such as *n*-alkanes, will tend to partition from the aqueous environment into the lipid regions of cell membranes. An understanding of membrane-bound activities will be useful in practical applications of microbial cells and their enzymes. For example, intact microbial cells are better than disrupted cells for the accumulation of primary alcohols (Benson and Shapiro, unpublished data), and the knowledge that an enzyme is at least partly bound to membrane can help in the design of artificial liposomes to carry out specific biochemical reactions.

We know that $AlkB^+$ and $AlkC^+$ activities sediment with membranes (3, 5). To prove that these results were not artifactual (perhaps due to the hydrophobicity of the substrates) and to localize these activities in the envelope of *P. putida,* we undertook a combined biochemical and genetic approach. Physical separation of *P. putida* inner (cytoplasmic) and outer membranes showed that alkane hydroxylase activity is located in the cytoplasmic membrane, as predicted from its biochemical properties and cofactor requirements. Gel electrophoresis identified two inducible peptides of about 40,000 and 47,000 daltons which are located in the inner membrane (Benson, Fennewald, and Shapiro, in preparation). Interestingly, both of these induced peptides are easily visualized by staining with Coomassie blue, and the 40,000-dalton peptide is the major inner membrane protein after induction. Electrophoresis of total membrane proteins from cells containing *alkB*::Tn7 or *alkA*::Tn7 mutant plasmids and their $AlkC^+$ revertants has shown a strict correlation between (i) presence of the 47,000-dalton peptide and $AlkC^+$ activity and (ii) presence of the 40,000-dalton peptide and the $AlkB^+$ activity. Thus, there is no doubt that initial oxidation of alkanes and their primary alcohol products occurs in the cytoplasmic membrane of alk^+ *P. putida.*

DISCUSSION

Genetic analysis and manipulation in *Pseudomonas* and other hydrocarbon-oxidizing bacteria present difficulties which have long since been overcome in *Escherichia coli* and other standard laboratory strains. Nonetheless, we believe that our work with the *Pseudomonas* alkane system illustrates some of the ways that basic genetic methods can be applied to organisms of industrial interest. Basically, two in vivo genetic tools made our work possible: (i) systems of gene transfer by conjugative plasmids and transducing phage and (ii) the use of insertion elements as mutagens. There is every reason to believe that both of these tools can be obtained for any bacterial species. We should emphasize that the use of insertion elements has especially great potential for in vivo genetic manipulation to create more useful organisms. Not only do these remarkable elements cause mutations but they can also be used to transpose and fuse genes (reference 7, p. 487–580; papers presented at the Third International Symposium on the Genetics of Industrial Microorganisms, Madison, Wis., 1978).

Bacteria display an extraordinary genetic plasticity and over the past three decades we have learned how to exploit it for basic research. Recombinant DNA technologies increase our ability to modify the bacterial genome. In a relatively brief period, we have seen our capacity for genetic modification of a specific hydrocarbon oxidation system increase manyfold, and we expect that this experience can be duplicated with many other systems. (Insertion element mutagenesis has already been applied to TOL plasmids [6; P. H. Williams, personal communication].) Thus, the prospects are good for developing a biopetrochemical industry based on the microbial conversion of hydrocarbon substrates by specifically constructed bacterial strains.

ACKNOWLEDGMENTS

We thank George Jacoby for countless strains and many helpful discussions, Pat Clarke and Bruce Holloway for strains of *P. aeruginosa* and transducing phages, William Prevatt and Richard Meyer for work on IncP-2 plasmids, and Gary Boch for technical assistance.

This research was supported by grants from the National Science Foundation (BMS 75-08951), the trustees of the Petroleum Research Fund administered by the American Chemical Society, and the Louis Block Fund of the University of Chicago. J.S. was the recipient of Public Health Service Research Career Development Award 1K04 AI-00118-1 and M.F. and S.B. were the recipients of Public Health Service predoctoral traineeships GM-07197 and GM-00090, respectively.

ADDENDUM IN PROOF

Further genetic studies have confirmed some of the conclusions stated above and forced modification of others. More regulatory mutations and a second *alkD* mutation map in the cluster near Tn7_{320} (Fig. 2). Polar *alkB*::Tn*7* and *alkA*::Tn*7* insertion mutations map under the ΔAB1145 deletion in the cluster to the right of Tn7_{322} (Fig. 2), as expected. However, *alkC* point mutations and *alkC*::Tn*7* insertion mutations are not linked by F116L transduction to either of the clusters shown in Fig. 2. This result, together with phenotypic characterization of polar insertion mutants and *alkC* mutants, indicates the existence of a second alcohol dehydrogenase gene, designated *alkE*, in the same transcription unit as *alkB* and *alkA*. We refer to this as the *alkBAE* operon (Fennewald, Benson, and Shapiro, in preparation).

REFERENCES

1. **Barth, P. T., N. Datta, R. W. Hedges, and N. J. Grinter.** 1976. Transposition of a deoxyribonucleic acid sequence encoding trimethoprim and streptomycin resistances from R483 to other replicons. J. Bacteriol. **125**:800–810.
2. **Benedik, M., M. Fennewald, and J. Shapiro.** 1977. Transposition of a beta-lactamase locus from RP1 into *Pseudomonas putida* degradative plasmids. J. Bacteriol. **129**:809–814.
3. **Benson, S., M. Fennewald, J. Shapiro, and C. Huettner.** 1977. Fractionation of inducible alkane hydroxylase activity in *Pseudomonas putida* and characterization of hydroxylase-negative plasmid mutations. J. Bacteriol. **132**:614–621.
4. **Benson, S., and J. Shapiro.** 1975. Induction of alkane hydroxylase proteins by unoxidized alkane in *Pseudomonas putida.* J. Bacteriol. **123**:759–760.
5. **Benson, S., and J. Shapiro.** 1976. Plasmid-determined alcohol dehydrogenase activity in alkane-utilizing strains of *Pseudomonas putida.* J. Bacteriol. **126**:794–798.

6. **Benson, S., and J. Shapiro.** 1978. TOL is a broad-host-range plasmid. J. Bacteriol. **135:**278–280.
7. **Bukhari, A. I., J. A. Shapiro, and S. L. Adhya (ed.).** 1977. DNA insertion elements, plasmids and episomes. Cold Spring Harbor Laboratory, Cold Spring Harbor, N.Y.
8. **Chakrabarty, A. M.** 1973. Genetic fusion of incompatible plasmids in *Pseudomonas.* Proc. Natl. Acad. Sci. U.S.A. **70:**1641–1644.
9. **Chakrabarty, A. M., G. Chou, and I. C. Gunsalus.** 1973. Genetic regulation of octane dissimilation plasmid in *Pseudomonas.* Proc. Natl. Acad. Sci. U.S.A. **70:**1137–1140.
10. **Fennewald, M., S. Benson, and J. Shapiro.** 1978. Plasmid-chromosome interactions in the *Pseudomonas* alkane system, p. 170–173. *In* D. Schlessinger (ed.), Microbiology—1978. American Society for Microbiology, Washington, D.C.
11. **Fennewald, M., W. Prevatt, R. Meyer, and J. Shapiro.** 1978. Isolation of Inc P-2 plasmid DNA from *Pseudomonas aeruginosa.* Plasmid **1:**164–173.
12. **Fennewald, M., and J. Shapiro.** 1977. Regulatory mutations of the *Pseudomonas* plasmid *alk* regulon. J. Bacteriol. **132:**622–627.
13. **Grund, A., J. Shapiro, M. Fennewald, P. Bacha, J. Leahy, K. Markbrieter, M. Nieder, and M. Toepfer.** 1975. Regulation of alkane oxidation in *Pseudomonas putida.* J. Bacteriol. **123:**546–556.
14. **Hansen, J. B., and R. H. Olsen.** 1978. Isolation of large bacterial plasmids and characterization of the P2 incompatibility group plasmids pMG1 and pMG5. J. Bacteriol. **135:**227–238.
15. **Holloway, B., and V. Krishnapillai.** 1975. Bacteriophages and bacteriocins, p. 99–132. *In* P. H. Clarke and M. H. Richmond (ed.), Genetics and biochemistry of *Pseudomonas.* John Wiley & Sons, London.
16. **Jacoby, G., and J. Shapiro.** 1977. Plasmids studied in *Pseudomonas aeruginosa* and other pseudomonads, p. 639–656. *In* A. I. Bukhari, J. A. Shapiro, and S. L. Adhya (ed.), DNA insertion elements, plasmids and episomes. Cold Spring Harbor Laboratory, Cold Spring Harbor, N.Y.
17. **Shapiro, J., S. Benson, M. Fennewald, A. Grund, and M. Nieder.** 1976. Genetics of alkane utilization, p. 568–571. *In* D. Schlessinger (ed.), Microbiology—1976. American Society for Microbiology, Washington, D.C.

Plasmids Involved in the Catabolism of Aromatic Hydrocarbons

PETER A. WILLIAMS

Department of Biochemistry and Soil Science, University College of North Wales, Bangor, Gwynedd LL57 2UW, Wales

Among natural populations of saprophytic microorganisms, members of the genus *Pseudomonas* are ubiquitous and exhibit a wide versatility in their degradative ability. The genetic basis for this in most cases has not been examined, but in a small number of instances it has been shown that the genes for the catabolic pathway enzymes are plasmid-coded; two of these plasmids were found in strains of *P. putida* and code for the degradation of aromatic hydrocarbons, naphthalene (3), toluene, and *m*- and *p*-xylenes (8, 10, 12). The first recognition of the extrachromosomal nature of the genes depended (i) upon the segregation of derivatives which had lost the ability to utilize the substrate either spontaneously or as a result of the presence of sublethal concentrations of mitomycin C during growth and (ii) upon the ability of the plasmid to determine its own transfer between genetically distinguishable strains. Subsequently, this genetic evidence has been reinforced by isolation of the DNA, correlation of its presence with the degradative ability, and restoration of that ability when the plasmid DNA is reintroduced into a plasmid-free strain by in vitro transformation (2, 6).

Both the NAH and TOL plasmids code for a sequence of at least 11 enzymes together with the associated regulatory apparatus, and they share part of this sequence, from catechol (a metabolite of both toluene and naphthalene) to pyruvate and acetaldehyde by way of the *meta* (or α-keto acid) pathway (Fig. 1).

As one possible way of assessing the frequency with which plasmid-specified catabolism occurs, we selectively enriched a number of bacteria with the same pathway as specified by TOL from different soil samples without any prejudice as to the nature of the isolate (9). All of them, together with several since isolated from both soils and marine sediments, were either *P. putida* or very similar strains, and all showed at least some of the characteristics of plasmid coding for the toluene-xylene pathway. In many cases the plasmid DNA has been isolated and characterized (2). Extrapolating from this result it might be tentatively postulated that plasmids are common vectors of catabolism in *Pseudomonas,* at least for some of the more exotic substrates. This could account, in part, for the catabolic versatility of the genus and would be a means whereby the population as a whole could conserve biosynthetic energy since the particular genes need only be maintained in a few members but with the potential of infectious spread throughout a wider spectrum of strains.

The importance and usefulness of degradative plasmids for possible industrial application is that they harbor on naturally occurring extrachromosomal elements a regulated biochemical sequence capable of converting a growth substrate to a metabolite(s) which, for the most part, could be used for energy and carbon by a wide range of bacteria of different genera (e.g., toluene to acetaldehyde and pyruvate). Consideration of the size of the DNA required (probably of molecular weight about 10^7) makes it seem unlikely that such units could be cloned from a chromosome even if all the genes were contiguous. If such pathways have potential use, either for degradation, for conversion of waste materials to biomass, or for accumulation of

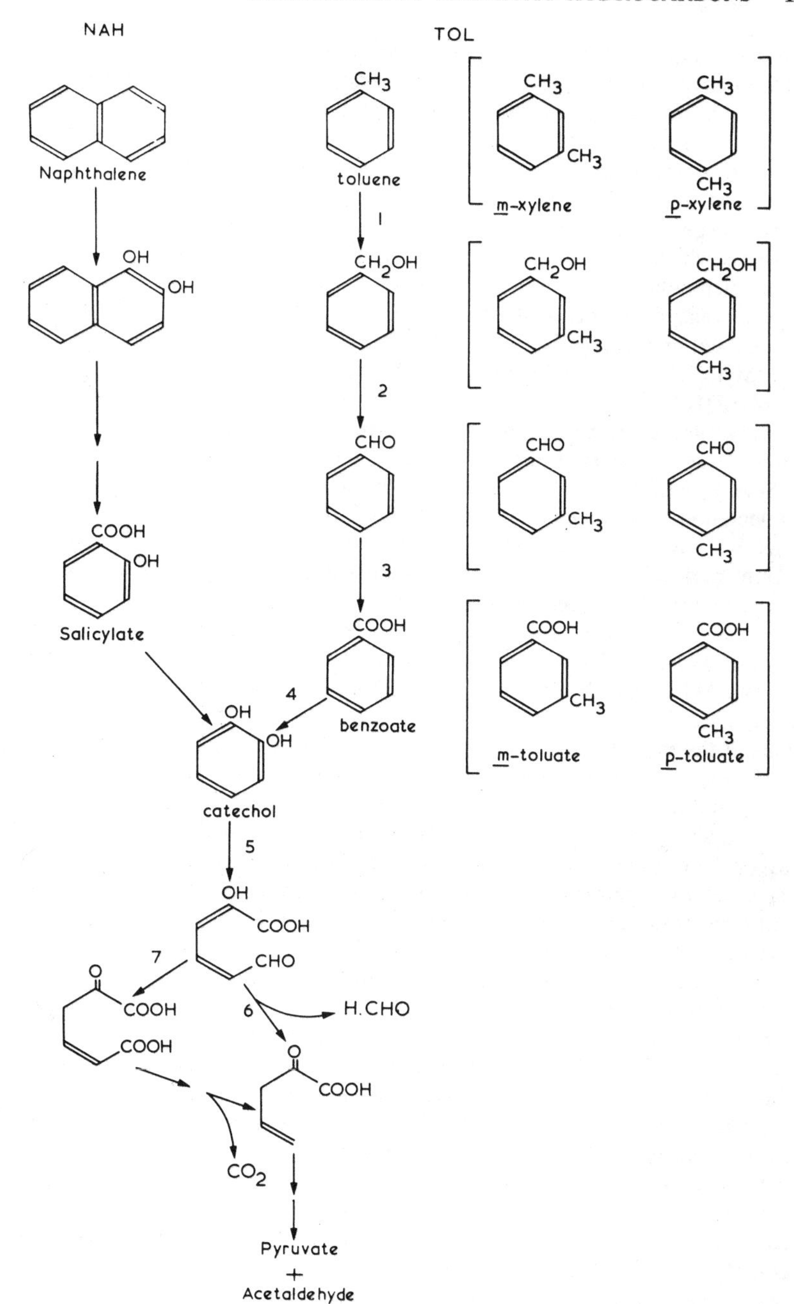

FIG. 1. *Metabolic pathways coded by NAH and TOL plasmids. The TOL plasmid confers the ability to utilize toluene, benzyl alcohol, benzaldehyde, and benzoate via catechol and* m- *and* p-*xylene and their corresponding metabolites (in parentheses) via 3- and 4-methyl catechol, respectively. The enzymes numbered (with abbreviations) are as follows: 1, xylene oxidase (XO); 2, benzyl alcohol dehydrogenase (BADH); 3, benzaldehyde dehydrogenase (BZDH); 4, toluate oxidase (TO); 5, catechol 2,3-oxygenase (C23O); 6, 2-hydroxymuconic semialdehyde hydrolase (HMSH); 7, 2-hydroxymuconic semialdehyde dehydrogenase (HMSD).*

useful by-products, it is necessary to understand (i) the properties of the plasmid DNA, to assess whether the whole catabolic unit could be maneuvered into strains which might prove more amenable to industrial processes, and (ii) the biochemistry and regulation of the pathways themselves and their interaction with chromosomal pathways, to assess how best they could be modified or mutated to useful ends. In this context, some of the recent work on the TOL plasmids will be discussed.

The independently isolated isofunctional TOL plasmids differ in a number of properties which include their transmissibility and their ability to utilize *p*-xylene and its metabolites (9). The molecular weights of those which have been isolated cover a range from the smallest at 52×10^6 up to 170×10^6 (2), although an isofunctional nonconjugative plasmid, XYL, appears to be smaller than any of these (J. R. Mylroie, D. A. Friello, and A. M. Chakrabarty, General Electric Co. Report no. 76CRD161).

As a structural probe, the purified plasmids have been subjected to restriction endonuclease hydrolysis and the fragments have been compared by electrophoresis (2). Several plasmids from apparently independent isolates were identical in size and fragmentation pattern, including the original TOL plasmid from *P. putida* mt-2 (which we have named pWW0). However, even some of the plasmids with quite different molecular weights show some similarities in fragmentation which might indicate that they share regions of homology.

We have investigated the regulation of the TOL-determined pathway in two strains: *P. putida* mt-2, carrying pWW0 with a molecular weight of 78×10^6 (2), and *P. putida* MT20, which contains plasmid DNA as yet structurally uncharacterized and is called TOL20.

One of the important techniques in studying the TOL plasmids relies on the fact that benzoate, which is a metabolite of toluene on the plasmid-coded pathway, can also be catabolized by an alternative chromosomal pathway (the *ortho* or β-ketoadipate pathway) which appears to be ubiquitous in saprophytic fluorescent *Pseudomonas* spp. The latter pathway supports a faster growth rate on benzoate than does the TOL-coded *meta* pathway, but is not expressed when the plasmid is present because of differences in the regulation of the two pathways (8). Consequently, growth on benzoate minimal medium selects for derivatives which have lost the *meta* pathway and were assumed to have lost the plasmid (8). *P. putida* MT20 is one of three isolates which not only segregate derivatives with the typical cured phenotype (loss of ability to grow on all the substrates except benzaldehyde and benzoate) but also segregate derivatives which use the *ortho* pathway for benzoate, no longer grow on *m*-toluate but retain the ability to grow on the hydrocarbons toluene and *m*-xylene, which they metabolize by the plasmid pathway (9). Examination of one of these, MT20-B3, led us to the conclusion that the TOL20 plasmid had suffered a deletion of some of its DNA, which included a regulatory gene involved in the induction by *m*-toluate and benzoate of at least some of the enzymes responsible for their dissimilation (13). A model for the regulation of the pathway enzymes was proposed, but further study was hampered by failure to obtain any structural gene mutants.

We therefore returned to *P. putida* mt-2, in which structural mutants of the pWW0 plasmid can be readily made. Analysis of a number of these mutants together with a regulatory mutant and its revertants led to a model (14) which is essentially the same as had been proposed for the TOL20 pathway (Fig. 2). According to this model, the deletion which gave rise to the particular phenotype of MT20-B3 included the *xylS* gene but none of the other pathway genes. One possibility which arises from

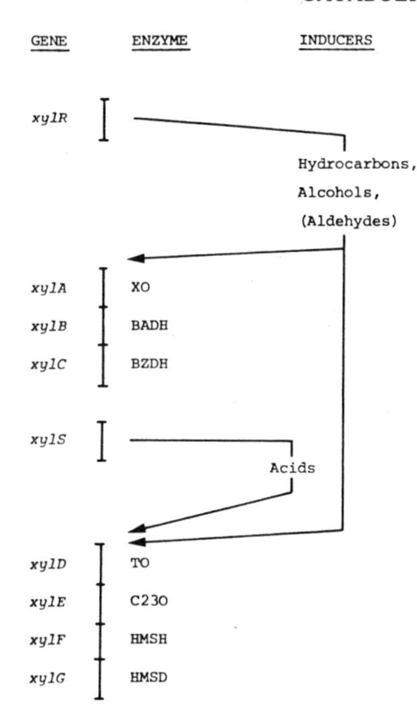

FIG. 2. *Model for the regulation of the early enzymes of the TOL-specified degradative pathway on pWW0 (TOL) and TOL20. The gene product of* xylR *combines with any of the hydrocarbon, alcohol, and probably aldehyde substrates to induce both blocks of enzymes. The product of a second regulator gene* xylS *combines only with the carboxylate substrates to induce only the second block. No particular gene order is implied. For abbreviations of the enzymes, see Fig. 1. In the study of TOL20, xylene oxidase was not assayed.*

these regulatory studies is that the induction by both the *xylR* and the *xylS* gene products may contain elements of positive control, but this aspect and the details of the model need further experimental verification. The similarity of the regulatory models for what appear to be very different plasmids was surprising and is reminiscent of the identity of some other genetic elements found on plasmids of different structures, for example, β-lactamases. This analogy with R plasmids has been given some added weight by studies in different laboratories.

As stated above, it had been assumed that derivative strains which had lost the ability to utilize all the plasmid-coded growth substrates except benzoate and benzaldehyde (both metabolized by the *ortho* pathway) and which did not express any of the plasmid-coded enzymes had totally lost the plasmid. However, Bayley et al. (1) found that the assumption was true in only about half of such derivatives of *P. putida* mt-2; in the others the pWW0 plasmid had lost a specific fragment with a molecular weight of about 30×10^6, leaving a residual plasmid with a molecular weight of 38×10^6 with no apparent function. The biochemical difference between the strains carrying the wild-type pWW0 plasmid and its smaller derivative would suggest that the Tol degradative function was carried on the fragment (30×10^6) deleted. The frequency of loss of the segment indicates that it might have terminal sequences of DNA which facilitate its ready excision. By inserting a Cb^r transposon, Tn401, onto the smaller plasmid, we have since demonstrated that it carries the *tra* function and moves between *P. putida* strains at about the same frequency as pWW0 itself (unpublished data).

The implications of this dissociation from TOL have recently been confirmed in two laboratories. Jacoby et al. (4) have demonstrated that transfer of TOL out of *P. aeruginosa* PAC containing both TOL and the IncP-1 plasmid RP4 ($Cb^rKm^rTc^r$) resulted in a few transconjugants which were $Cb^rKm^rTc^s$ and grew on all the TOL substrates. These contained a single plasmid which retained all the properties of RP4 except Tc^r but had gained an additional sequence of 30×10^6 to 40×10^6. The TOL function could be further transposed from this cointegrate into the IncP-2 plasmid pMG5, which in the process gained a similar-sized fragment of DNA, but the dependence of the transposition on the *rec* system was not tested. A similar set of experiments has been independently carried out by A. M. Chakrabarty, D. A. Friello, and L. H. Bopp (abstracts of EMBO workshop on plasmids, Berlin, April 1978), who have additionally shown that in *P. aeruginosa* PAO the plasmid with a molecular weight of 78×10^6 can dissociate to leave a residual plasmid of 30×10^6 which specifies the complete degradative pathway but is transfer deficient. Recombinants between TOL and R plasmids have also been obtained by White and Dunn (7) with R91 and by Nakazawa et al. (5) with RP4, both of which might also be insertions of part of TOL into the R plasmid.

There is a strong implication that the element transposed in these experiments is the same fragment which is so readily lost during growth on benzoate and that TOL itself is the result of recombination between a 48×10^6 sex factor and a 30×10^6 segment carrying the degradative pathway genes, which may be bounded by sequences analogous to insertion sequences. It is interesting to speculate that the other isofunctional TOL plasmids may also share similar or even the same sequence.

Transposition of degradative genes into carefully chosen vector plasmids increases the possible range of host bacteria into which they can be introduced. The toluene-xylene genes from TOL have been introduced into *Escherichia coli* on promiscuous R plasmids (4, 7). Although in this instance there appears to be little or no gene expression (for reasons which have yet to be ascertained), the procedure could be an important way of constructing potentially useful bacteria.

An alternative approach to genetic manipulation for deriving useful strains from the degradative plasmids involves making use of the overlap which the plasmid-coded pathways often have in common with chromosomal pathways. For example, a mutant of *P. putida* mt-2 lacking the plasmid-coded catechol 2,3-oxygenase (14) does not grow on *m*- or *p*-xylene since these can only be metabolized by the *meta* pathway, but it does grow on toluene. Toluene is metabolized to benzoate by the plasmid-determined enzymes and further by the chromosomal *ortho* pathway. An alternative approach to the construction of hybrid plasmid-chromosomal pathways has been used by Wong and Dunn (11). *P. putida* PPl-2 metabolizes phenol via catechol by a phenol hydroxylase and subsequent *ortho* sequence, which are all chromosomally determined. When TOL was introduced into a mutant of PPl-2 unable to grow on phenol because of a nonfunctional catechol 1,2-oxygenase, the resultant transconjugants grew very slowly on phenol because the catechol produced by the phenol hydroxylase could be slowly metabolized by low uninduced levels of the plasmid *meta* pathway. However, mutant strains with greatly improved growth rates on phenol were selected, and in these the regulation of the plasmid had been modified such that phenol acted as an inducer.

With a detailed knowledge of the biochemistry and the regulation of the pathways involved, it should be possible to construct novel hybrid pathways, part chromosomal and part plasmid in their coding, particularly if promiscuous recombinant plasmids were used to introduce the degradative function into new hosts. The net result could be important in degrading otherwise recalcitrant molecules or, if genetic blocks were introduced in the hybrid pathway, in accumulating valuable intermediates from cheap, readily available substrates.

REFERENCES

1. **Bayley, S. A., C. J. Duggleby, M. J. Worsey, P. A. Williams, K. G. Hardy, and P. Broda.** 1977. Two modes of loss of the Tol function from *Pseudomonas putida* mt-2. Mol. Gen. Genet. **154:**203–204.
2. **Duggleby, C. J., S. A. Bayley, M. J. Worsey, P. A. Williams, and P. Broda.** 1977. Molecular sizes and relationships of TOL plasmids in *Pseudomonas.* J. Bacteriol. **130:**1274–1280.
3. **Dunn, N. W., and I. C. Gunsalus.** 1973. Transmissible plasmid coding early enzymes of naphthalene oxidation in *Pseudomonas putida.* J. Bacteriol. **144:**974–979.
4. **Jacoby, G. A., J. E. Rogers, A. E. Jacob, and R. W. Hedges.** 1978. Transposition of *Pseudomonas* toluene-degrading genes and expression in *Escherichia coli.* Nature (London), **274:**179–180.
5. **Nakazawa, T., E. Hayashi, T. Yokota, Y. Ebina, and A. Nakazawa.** 1978. Isolation of TOL and RP4 recombinants by integrative suppression. J. Bacteriol. **134:**270–277.
6. **Palchaudhuri, S., and A. Chakrabarty.** 1976. Isolation of plasmid deoxyribonucleic acid from *Pseudomonas putida.* J. Bacteriol. **126:**410–416.
7. **White, G. P., and N. W. Dunn.** 1977. Apparent fusion of the TOL plasmid with the R91 drug resistance plasmid in *Pseudomonas aeruginosa.* Aust. J. Biol. Sci. **30:**345–355.
8. **Williams, P. A., and K. Murray.** 1974. Metabolism of benzoate and methylbenzoates by *Pseudomonas putida* (*arvilla*) mt-2: evidence for the existence of a TOL plasmid. J. Bacteriol. **120:**416–423.
9. **Williams, P. A., and M. J. Worsey.** 1976. Ubiquity of plasmids in coding for toluene and xylene metabolism in soil bacteria: evidence for the existence of new TOL plasmids. J. Bacteriol. **125:**818–828.
10. **Wong, C. L., and N. W. Dunn.** 1974. Transmissible plasmid coding for the degradation of benzoate and *m*-toluate in *Pseudomonas arvilla* mt-2. Genet. Res. **23:**227–232.
11. **Wong, C. L., and N. W. Dunn.** 1976. Combined chromosomal and plasmid encoded control for the degradation of phenol in *Pseudomonas putida.* Genet. Res. **27:**405–412.
12. **Worsey, M. J., and P. A. Williams.** 1975. Metabolism of toluene and xylenes by *Pseudomonas putida* (*arvilla*) mt-2: evidence for a new function of the TOL plasmid. J. Bacteriol. **124:**7–13.
13. **Worsey, M. J., and P. A. Williams.** 1977. Characterization of a spontaneously occurring mutant of the TOL20 plasmid in *Pseudomonas putida* MT20: possible regulatory implications. J. Bacteriol. **130:** 1149–1158.
14. **Worsey, M. J., F. C. H. Franklin, and P. A. Williams.** 1978. Regulation of the degradative pathway enzymes coded for by the TOL plasmid (pWW0) from *Pseudomonas putida* mt-2. J. Bacteriol. **134:** 757–764.

Genetics of *Saccharomycopsis lipolytica,* with Emphasis on Genetics of Hydrocarbon Utilization

JOHN BASSEL AND DAVID M. OGRYDZIAK

Donner Laboratory, University of California, Berkeley, California 94720, and Institute of Marine Resources, University of California, Davis, California 95616

The dimorphic yeast *Saccharomycopsis lipolytica* is of special interest because of its ability to utilize hydocarbons. The ability to manipulate *S. lipolytica* genetically will enhance efforts to develop strains with superior growth characteristics, metabolite production, and nutritional value.

INITIAL DEVELOPMENTS

The development of the genetics of *S. lipolytica* began in 1945 when several strains of *Candida (Saccharomycopsis) lipolytica* obtained from a corn processing plant were isolated at the Northern Regional Research Laboratory (NRRL). It was not until 1969 that the first report on sexuality in these strains was published. Wickerham, Kurtzman, and Herman found that one strain, NRRL YB-423, produced ascospores (13). The spores varied in size and shape, one to four spores per ascus were produced, and spore viability was extremely low (<0.1%). Segregants of YB-423 were mated with each other and with other haploid isolates. Most crosses again produced spores of varying size and shape, and spore viability remained very low. However, when YB-421 was crossed with YB-423-12, an abundance of uniformly shaped ascospores (shallow bowls with projecting rim) was produced, and spore viability increased to 12%.

The NRRL group established that *S. lipolytica* was heterothallic (Fig. 1), and they published procedures for mating, sporulation, and stabilization of haploids and diploids (9, 13, 14).

STRAIN DEVELOPMENT IN OTHER LABORATORIES

In 1970, a study was initiated with YB-421 and YB-423-12, obtained from Wickerham, for the purpose of developing genetic procedures and establishing the basic genetics of *S. lipolytica.* Genetic markers were introduced into these strains, and complementation and genetic recombination were demonstrated (2). From the beginning, genetic studies of *S. lipolytica* were hampered by low sporulation frequencies and low ascospore viability. Two explanations could be offered for the initial low spore viability. (i) The original parental strains differed by chromosome translocations or inversions which would lead to the generation of duplications and deletions in meiosis. (ii) The original strains might have been aneuploid, and the resultant cross would yield an assortment of aneuploid meiotic products which could have low viability due to chromosomal imbalance. In both cases, inbreeding should eventually lead to homogeneous haploid parents and normal viability.

The inbreeding program was successful; during 10 rounds of selection ascospore viability increased from approximately 15% to 85% (11). Spore viability was determined by dissecting the asci with a micromanipulator after first digesting the ascus walls with the enzyme glusulase. In the later rounds of inbreeding, only spores from

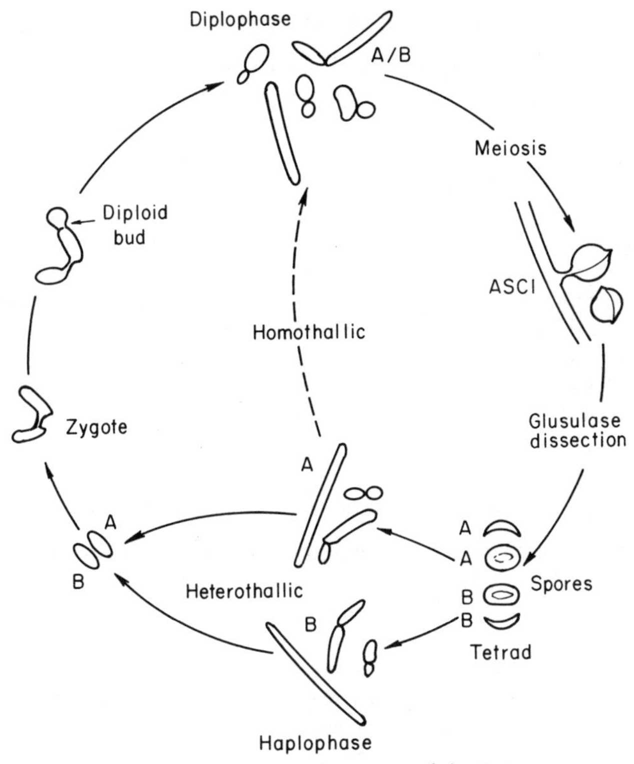

FIG. 1. *Life cycle of* Saccharomycopsis lipolytica.

asci that yielded four viable products were used. Sporulation frequency and the percentage of four-spored asci increased in the inbred strains, and tetrad analysis was finally practical.

Gaillardin, Charoy, and Heslot followed a similar plan for developing the genetics of *S. lipolytica* (5). Strain YB-423-12 obtained from Wickerham and strain W isolated from soil were used as parental strains. Genetic markers were introduced, and three rounds of inbreeding were performed. Spores for inbreeding were obtained as random spores by using paraffin oil to separate spores from vegetative cells. The inbreeding improved the segregation ratios of genetic markers determined by random spore analysis, increased the percentage of four-spored asci, and improved spore viability to a reported 90% based on a comparison of microscopic counts and plate counts of spores obtained by the paraffin oil technique. Our results with Heslot's strains (obtained by using tetrad dissection to determine spore viability) indicate that estimates of spore viability determined by the paraffin oil method were high by a factor of two to four (unpublished data).

Esser and Stahl reported on the life cycle of *S. lipolytica* (4). They crossed a wild strain with one of the Heslot's inbred strains and then crossed a segregant of this cross with a second wild-type strain. They found a high percentage of asci containing two binucleate spores and very few four-spored asci. Spore viability of the binucleate spores isolated by micromanipulation was only 25%. Cytological and genetic evidence indicated that the variability in ascospores per ascus was caused by the absence of correlation between the meiotic division and spore wall formation.

SUITABILITY OF WILD STRAINS FOR GENETIC ANALYSIS

The results obtained from the four laboratories involved in *S. lipolytica* strain development (i.e., the spore viabilities found by the NRRL group with the original *S. lipolytica* sexual strain and with crosses between different wild strains; the positive effects of inbreeding on spore viability and the percentage of four-spored asci found by Mortimer's and Heslot's groups; and the low spore viability and low percentage of four-spored asci found by Esser's group in a cross of an inbred strain with a wild strain) suggest that one cannot isolate *S. lipolytica* strains from nature and expect immediately to do genetic studies with these strains. Spore viability and the percentage of four-spored asci would probably be so low as to make genetic analysis impossible. As shown by Esser and Stahl (4), even crossing the wild strain with an inbred strain does not eliminate the need for further inbreeding. In fact, crosses between Heslot's and Mortimer's inbred strains (which have YB-423-12 as a common parental strain) show incompatibility, as spore viability in these crosses dropped to about 10% (unpublished data).

STATE OF DEVELOPMENT OF GENETIC TECHNIQUES

The NRRL group established procedures for mating, sporulation, and tetrad dissection in *S. lipolytica.* We, as well as several other investigators (3, 5, 9, 13), have investigated the influence of various combinations of factors on copulation but have had limited success in increasing the very low mating frequencies (best results are 2 to 4% zygotes for crosses between prototrophs and a frequency of 10^{-3} to 10^{-4} for crosses between auxotrophs). This low mating frequency is a major obstacle to the rapid development of *S. lipolytica* genetics. All crosses must be forced by using complementary genetic markers. Therefore, complementation analysis, scoring of mating type, and scoring involving complementation are quite laborious. Improved mating procedures or strains exhibiting much improved mating frequency would greatly facilitate genetic studies.

Inbreeding has improved sporulation, the percentage of four-spored asci, and spore viability sufficiently so that tetrad analysis is now possible. Tetrad dissection with *S. lipolytica* is more difficult than with *Saccharomyces cerevisiae,* but with a good spore preparation 15 to 20 four-spored tetrads can be dissected in 1 h.

Random spore analysis has been useful in the case of *S. lipolytica,* but the interpretation of random spore results is often complicated by diploids which survive and by selection against certain markers and combinations of markers. Evidence for gene conversion and reciprocal mitotic recombination have been obtained. Colonies sectored for the red porphyrin-excreting marker were used for analysis, and preliminary results indicate that mitotic genetic analysis should be feasible.

GENETIC MARKERS, THE GENETIC MAP, AND GENETIC STUDIES

The current genetic markers and the genetic map developed by Mortimer's group can be summarized as follows:

(i) Eighty-seven mutations in 63 genes have been shown to be mutations in single nuclear genes.

(ii) Twenty-four cases of linkage in 300 gene pairs investigated—8% linkage suggests an estimated map length of 450 map units based on 1% linkage and about 3,600 total map units in *Saccharomyces cerevisiae.*

(iii) Six linkage fragments—no centromere markers yet identified.

About 20 different genes have been described by Heslot's group, so over 80 single nuclear genes are now available for genetic studies in *S. lipolytica.* Most of the mutants were isolated by use of ethyl methane sulfonate mutagenesis, but X rays and other chemical mutagens have also been used successfully (11).

The high percentage of linked markers compared to the percentage in *S. cerevisiae* indicates that the *S. lipolytica* genetic map is probably substantially shorter.

Six linkage groups have been found, but, since no centromere markers have yet been identified, the genetic studies so far have yielded little information on the number of chromosomes. Centromere linkage can only be detected by tetrad analysis, and both genes of the gene pair must be centromere-linked. We have tetrad data on 215 gene pairs, but 200 of the gene pairs included at least one of six markers which do not seem to be centromere-linked. The fact that the first six markers, for which there are sufficient data, show no centromere linkage indicates that *S. lipolytica* may have a few long chromosomes (11).

There are at least nine studies in which genetic analysis (including confirmation of mutations as single nuclear gene mutations and determination of dominance, complementation, and linkage relationships) has been used in *S. lipolytica.* These studies include those by Heslot's group on lysine excretion (7) and catabolism (6) and those by Mortimer's group on mating type and alkane utilization (1), extracellular protease production (12), adenine biosynthesis, porphyrin production, mutants temperature sensitive for RNA synthesis, and radiation sensitivity (all unpublished data).

HYDROCARBON METABOLISM IN *S. LIPOLYTICA*

We have been isolating mutants of *S. lipolytica* unable to utilize *n*-alkanes as a carbon source for the purpose of carrying out an integrated genetic-biochemical study of hydrocarbon metabolism in this yeast. The mutants were induced by exposing cells to UV light (ca. 30% survival). A preliminary report on the isolation of hydrocarbon-negative mutants has been published (1).

n-Alkane metabolism in yeast is thought to proceed via an initial hydroxylation step which converts the *n*-alkane to a primary alcohol. Two successive steps oxidize the alcohol to an aldehyde and finally to the corresponding fatty acid (8, 10). Fatty acids are metabolized further by β-oxidation, a series of five reactions converting fatty acids into two-carbon (acetate) moieties. Oxidation of the acetate molecules is carried out via the Krebs cycle and the glyoxylate shunt. We have isolated 28 mutants unable to utilize *n*-decane as a carbon source. Table 1 summarizes substrate utilization

TABLE 1. *Substrate utilization tests of alkane-negative mutants of* S. lipolytica

Phenotypic designation	Growth[a] on carbon source:						No. of mutants
	Hydrocarbon	Alcohol	Aldehyde	Fatty acid	Acetate	Glucose	
A	−	+	+	+	+	+	6
B	−	−	+	+	+	+	0
C	−	−	−	+	+	+	0
D	−	−	−	−	+	+	14
E	−	−	−	−	−	+	8
Total							28

[a] +, Growth; −, no growth.

TABLE 2. *Complementation between phenotype A mutants of* S. lipolytica[a]

trp alk B mutant no.	*ade alk A* mutant no.						Loci
	207	208	222	238	240	245	
207	−	+	+	+	+	+	*alk1* = 208, 238
208		−	+	−	+	+	*alk2* = 207
222			−	+	+	+	*alk3* = 222
238				−	+	+	*alk4* = 240
240					−	+	*alk5* = 245
245						−	

[a] The *ade alk A* and *trp alk B* strains were selected from crosses of the original mutants to wild type. The symbols + and − indicate growth or no growth of intermutant hybrids on synthetic minimal medium with decane as the carbon source.

TABLE 3. *Linkage relationships of phenotype A mutants of* S. lipolytica *detected by the random spore analysis of intermutant hybrids*[a]

trp alk B mutant no.	*ade alk A* mutant no.						Loci
	207	208	222	238	240	245	
207		5:17		30:84	8:79[b]	17:66	*alk1* = 208, 238
208		0:28	28:44	0:56	16:42	20:58	*alk2* = 207
222							*alk3* = 222
238				0:32	24:58	20:77	*alk4* = 240
240						18:49	*alk5* = 245
245							

[a] The ratio of wild-type spores to *alk* spores in random spore preparations is shown. Linkage is indicated by a ratio significantly less than 1:3. Ratios greater than 1:3 presumably reflect a selective advantage of the wild-type allele.

[b] Linkage significantly less than 25% alk^+ spores ($P < 0.01$).

studies designed to identify tentatively the genetic blocks associated with these mutants.

The six phenotype A mutants which are unable to utilize hydrocarbons but can grow at the expense of the other products of alkane metabolism have been analyzed genetically. All six phenotype A mutants are apparently chromosomal. In crosses to wild type, these mutants segregated and, with the exception of 222, showed 1:1 segregation ratios in whole tetrads and in random spores. A large excess of wild-type spores were recovered from crosses of 222 to wild type, suggesting a selection against this mutant in spore clones. Table 2 presents a complementation analysis of the six mutants, indicating that they represent five genetic loci. The mutants were also crossed in pairwise combinations, and random spores from the resulting diploids were analyzed (Table 3). Significant linkage was detected in one case between *alk2-207* and *alk4-240.* Because of the selection against 222, this mutant was not included in the analysis.

Many mutants representing the other phenotypic classes of alkane-negative mutant (phenotypes D and E, Table 1) have been put into crosses. All the D and E mutants tested so far segregate normally, indicating that they too are chromosomal.

ACKNOWLEDGMENT

This research was supported by National Science Foundation grant PCM76-11663.

REFERENCES

1. **Bassel, J., and R. Mortimer.** 1973. Genetic analysis of mating type and alkane utilization in *Saccharomycopsis lipolytica.* J. Bacteriol. **114:**894–896.
2. **Bassel, J., J. Warfel, and R. Mortimer.** 1971. Complementation and genetic recombination in *Candida lipolytica.* J. Bacteriol. **108:**609–611.
3. **Bojnanská, A.** 1977. Determination of mating types and frequency of zygotes in strains of the species of *Candida lipolytica.* Acta Fac. Rerum Nat. Univ. Commenianae Genet. **8:**55–65.
4. **Esser, K., and U. Stahl.** 1976. Cytological and genetic studies of the life cycle of *Saccharomycopsis lipolytica.* Mol. Gen. Genet. **146:**101–106.
5. **Gaillardin, C. M., V. Charoy, and H. Heslot.** 1973. A study of copulation, sporulation and meiotic segregation in *Candida lipolytica.* Arch. Microbiol. **92:**69–83.
6. **Gaillardin, C., P. Fournier, G., Sylvestre, and H. Heslot.** 1976. Mutants of *Saccharomycopsis lipolytica* defective in lysine catabolism. J. Bacteriol. **125:**48–57.
7. **Gaillardin, C. M., G. Sylvestre, and H. Heslot.** 1975. Studies on an unstable phenotype induced by UV irradiation. Arch. Microbiol. **104:**89–94.
8. **Gallo, J., J. C. Bertrand, B. Roche, and E. Azoulay.** 1973. Alkane oxidation in *Candida tropicalis.* Biochim. Biophys. Acta **296:**624–638.
9. **Herman, A. I.** 1971. Mating responses in *Candida lipolytica.* J. Bacteriol. **107:**371.
10. **Liu, C. M., and M. Johnson.** 1971. Alkane oxidation by a particulate preparation from *Candida.* J. Bacteriol. **106:**830–834.
11. **Ogrydziak, D., J. Bassel, R. Contopulou, and R. Mortimer.** 1978. Development of genetic techniques and the genetic map of the yeast *Saccharomycopsis lipolytica.* Mol. Gen. Genet., **163:**229–239.
12. **Ogrydziak, D. M., and R. K. Mortimer.** 1977. Genetics of extracellular protease production in *Saccharomycopsis lipolytica.* Genetics **87:**621–632.
13. **Wickerham, L. J., C. P. Kurtzman, and A. I. Herman.** 1969. Sexuality in *Candida lipolytica,* p. 81–92. *In* Recent trends in yeast research. D. G. Ahearn (ed.), vol. 1. Spectrum (Georgia State University), Atlanta.
14. **Wickerham, L. J., C. P. Kurtzman, and A. I. Herman.** 1970. Sexual reproduction in *Candida lipolytica.* Science **167:**1141.

Occurrence and Function of Aminoglycoside-Modifying Enzymes

J. DAVIES, C. HOUK, M. YAGISAWA,[1] AND T. J. WHITE [2]

Department of Biochemistry, University of Wisconsin, Madison, Wisconsin 53706

Enzymes modifying the aminoglycoside-aminocyclitol antibiotics have been isolated from a variety of bacterial sources. These enzymes were originally described as determinants of antibiotic resistance in bacteria isolated clinically (2). Since the time of their original description, the distribution and types (substrate range) of modifying enzymes have increased markedly. They can be found in most gram-negative and gram-positive organisms that inhabit hospitals (Table 1). Following the suggestion of Benveniste and Davies that the source of these resistance determinants might be the organisms that produce the aminoglycoside antibiotics (1, 3), we have found that the aminoglycoside-modifying enzymes are of wide distribution among bacterial genera known to produce antibiotics (Table 1). In this brief review we consider the distribution of aminoglycoside-modifying enzymes and their possible role(s) in the antibiotic-producing organisms.

Most aminoglycoside-producing organisms possess enzymatic activities which modify the antibiotics they produce. In several instances the antibiotic produced is the best substrate for the enzyme present in that particular strain (3; Table 2). Thus, there appear to be multiple forms of the same enzymatic activity which differ in their substrate range for the aminoglycosides. This is especially true for the 3′-*O*-phosphotransferases, which form at least four distinct classes. A recent example is *Streptomyces lividus,* which produces lividomycin and a lividomycin phosphotransferase. Enzymes with this substrate range (modification of lividomycin, neomycin, and kanamycin) have been found in certain clinical isolates but in no other actinomycete so far examined.

In contrast, the organisms which produce gentamicin and the nonaminoglycoside spectinomycin appear to have no mechanism for modifying their antibiotic product that is related to any of the resistance mechanisms in clinical isolates. Although plasmids determining the adenylylation (or nucleotidylation) of aminoglycoside antibiotics appear to be fairly common in clinical isolates, such activities have not been detected in any of the antibiotic-producing strains so far examined.

Organisms producing antibiotics of the neomycin type (4,5-disubstituted 2-deoxystreptamines) contain a 3′-*O*-phosphotransferase and a presumed 3-*N*-acetyltransferase (Table 3). The latter enzyme is found in both producing and nonproducing strains (Table 4) and might not have a biochemical role in aminoglycoside production. Surprisingly, this acetyltransferase can utilize the unusual aminoglycoside fortimycin as a substrate (Table 5). In this respect, it resembles aminoglycoside acetyltransferase (3)-I, and we suggest that this enzyme may be present in the *Micromonospora* strain that produces fortimycin (6).

Neomycin-producing strains apparently require a 3′-*O*-phosphotransferase. The link between production of the neomycin-like antibiotics and the presence of a 3′-*O*-phosphotransferase is emphasized when we realize that *M. chalcea,* a producer of

[1] Present address: Kyowa Hakko Co., Tokyo, Japan.
[2] Present address: Cetus Corp., Berkeley, CA 94710.

TABLE 1. *Sources of aminoglycoside-aminocyclitol–modifying enzymes*

Enzyme[a]	Substrate[b]	Gram-negative	Gram-positive	Producing organism
N-acetyltransferases				
AAC(3)	Gen, Tob	+	?	+
AAC(2′)	Gen	+	−	+
AAC(6′)	Kan	+	+	+
O-phosphotransferases				
APH(6)	Str	+	−	+
APH(3′)	Kan	+	+	+
APH(2″)	Gen, Tob	−	+	−
APH(3″)	Str	+	+	+
APH(5″)	Rib	+	−	−
O-nucleotidyltransferases				
AAD(6)	Str	−	+	−
ANT(4′, 4″)	Tob	−	+	−
AAD(2″)	Gen	+	?	−
AAD(3″, 9)	Str, Spc	+	+	−

[a] AAC, Aminoglycoside acetyltransferase; APH, aminoglycoside phosphotransferase; AAD, aminoglycoside adenylyltransferase; ANT, aminoglycoside nucleotidyltransferase.
[b] Gen, Gentamicin; Tob, tobramycin; Kan, kanamycin; Str, streptomycin; Rib, ribostamycin; Spc, spectinomycin.

TABLE 2. *Comparisons of substrate ranges of aminoglycoside-3′-phosphotransferases [APH(3′)]*

Source[a]	Substrates[b]					
	Neo	KanA	Liv	But	Tob	Ami
S. fradiae (neomycin)	++	+	−	+	−	−
S. lividus (lividomycin)	+	+	+	+	−	−?
S. rimosus (paromomycin)	++	+	−	+	−	−
S. ribosidificus (ribostamycin)	++	+	−	+	−	−
APH(3′)-I	+	+	+	−	−	−
APH(3′)-II	+	+	−	+	−	−
APH(3″)-III	+	+	+	+	−	+

[a] The antibiotics in parentheses are those produced by the strains listed
[b] Neo, Neomycin; KanA, kanamycin A; Liv, lividomycin; But, butirosin; Tob, tobramycin; Ami, amikacin; ++, good substrate; +, poor substrate; −, nonsubstrate.

TABLE 3. *Antibiotic-producing strains with the aminoglycoside-aminocyclitol–modifying enzymes aminoglycoside-3′-phosphotransferase and aminoglycoside-3-acetyltransferase*

Strain	Antibiotic produced
S. fradiae	Neomycin
S. ribosidificus	Ribostamycin
S. lividus	Lividomycin
Bacillus circulans	Butirosin

TABLE 4. *Some non-aminoglycoside-producing strains with aminoglycoside-3-acetyltransferase*

Strain	Antibiotic
S. fradiae	Tylosin
S. fradiae	Phosphonomycin
S. crysomalleus	Actinomycin
S. glaucescens	?
S. griseus	Streptomycin

neomycin, appears to possess the same phosphotransferase and acetyltransferase activities as the generically unrelated *S. fradiae,* the type neomycin producer (Table 6). (Related *S. fradiae* that do not produce neomycin do not have these enzymes.) This implies that a linked set of characteristics (genes) associated with neomycin production in one organism may be transferred to another organism, and, acting in concert with the primary metabolic pathways of the new organism, produce neo-

TABLE 5. *Substrate range of* "Streptomyces-*acetyltransferase*"

Compound	Use as substrate[a]
Kanamycin A	+
Kanamycin B	++
Kanamycin C	++
Paromomycin	++
Amikacin	0
Tobramycin	++
Gentamicin C_{1a}	++
Gentamicin C_1	++
Fortimycin A	+

[a] ++, Good substrate; +, poor substrate; 0, nonsubstrate.

TABLE 6. *Aminoglycoside-aminocyclitol–modifying enzymes in neomycin-producing strains and in non-neomycin-producing* S. fradiae

Strain	Modifying enzyme[a]
S. fradiae	APH(3′); AAC(3)
S. albogriseolus	APH(3′); AAC(3)
M. chalcea	APH(3′); AAC(3)
S. fradiae (tylosin)	AAC(3)
S. fradiae (phosphonomycin)	AAC(3)

[a] APH, Aminoglycoside phosphotransferase; AAC, aminoglycoside acetyltransferase.

mycin. The possibility that these linked genes may be on a plasmid is obvious and implies that a plasmid associated with the production of a particular antibiotic may direct the synthesis of that antibiotic whenever the plasmid is present in a strain with the proper metabolic background.

The role of plasmids in the biosynthesis of aminoglycoside antibiotics has received much support (5–7). If a plasmid were required for neomycin biosynthesis, we might expect that it would carry a gene for a phosphotransferase. When neomycin-producing *S. fradiae* was treated with any one of a variety of plasmid-curing agents (acridine orange, rifampin, or rubradirin), a number of "cured" derivatives were obtained that were no longer capable of neomycin production. Analysis of some of the cured derivatives for the presence of modifying enzymes gave the results shown in Table 7. It can be seen that some of the nonproducing variants maintained their capacity to produce the 3′-phosphotransferase (AO 144) and others did not (AO 80, AO 83). All retained the capacity to make the acetyltransferase. Interestingly, when these strains were grown in the presence of 2-deoxystreptamine, only AO 144 produced neomycin (Table 8). Repeated subculture of strain AO 144 led to the spontaneous production, at high frequency, of variants (AO 144-1) that no longer possessed 3′-phosphotransferase and were no longer capable of producing neomycin when grown in the presence of 2-deoxystreptamine.

These results are consistent with, but do not prove, the notion that the 3′-phosphotransferase is required for neomycin production. It is possible that AO 80, AO 83, and AO 144-1 have lost some other function required for neomycin synthesis. However, the simplest interpretation is that two plasmids are present in *S. fradiae* and are required for neomycin production. One plasmid carries the gene for the 3′-phosphotransferase and the other carries genes for regulatory or biosynthetic steps for neomycin production. Why is the 3′-phosphotransferase required for neomycin production? It seems likely that it is needed to protect the producing strain against suicide from its own antibiotic or as a required modification (blocking or activation) in biosynthesis, or both. Studies of the requirement for plasmid-coded genes in the biosynthesis of neomycin by *S. fradiae* are described elsewhere (8). Similar results with *S. rimosus* forma. *paromomycinus,* the organism that produces paromomycin, have also been obtained (T. J. White and G. S. Gray, unpublished data).

In conclusion, studies of antibiotic-modifying enzymes in resistant bacteria have pointed to a possible role for related enzymes in antibiotic-producing organisms. It seems likely that for aminoglycoside-aminocyclitol antibiotics, high-yielding strains always possess at least one activity that modifies (detoxifies) the antibiotic produced.

TABLE 7. *Enzyme assays in cured strains of* S. fradiae

Strain	Phosphotransferase	Acetyltransferase
Type strain	+	+
CMP 487	+	+
AO 80	−	+
AO 83	−	+
AO 144	+	+
AO 144-1	−	+

TABLE 8. *Feeding of cured strains of* S. fradiae *with 2-deoxystreptamine*

Strain	Phenotype	Production of antibiotic on feeding
Type strain	neo^+ neo^R APH^+ AAC^+	Producer
AO 80	neo^- neo^S APH^- AAC^+	−
AO 83	neo^- neo^S APH^- AAC^+	−
AO 144	neo^- neo^R APH^+ AAC^+	+
CMP 487	$neo^\pm$ neo^R APH^+ AAC^+	±
AO 144–1	neo^- neo^S APH^- AAC^+	−

This suggests that one route to increasing the yield of antibiotics of this class would be to introduce modifying enzymes (by use of plasmids from resistant strains) into the producing strain. An alternative possibility is to isolate plasmids associated with antibiotic production (assuming that they do exist) and to introduce them into an organism with similar primary metabolic pathways but possessing enzyme activities that can modify the antibiotic product. It should be of interest, therefore, to determine, for example, whether gentamicin production would be altered if the necessary biosynthetic and regulatory genes (on a plasmid?) were introduced into a strain of *S. fradiae* carrying 3-*N*-acetyltransferase, an enzyme known to modify gentamicin.

ACKNOWLEDGMENTS

This work was supported by research grants from the National Institutes of Health (AI 10076) and the National Science Foundation (BMS 7202264). T.J.W. was recipient of a National Institutes of Health postdoctoral fellowship.

REFERENCES

1. **Benveniste, R., and J. Davies.** 1973. Aminoglycoside-antibiotic inactivating enzymes in Actinomycetes similar to those present in clinical isolates of antibiotic-resistant bacteria. Proc. Natl. Acad. Sci. U.S.A. **70:**2276–2280.
2. **Benveniste, R., and J. Davies.** 1973. Mechanisms of antibiotic resistance in bacteria. Annu. Rev. Biochem. **42:**471–506.
3. **Dowding, J., and J. Davies.** 1975. Mechanisms and origins of plasmid-determined antibiotic resistance, p. 179–186. *In* D. Schlessinger (ed.), Microbiology—1974. American Society for Microbiology, Washington, D.C.
4. **Hopwood, D. A., and M. J. Merrick.** 1977. Genetics of antibiotic production. Bacteriol. Rev. **41:**595–635.
5. **Hotta, K., Y. Okami, and H. Umezawa.** 1977. Elimination of the ability of a kanamycin-producing strain to biosynthesize deoxystreptamine moiety by acriflavine. J. Antibiot. **30:**1146–1149.
6. **Sato, S., T. Iida, R. Okachi, K. Shirahata, and T. Nara.** 1977. Enzymatic acetylation of fortyimycin A and seldomycin factor 5 by aminoglycoside 3-acetyltransferase I: [AAC(3)-I] of *E. coli* KY8348. J. Antibiot. **30:**1025–1027.
7. **Shaw, P. D., and J. Piwowarski.** 1977. Effect of ethidium bromide and acriflavine on streptomycin production by *Streptomyces bikiniensis.* J. Antibiot. **30:**404–408.
8. **Yagisawa, M., T-S. R. Huang, and J. Davies.** 1978. The possible role of plasmids in neomycin biosynthesis and modification. J. Antibiot., in press.

Bacterial β-Lactamases

R. B. SYKES

The Squibb Institute for Medical Research, Princeton, New Jersey 08540

β-Lactamase activity was first recorded in 1940, ironically, if not inappropriately, by a co-discoverer of penicillin (2). Since then, however, in contrast to the ever-widening range of beneficial applications of its most distinguished substrates in all their various forms, the enzyme has grown in danger and importance.

β-Lactamases hydrolyze the cyclic amide bond in the β-lactam ring of enzyme-susceptible penicillins and cephalosporins, rendering the compounds antibacterially inactive. They are produced by gram-positive and gram-negative bacteria (32, 37), actinomycetes (27, 28), yeasts (21), and blue-green algae (18). The enzymes exhibit a considerable specificity, in that their only substrates are compounds containing a reactive β-lactam ring, e.g., penicillins and cephalosporins. They provide the only penicillin and cephalosporin resistance mechanism which is properly understood at the molecular level, and their importance as antibiotic resistance mechanisms cannot be overemphasized.

With the introduction of penicillin in the early 1940s, selection pressure was immediately applied to a specific section of the microbial population, i.e., the gram-positive cocci. The response to this pressure was the appearance in hospitals of β-lactamase-producing, penicillin-resistant strains of *Staphylococcus aureus*. The emergence of such strains in turn applied pressure to antibiotic research programs, which responded by the identification of new antibiotics. In turn, the introduction of new antibiotics applied more pressure on the microbial population (Fig. 1), completing the cycle. The driving force for this cycle of events has been provided by the β-lactamases and, although it has been going on for almost 40 years, the cycle has recently gained momentum. The β-lactamase calendar shown in Fig. 2 depicts some of the important β-lactamase–related events that have taken place since 1940.

By 1948, as many as 50% of strains in hospitals were found to be penicillin resistant, and by the mid-1950s the incidence in some hospitals had risen to as high as 80% (5, 33, 39). A number of approaches were taken in an attempt to overcome the resistance problem, including modification of the penicillin molecule and the search for β-lactamase inhibitors and for new agents having stability to the β-lactamase.

The first real advance came in 1955 with the isolation of cephalosporin C by Newton and Abraham (24). The antibacterial activity of cephalosporin C was low, but the apparent relationship of the substance to the penicillin family, coupled with its resistance to staphylococcal β-lactamase, made it of immediate clinical interest. Consequently, an enormous effort was put into the development of cephalosporin C by a number of pharmaceutical companies under the auspices of the National Research Development Corporation (1).

Modification of the penicillin molecule had been the aim of organic chemists ever since its structure was established in 1943 (9). However, it was the observation in 1959 that organisms produce amidases which split off the side chains of various penicillins to give the penicillin nucleus, 6-aminopenicillanic acid (6), that led the way to the synthesis of the semisynthetic penicillins.

By 1960, the problem of resistant staphylococci in hospitals was made even more

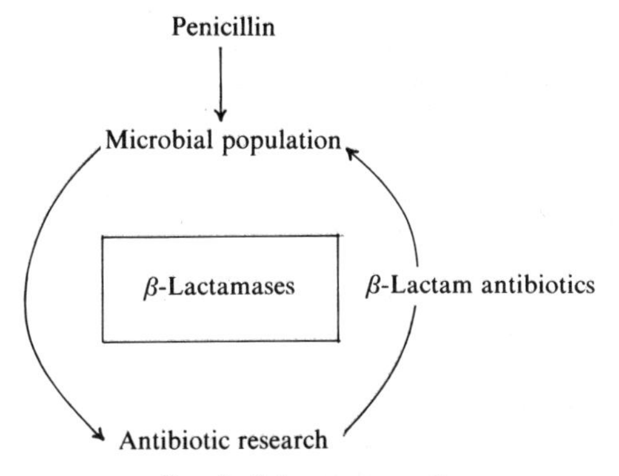

FIG. 1. *β-Lactamase cycle.*

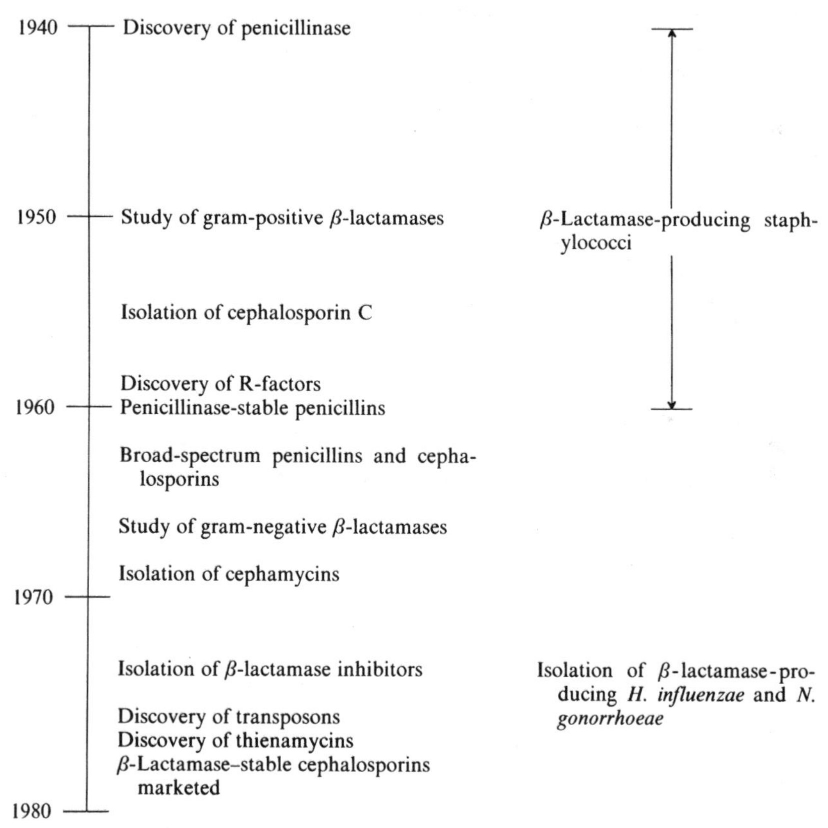

FIG. 2. *β-Lactamase calendar.*

difficult by the high incidence of strains resistant not only to penicillin G but also to other antibiotics including tetracycline, streptomycin, erythromycin, chloramphenicol, and novobiocin. Into this arena was introduced the first of the β-lactamase–stable penicillins, methicillin (35), soon to be followed by nafcillin, oxacillin, cloxacillin, dicloxacillin, and flucloxacillin. All of these compounds showed a high degree of stability to staphylococcal β-lactamase and good activity against gram-positive cocci; thus, they eliminated the need for cephalosporin C as a therapeutic agent.

With cephalosporin C out of the running for a place in clinical medicine, emphasis was then placed on the preparation of cephalosporin analogs with broad-spectrum antimicrobial activity. In addition, one of the original objectives of the work on derivatives of 6-aminopenicillanic acid was to obtain penicillins with a broader spectrum of activity than that shown by penicillin G. These studies resulted in the discovery of the broad-spectrum β-lactams ampicillin (34), cephalothin (12), and cephaloridine (22).

With the introduction of the β-lactamase–stable penicillins, the problem of the penicillin-resistant staphylococci was reduced considerably. However, the introduction of the broad-spectrum β-lactam antibiotics soon revealed another hazard: β-lactamase–producing gram-negative organisms resistant to penicillins and cephalosporins. A new goal was now identified: to find new penicillins and cephalosporins with stability to β-lactamases produced by gram-negative bacteria and with activity against the organisms that produce these enzymes.

In 1971, Nagarajan and his colleagues (23) described the isolation of a new family of antibiotics, the naturally occurring 7α-methoxy cephalosporins or cephamycins. These compounds were shown to possess broad gram-negative activity along with high stability to inactivation by β-lactamases (10, 36).

Although the search for β-lactamase inhibitors had started in the early days of penicillin (7, 16), the first significant breakthroughs came recently with the isolation from natural sources of clavulanic acid (17) and olivanic acids (8). These β-lactam–containing compounds are highly effective in enhancing the activity of enzyme-susceptible penicillins and cephalosporins against gram-negative bacteria.

This burst of activity in the discovery of novel β-lactam–containing compounds included the discovery of the nocardicins (3) and the thienamycins, which show a high degree of stability to β-lactamases.

Finally, 17 years after the introduction of the first broad-spectrum β-lactam antibiotic, came the first β-lactamase–stable cephalosporins, cefuroxime (25) and cefoxitin (29), which began to be marketed in Europe at the beginning of 1978.

In response to all this activity, the microbial population has not remained constant. A most serious consequence has been the emergence of β-lactamase–producing strains of *Haemophilus influenzae* (38), and *Neisseria gonorrhoeae* (30).

During the past 20 years, two important discoveries were made which changed our concept of the development of antibiotic resistance in bacteria. In 1959 came the discovery of R-factors by Ochiai and his co-workers (26), which provided the first proof of transferable nonchromosomal replicating genetic material from one bacterial cell to another. Then, in 1974, Hedges and his colleagues (14) advanced the concept of transposition, the ability of certain pieces of genetic material to be translocated from one replicon to another. The term transposon has been proposed for such mobile R-factor antibiotic resistance genes (15).

The possible locations and genetic events that may take place with respect to the genes for β-lactamase production are shown in Fig. 3. β-Lactamase genes may be chromosomally or extrachromosomally mediated and may be mobilized intragenetically by a recombination/transposition event and intergenetically by one of the transfer mechanisms such as transformation, transduction, or conjugation. β-Lactamase genes thus have the potential for a high degree of mobility, explaining in part why they have been so successful as antibiotic resistance agents.

At this point, it is instructive to look at the properties of β-lactamases produced by organisms of clinical importance, especially at those enzymes which have the potential to play a role in resistance to antibiotic therapy. The substrate profile of staphylococcal β-lactamases presented in Table 1 shows the hydrolytic activity of the enzyme against selected penicillins and cephalosporins. Although the β-lactamases from *S. aureus* can be serologically divided into four types (31), the enzymes are otherwise indistinguishable. They are predominantly active against penicillins (with the exception of the β-lactamase–stable penicillins), showing little activity against cephalosporins with the exception of cephaloridine. In all but a few instances (4), the enzymes have been shown to be plasmid-mediated and the cell-to-cell transfer is mediated by transduction (13).

Staphylococcal β-lactamase is an inducible enzyme (11) which is secreted from the cells in large quantities. It has a high affinity for penicillin substrates and is very effective in conferring penicillin resistance.

Gram-negative bacteria produce a plethora of β-lactamases which on a genetic basis can be classified as either chromosomally or extrachromosomally mediated. Table 2 lists some of the properties of chromosomally mediated enzymes and the organisms which produce them. Enzymes that show a predominant activity against cephalosporins are relatively common among the *Enterobacteriaceae* and in many instances are characteristic of the producing organism (19). Many of these enzymes are inducible, an uncommon property among gram-negative β-lactamases. Unlike the "cephalosporinases," enzymes showing selective activity against penicillins are relatively rare. Among the broad-spectrum β-lactamases, by far the most important from a resistance point of view are those produced by *Klebsiella* species.

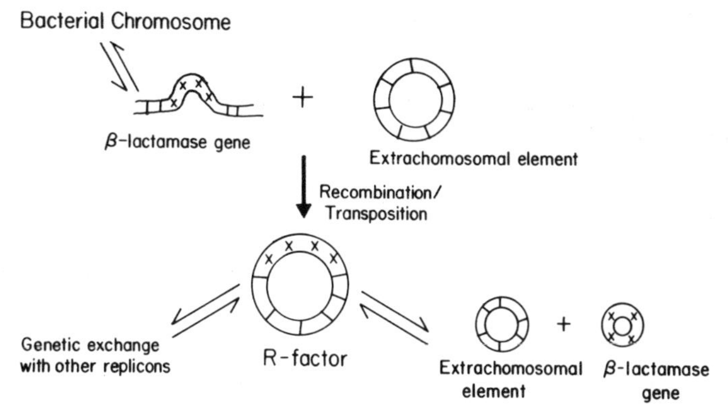

FIG. 3. *Possible locations and genetic events that may take place with respect to the genes for β-lactamase production.*

TABLE 1. *Substrate profile of staphylococcal β-lactamases*

Substrate	Hydrolytic activity
Penicillin G	100[a]
Ampicillin	120
Cloxacillin	2
Cephradine	2
Cephaloridine	10

[a] Arbitrary value for comparison purposes.

The plasmid-mediated β-lactamases are shown in Table 3 and are all constitutive. Among the R-factor–mediated cephalosporinases are listed those enzymes produced by certain strains of *Bacteroides fragilis.* It was believed until quite recently that the genes for β-lactamase production present in these strains were chromosomally mediated, but recent studies have proved differently (Frank Young, personal communication).

TABLE 2. *Properties of chromosomally mediated β-lactamases (37)*

Substrate profile				Inducible or constitutive	Enzyme production	
Penicillin G	Ampicillin	Cephaloridine	Cefoxitin		Organisms	Occurrence
100[a]	<5	200–10,000	<5	Inducible	*Acinetobacter*	Frequent
					Citrobacter	Frequent
					Enterobacter	Frequent
					Indole positive	
					Proteus	Frequent
					Pseudomonas	Frequent
					Serratia	Frequent
100	<5–10	300–10,000	<5	Constitutive	*Citrobacter*	Infrequent
					Enterobacter	Infrequent
					E. coli	Frequent
					Pseudomonas	Infrequent
					Salmonella	Frequent
					Shigella	Frequent
100	100	2–50	<5	Constitutive	*Aeromonas*	Infrequent
					Bacteroides	Infrequent
					Proteus	Infrequent
					Pseudomonas	Infrequent
100	20–200	70–150	<5	Constitutive	*Bacteroides*	Infrequent
					Branhamella	Infrequent
					Klebsiella	Frequent

[a] Arbitrary value for comparison purposes.

TABLE 3. *Properties of plasmid-mediated β-lactamases (37)*

Substrate profile					Enzyme production	
Penicillin G	Ampicillin	Methicillin	Cloxacillin	Cephaloridine	Organism	Occurrence
100[a]	5–100	<5	<5	200–1,000	*Bacteroides*	Frequent
100	200	20–400	200	30	*Bordetella*	Infrequent
					Klebsiella	Infrequent
					Proteus	Infrequent
					Providencia	Infrequent
					Salmonella	Infrequent
100	150	<5	<5	120	*E. coli*	Frequent
					Haemophilus	Infrequent
					Klebsiella	Infrequent
					Neisseria	Infrequent
					Proteus	Frequent
					Pseudomonas	Infrequent
					Salmonella	Infrequent
					Serratia	Infrequent
					Shigella	Frequent

[a] Arbitrary value for comparison purposes.

The oxacillin-hydrolyzing β-lactamases are distinguished by the fact that they hydrolyze the "penicillinase-resistant" isoxazoyl penicillins more rapidly than benzylpenicillin. These enzymes are relatively rare, accounting for less than 10% of R-factor–mediated lactamases (20).

The R-factor–mediated broad-spectrum β-lactamases, or TEM-type lactamases, are by far the most common enzymes in this class. In 1974, Hedges and his colleagues (14) compared the β-lactamases mediated by 29 different plasmids and found that one enzymatic type (the TEM-type β-lactamase) was determined by plasmids from among a broad taxonomic range of bacteria isolated on four different continents. From these studies (14), it was concluded that the ubiquity of the structural gene for this type of β-lactamase could only be explained by the gene's possessing a special ability to be translocated from one replicon to another. This special ability has provided the TEM-type lactamases with a very broad host range, the more important members of which are shown in Table 3.

From the data presented, one can understand some of the factors which have made β-lactamases such highly successful mechanisms for bestowing antibiotic resistance. The variability among the gram-negative lactamases is astounding, and the ability of certain lactamase genes to integrate into homologous and nonhomologous genetic systems provides an infinite resource. From the lactamase point of view, the driving force is the selective pressure associated with antibiotic therapy. It is predictable that we have not seen the last of the effects of β-lactamases on the therapy of infectious disease.

REFERENCES

1. **Abraham, E. P.** 1967. The cephalosporin C group. Q. Rev. Chem. Soc. **21:**231–248.
2. **Abraham, E. P., and E. Chain.** 1940. An enzyme from bacteria able to destroy penicillin. Nature (London) **146:**837.
3. **Aoki, H., H. Sakai, M. Kohsaka, T. Konomi, J. Hosoda, Y. Kubochi, E. Iguchi, and H. Imanaka.** 1976. Nocardicin A, a new monocyclic β-lactam antibiotic. J. Antibiot. **29:**492–500.
4. **Asheshov, E. H.** 1966. Chromosomal location of the genetic elements controlling penicillinase production in a strain of *Staphylococcus aureus*. Nature (London) **210:**804–806.
5. **Barber, M., and M. Rozwadowska-Dowzenko.** 1948. Infection by penicillin resistant staphylococci. Lancet **2:**641–644.
6. **Batchelor, F. R., F. P. Doyle, J. H. C. Nayler, and G. N. Rolinson.** 1959. Synthesis of penicillin: 6-amino penicillanic acid in penicillin fermentations. Nature (London) **183:**257–258.
7. **Behrens, O. K., and L. Garrison.** 1950. Inhibitors for penicillinase. Arch. Biochem. Biophys. **27:**94–98.
8. **Brown, A. G., D. F. Corbett, A. J. Eglington, and T. T. Howarth.** 1977. Structures of olivanic acid derivatives MM4550 and MM13902; two new, fused β-lactams isolated from *Streptomyces olivaceus*. J. Chem. Soc. Chem. Commun., p. 523–525.
9. **Clarke, H. T., J. R. Johnson, and R. Robinson.** 1949. The chemistry of penicillin. Princeton University Press, Princeton, N.J.
10. **Daoust, D. R., H. R. Onishi, H. Wallick, D. Hendlin, and E. O. Stapley.** 1973. Cephamycins, a new family of β-lactam antibiotics: antibacterial activity and resistance to β-lactamase degradation. Antimicrob. Agents Chemother. **3:**254–261.
11. **Geronimus, L. H., and S. Cohen.** 1957. Induction of staphylococcal penicillinase. J. Bacteriol. **73:**28–34.
12. **Griffith, R. S., and H. R. Black.** 1964. Cephalothin, a new antibiotic. J. Am. Med. Assoc. **189:**823–828.
13. **Harmon, S. A., and J. N. Baldwin.** 1964. Nature of the determinant controlling penicillinase production in *Staphylococcus aureus*. J. Bacteriol. **87:**593–597.
14. **Hedges, R. W., N. Datta, P. Kontomichalou, and J. T. Smith.** 1974. Molecular specificities of R factor-determined beta-lactamases: correlation with plasmid compatibility. J. Bacteriol. **117:**56–62.
15. **Hedges, R. W., and A. E. Jacob.** 1974. Transposition of ampicillin resistance from RP4 to other replicons. Mol. Gen. Genet. **132:**31–40.
16. **Housewright, R. D., and R. J. Henry.** 1947. Studies on penicillinase. III. The effect of antipenicillinase on penicillin-resistant organisms. J. Bacteriol. **53:**241–247.
17. **Howarth, J. T., and A. G. Brown.** 1976. Clavulanic acid, novel β-lactam isolated from *Streptomyces clavuligerus*; x-ray crystal structure analysis. J. Chem. Soc. Chem. Commun., p. 266–267.
18. **Kushner, D. J., and C. Breil.** 1977. Penicillinase formation by blue-green algae. Arch. Microbiol. **112:** 219–223.

19. **Matthew, M., and A. M. Harris.** 1976. Identification of β-lactamases by analytical isoelectric focusing: correlation with bacterial taxonomy. J. Gen. Microbiol. **94:**55–67.
20. **Matthew, M., and R. W. Hedges.** 1976. Analytical isoelectric focusing of R factor-determined β-lactamases: correlation with plasmid compatibility. J. Bacteriol. **125:**713–718.
21. **Mehta, R. J., and C. H. Nash.** 1978. β-Lactamase activity in yeast. J. Antibiot. **31:**239–240.
22. **Muggleton, P. W., C. H. O'Callaghan, and W. K. Stevens.** 1964. Laboratory evaluation of a new antibiotic—cephaloridine (ceporin). Br. Med. J. **2:**1234–1237.
23. **Nagarajan, R., L. D. Boeck, M. Gorman, R. L. Hamill, C. E. Higgens, M. M. Hoehn, W. H. Stark, and J. G. Whitney.** 1971. β-Lactam antibiotics from Streptomyces. J. Am. Chem. Soc. **93:**2308–2310.
24. **Newton, G. G. F., and E. P. Abraham.** 1956. Isolation of cephalosporin C, a penicillin-like antibiotic containing D-α-aminoadipic acid. Biochem. J. **62:**651–658.
25. **O'Callaghan, C. H., R. B. Sykes, D. M. Ryan, R. D. Foord, and P. W. Muggleton.** 1975. Cefuroxime, a new cephalosporin antibiotic. J. Antibiot. **29:**29–37.
26. **Ochiai, K., T. Yamanaka, K. Kimura, and O. Sawada.** 1959. Studies on inheritance of drug-resistance between Shigella strains and *Escherichia coli* strains. Nihon Iji Shimpo **861:**34–46.
27. **Ogawara, H.** 1975. Production and property of β-lactamases in *Streptomyces.* Antimicrob. Agents Chemother. **8:**402–408.
28. **Ogawara, H., S. Horikawa, S. Shimada-Miyoshi, and K. Yasuzawa.** 1978. Production and property of beta-lactamases in *Streptomyces*: comparison of the strains isolated newly and thirty years ago. Antimicrob. Agents Chemother. **13:**865–870.
29. **Onishi, H. R., D. R. Daoust, S. B. Zimmerman, D. Hendlin, and E. O. Stapley.** 1974. Cefoxitin, a semisynthetic cephamycin antibiotic: resistance to beta-lactamase inactivation. Antimicrob. Agents Chemother. **5:**38–48.
30. **Percival, A., J. Rowlands, J. E. Corkill, C. D. Alergent, O. P. Arya, E. Rees, and E. H. Annels.** 1976. Penicillinase producing gonococci in Liverpool. Lancet **2:**1379–1382.
31. **Richmond, M. H.** 1965. Wild type variants of exopenicillinase from *Staphylococcus aureus.* Biochem. J. **94:**584–593.
32. **Richmond, M. H., and R. B. Sykes.** 1973. The β-lactamases of gram-negative bacteria and their possible physiological role. Adv. Microb. Physiol. **9:**31–88.
33. **Ridley, M., D. Barrie, R. Lynn, and K. C. Stead.** 1970. Antibiotic-resistant *Staphylococcus aureus* and hospital antibiotic policies. Lancet **1:**230–233.
34. **Rolinson, G. N., and S. Stevens.** 1961. Microbiological studies on a new broad-spectrum penicillin "Penbritin." Br. Med. J. **2:**191–196.
35. **Rolinson, G. N., S. Stevens, F. R. Batchelor, J. Cameron-Wood, and E. B. Chain.** 1960. Bacteriological studies on a new penicillin—BRL 1241. Lancet **2:**564–567.
36. **Stapley, E. O., M. Jackson, S. Hernandez, S. B. Zimmerman, S. A. Currie, S. Mochales, J. M. Mata, H. B. Woodruff, and D. Hendlin.** 1972. Cephamycins, a new family of β-lactam antibiotics. I. Production by actinomycetes, including *Streptomyces lactamdurans* sp. n. Antimicrob. Agents Chemother. **2:** 122–131.
37. **Sykes, R. B., and M. Matthew.** 1976. The β-lactamases of gram-negative bacteria and their role in resistance to β-lactam antibiotics. J. Antimicrob. Chemother. **2:**115–157.
38. **Sykes, R. B., M. Matthew, and C. H. O'Callaghan.** 1975. R-factor mediated β-lactamase production by *Haemophilus influenzae.* J. Med. Microbiol. **8:**437–441.
39. **Thompson, R. E. M., J. W. Harding, and R. D. Simon.** 1960. Sensitivity of *Staphylococcus pyogenes* to benzylpenicillin and BRL 1241. Br. Med. J. **2:**708–709.

Epidemiology of Plasmid-Mediated Ampicillin Resistance in Pathogenic Microorganisms

LUCY S. TOMPKINS, MARILYN ROBERTS, JORGE H. CROSA, AND STANLEY FALKOW

Department of Microbiology and Immunology, University of Washington School of Medicine, Seattle, Washington 98195

Antibiotic-resistant organisms are posing an increasing public health problem throughout the world. This situation has been highlighted by the recent widespread outbreaks of enteric disease caused by resistant dysentery and typhoid organisms. The delayed recognition that these previously treatable infections were caused by antibiotic-resistant organisms resulted in unnecessary suffering and loss of life. A serious corollary has been the emergence within the hospital environment of antibiotic-resistant microorganisms that can opportunistically infect and cause disease in patients whose normal defense mechanisms have been weakened by disease, major surgery, or immunosuppression. Perhaps the most worrisome feature of recent investigations into the epidemiology of antibiotic resistance has been the recognition of R plasmids in *Haemophilus influenzae* and *Neisseria gonorrhoeae.* Heretofore, R plasmids have played a prominent role in epidemic enteric disease or in nosocomial infections in highly selective hospital environments, but gonococcal infections and most *H. influenzae* infections are acquired out-of-hospital and affect broad segments of the populations.

One particularly illustrative example of plasmid-mediated antibiotic resistance has been the marked increase in ampicillin resistance in the past decade. Identical ampicillin resistance genes have occurred in a variety of plasmids in many different bacterial species. There is now good evidence to explain the fact that plasmids of substantially diverse origin possess identical ampicillin resistance. This knowledge, which we shall review in the following sections, provides some new insights into plasmid evolution.

RESISTANCE GENES AS DISCRETE, MOBILE, GENETIC ELEMENTS

R plasmids of gram-negative bacteria are a very diverse group of genetic elements and are found in a variety of sizes and overall DNA composition. One of the most convenient methods for classifying plasmids is incompatibility grouping. The basis for this grouping is quite straightforward: if two plasmids are closely related, they cannot coexist in the same bacterial host. The plasmids are called incompatible and are said to be part of the same incompatibility group. Members of the same incompatibility group usually have a similar mass and DNA base composition; moreover, they share a significant proportion of their nucleotide sequences (8). Compatible plasmids, that is, plasmids which can coexist in the same bacterial host, are generally quite unrelated and rarely share more than 10% of their nucleotide sequences (8). Yet, despite their diversity, R plasmids of different incompatibility groups frequently determine resistance to the same antibiotics or combination of antibiotics. In the case of ampicillin resistance, identical β-lactamases are present in all incompatibility groups. This has been found to be the case for other antibiotic resistance enzymes as well. Such findings indicate, therefore, that there has been a sharing of resistance mechanisms between members of different unrelated plasmid groups.

Studies of the plasmid sequences which encode for antibiotic resistance have now established that most plasmid-mediated antibiotic resistance determinants reside upon discrete DNA segments, called transposons, which have the capacity for excising themselves from one chromosome and inserting themselves into another. In practice, this is easy to demonstrate. One introduces a plasmid carrying the structural gene for ampicillin resistance, Ap^r, into a cell carrying an easily recognizable compatible plasmid, for example, a ColE1 plasmid, which does not encode for antibiotic resistance. It is a simple matter, by use of appropriate molecular and genetic techniques, to demonstrate that within 30 generations 1 in 10^5 ColE1 plasmids has acquired the Ap^r gene(s) (11, 22). The frequency of ColE1 plasmids carrying Ap^r does not increase much beyond this level, suggesting that some mechanism controls the capacity for Ap^r genes to transpose from plasmid to plasmid or even from plasmid to chromosome (11, 22).

The examination of the precise DNA sequence of this ampicillin transposon, Tn*A*, has revealed that the majority are 3.2×10^6 daltons in mass and are bound by inverted repeated sequences some 100 to 150 base pairs in length (10). Though a variety of Tn*A* transposons have been described (Tn*1*, Tn*2*, Tn*3*, etc.), most are quite closely related (12; Table 1). Recent studies suggest that Tn*A* sequences are highly evolved units of DNA. That is, in addition to the structural gene for β-lactamase and its control, the Tn*A* sequences also carry genes which are required for transposition of the element. Transposon-mediated resistance to tetracycline (14), kanamycin (4), chloramphenicol (9), trimethoprim and streptomycin (3), and mercury ion (24) is also well documented. These transposons are distinct from Tn*A* and from one another. Yet, all share the very general common property of excision and integration through flanking repeated sequence.

The ability of transposons to "jump" from plasmid to plasmid occurs independently of the usual recombination processes of the bacterial cell. Since transposition occurs by illegitimate recombination, homology is not required between interacting DNA sequences; this surely explains why one can find the identical Tn*A* sequence

TABLE 1. *Relatedness of various plasmids carrying the transposon coding after TEM beta-lactamase*

Resistance plasmid	Incompatibility class	Percent guanine + cytosine	Plasmid mass (daltons)	Microbial origin of R plasmid	Relatedness of Tn*A* sequence (%)[a]
RI	FII	52	62×10^6	*Salmonella paratyphi*	100
RP4	P	59	34×10^6	*Pseudomonas aeruginosa*	82
R7K	W	62	22×10^6	*Proteus rettgeri*	90
R648	?	56	6.1×10^6	*S. typhimurium*	99
R6K	X	45	26×10^6	*Escherichia coli*	99
RSF007	?	40	30×10^6	*Haemophilus influenzae*	97

[a] The DNA of a ColE1 plasmid which had received by transposition a Tn*A* sequence (Tn*1*) originating from the R plasmid R1 was labeled with [^{3}H]thymine and hybridized with unlabeled DNA from each of the indicated R plasmids. ColE1 has no detectable sequences in common with any of the R plasmids; hence, the degree of DNA-DNA binding between the labeled ColE1::Tn*1* DNA and the unlabeled DNA was restricted to Tn*A* sequences. All reactions were normalized to the reaction between ColE1::Tn*1* and plasmid R1, which was set at 100%. (The actual degree of DNA-DNA binding was 47%.)

on plasmids that are totally different in other biological properties (Table 1). The behavior of transposons accounts for the fact that, when a new antibiotic is introduced, "new" R plasmids containing a gene conferring resistance to the new agent appear so rapidly. Thus, the R plasmids are generally not "new"; they are often simply preexistent R plasmids which have received the appropriate transposon.

R-PLASMID TRANSFER IN VIVO

It is well established that the incidence of organisms harboring R plasmids is increasing. Given the many ways by which resistance genes can be disseminated (transformation, conjugation, transduction, and now transposition) it might be imagined that R plasmids are transmitted with great ease in natural populations. From a practical standpoint, however, R-plasmid transmission in vivo does not occur on so large a scale as observed in laboratory experiments.

To be sure, R-plasmid transfer occurs in vivo in the absence of purposive antibiotic selection (17), but this transfer is clearly a rare event, and it appears that man's normal flora often acts as an effective prophylactic to plasmid transfer (18, 19). Not surprisingly, R plasmid transfer in vivo can be brought to a readily detectable level under the force of antibiotic selection (18, 19). In the developed countries of the world, the greatest problem related to the presence of R plasmids occurs in the hospital environment, where selective pressure is intense. Indeed, R-plasmid transfer occurs on the surface of wounds (13) and has even been demonstrated in the urine of patients (23).

Though R-plasmid transfer can be shown to occur within the hospital environment, "helter skelter" transfer of R plasmids residing in strains patients carry as part of their normal flora, producing multiple plasmid morphologies in many different host strains, is not a frequent event. Rather, one is often confronted with "special" hospital strains containing a single plasmid species. The point may be illustrated by the sudden emergence of a multiply resistant strain of *Serratia marcescens* which appeared in early 1977 in a Veterans Hospital in Seattle. A great many patients admitted to the Urology and General Surgery wards became colonized with this organism. In a few instances the organisms caused serious disease including pyelonephritis and bacteremia. Serological and bacteriocin testing revealed that the responsible *Serratia* strain was a single clone that had become widely distributed in the hospital. This clone harbored a plasmid conferring high-level carbenicillin resistance as a result of the presence of Tn*A* and low-level resistance to kanamycin, gentamicin, and tobramycin, but not amikacin. During the course of monitoring patients infected with this nosocomial strain, it was possible to demonstrate that the R plasmid of the *S. marcescens* strain had a high ability to be transmitted in vivo to other species of gram-negative microorganisms which also colonized these patients, including *Citrobacter* sp., *Escherichia coli*, and *Klebsiella pneumoniae* (Fig. 1).

In the hospital setting there is an enormous reservoir of R plasmids and resistant species (1). It should not be surprising to find, therefore, that transposition of resistance genes coupled with plasmid transfer mechanisms, such as conjugation, soon leads to one or two plasmid species carrying resistance to most of the commonly employed antibiotics in a given clinical setting. Moreover, these "most fit" plasmids find their way into the "most fit" organism within the clinical setting. It appears that nosocomial infection by resistant organisms is associated with intensive antibiotic usage in compromised hosts and that nosocomial infection problems will diminish when antibiotic selection is decreased (21). In general, this is true, but we should be

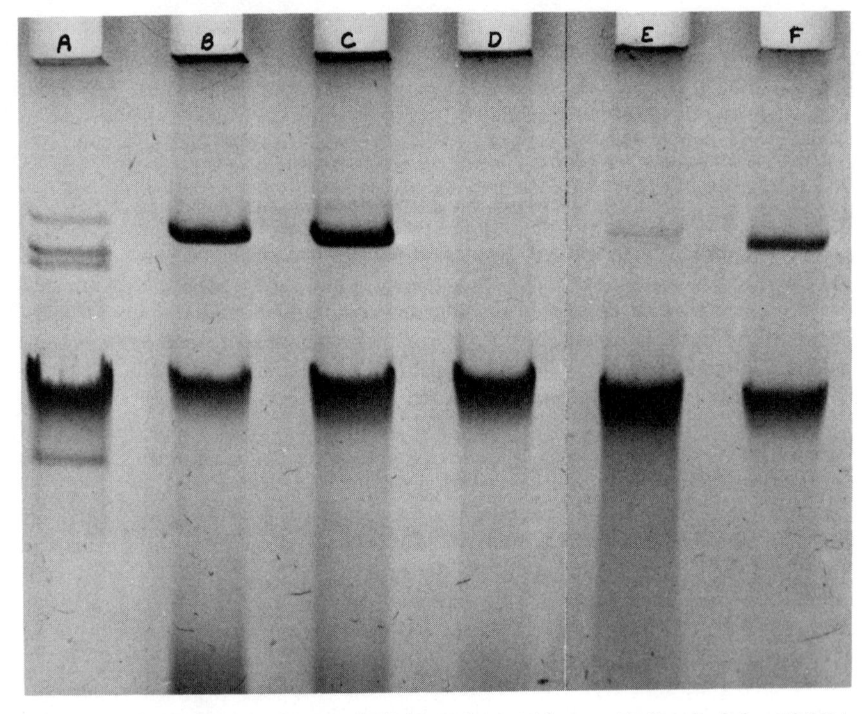

FIG. 1. *Agarose gel electrophoresis of DNA from bacterial lysates. (A) Standard plasmid DNAs with molecular weights of 62, 34, and 23 megadaltons, and chromosomal DNA. (B) Lysate from* Serratia marcescens *strain 50, Seattle Veterans Administration Hospital "epidemic" strain, resistant to carbenicillin, ampicillin, gentamicin, tobramycin, and kanamycin, and containing a single 45-megadalton plasmid. (C) Lysate from* Citrobacter freundii, *resistant to carbenicillin, ampicillin, gentamicin, tobramycin, and kanamycin, isolated from urine of a Veterans Administration Hospital patient, and containing a 45-megadalton plasmid. (D) Lysate from* S. marcescens, *susceptible to carbenicillin, ampicillin, gentamicin, tobramycin, and kanamycin, and lacking plasmid DNA. (E) Lysate from* Klebsiella pneumoniae, *resistant to carbenicillin, ampicillin, gentamicin, tobramycin, and kanamycin, isolated from urine of Veterans Administration Hospital patient BL, and containing a 45-megadalton plasmid. (F) Lysate from* S. marcescens *strain 50, isolated from urine of patient BL simultaneously with the* Klebsiella *isolate and containing a 45-megadalton plasmid.*

aware that continued selection may eventually result in strains that colonize as well when antibiotics are absent as when antibiotic selection is present. Certainly, we are still largely ignorant of the fate of the resistant bacteria and their R plasmids when patients are discharged from the hospital or of the extent to which this may contribute to the plasmid burden in the community at large. Whereas resistant opportunistic species like *Serratia* and their R plasmids may require antibiotics to be maintained, this is not true for the essential pathogens such as *Shigella dysenteriae* (5) or *Salmonella typhi* (2). As E. S. Anderson has remarked, if these essential pathogens acquire an R plasmid, so much the better for the organism, so much the worse for the patient (1). Once established, resistant epidemic pathogens will spread with or without antibiotic pressure, often with grim results.

Yet, as seems to be the case with local outbreaks of nosocomial infection, resistant pathogens often represent a single strain containing a single class of plasmid. It is an interlocked epidemiology in which one is dealing with epidemics of plasmids in bacteria as well as of bacteria in humans. As an example, *S. dysenteriae* type 1 was the causative organism in the 1970 epidemic of dysentery in Central America and Mexico (15, 16). The strain was resistant to chloramphenicol, tetracycline, streptomycin, and sulfonamides but uniformly susceptible to ampicillin (16). Subsequently, during 1972 through 1974, outbreaks of dysentery occurred in Mexico, Central America, and Bangladesh in which resistant Shiga strains were isolated. Crosa et al. (5) showed that the ampicillin resistance in these strains was associated with the acquisition of a 5.5×10^6-dalton plasmid (Fig. 2). They also showed that this plasmid species is quite distinctive and commonly found in other strains of the *Enterobacteriaceae* (Table 2). Moreover, this plasmid possessed the Tn*A* sequence. Indeed, Tn*A* comprised well over half of the plasmid genome, with the remaining plasmid genes dedicated to replicative and other essential functions (5). Presumably, Tn*A* could potentially find its way onto many plasmids of various classes. Why, then, does this particular plasmid appear to be so successful in being disseminated throughout the *Enterobacteriaceae*? One really does not know, although it appears that a particular plasmid may evolve through conjugation and transposition, and then, because of some unknown essential plasmid property, probably unrelated to antibiotic resistance per se, can be spread among and maintained in a wide variety of bacterial species. Of course, antibiotic usage is important in the sense that the more intense the selective pressure, and the longer it is applied, the higher is the probability that some "special" plasmid, perhaps containing virulence genes as well as antibiotic resistance markers, will arise.

R PLASMIDS IN *HAEMOPHILUS* AND *NEISSERIA*

The selection of "special" plasmids in part seems to explain the rather sudden emergence of R plasmids in both *H. influenzae* and *N. gonorrhoeae.* The situation in *H. influenzae* has been particularly dramatic, with the sequential appearance of plasmid-mediated ampicillin resistance (7), tetracycline resistance (6), chloramphenicol resistance, and kanamycin resistance (25). The examination of the plasmids mediating ampicillin resistance in both *Haemophilus* and the gonococcus has been particularly instructive. As shown in Table 2, both transferable and nontransferable plasmids have been found in *Haemophilus* (20). Only nonconjugative plasmids have been found in the gonococcus (20). All of these plasmids have a rather distinctive DNA base composition of about 40% guanine plus cytosine; yet, they possess all or part of the same Tn*A* sequences of plasmids found in enteric species (20). The small nonconjugative plasmids of gonococci and *H. influenzae* are very closely related, but are not at all related (except for Tn*A* sequences) to the epidemic 5.5×10^6-dalton Ap^r plasmid, RSF1030, of the enteric species described earlier (20).

In some basic biological properties, however, the special Ap^r plasmids of enteric species, *Haemophilus*, and *Neisseria* are quite similar. They all possess part or all of Tn*A*. Many of these plasmids have a strict dependence upon DNA polymerase I for replication (Table 2) and continue to replicate in the presence of chloramphenicol. The full implications, if any, of these common biological properties is unknown.

CONCLUSIONS

The Tn*A* transposon is a remarkable structure. It is a "special" element which has been conserved in a variety of hosts. In theory, *any* resident plasmid could gain a transposon and metamorphose from an innocuous molecular parasite one moment

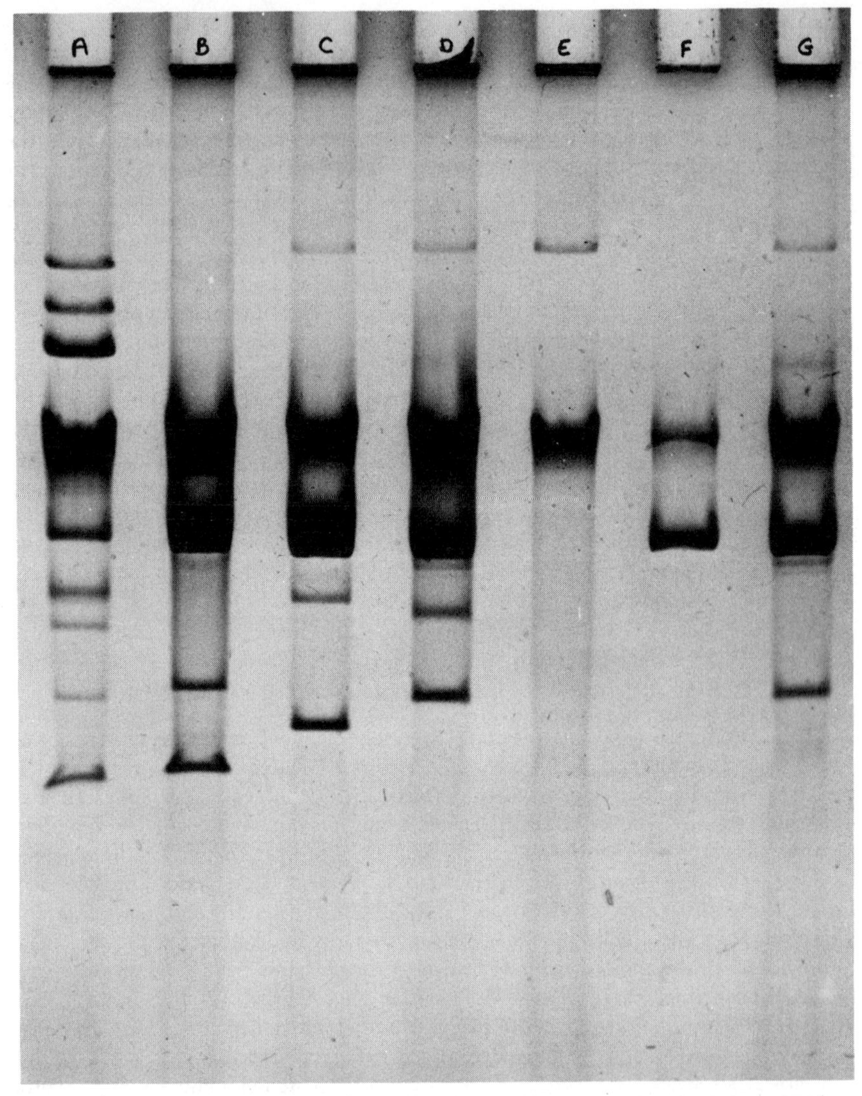

FIG. 2. *Agarose gel electrophoresis of DNA from bacterial lysates. (A) Standard plasmid DNAs varying from 62 megadaltons (uppermost band) to 1.8 megadaltons (lowest band). (B) Lysate of* Shigella dysenteriae *type 1 strain M6982 Ap^r from Bangladesh containing several plasmids, including a 5.5-megadalton plasmid (nonconjugative) as well as six others (three are barely visible). (C) Lysate of* S. dysenteriae *type 1 strain 762 Ap^r from Costa Rica, containing a 5.5-megadalton plasmid, an 80-megadalton plasmid, and five others. (D) Lysate of* S. dysenteriae *type 1 strain 51B Ap^r from a Mexican epidemic, containing a 5.5-megadalton plasmid and four others. (E) Lysate of* Escherichia coli *K-12 (exconjugant no. 5, obtained from a cross between* S. dysenteriae *type 1 [Mexico] and* E. coli *K-12 W1485-1 Nx^r), susceptible to ampicillin and containing the 80-megadalton conjugative plasmid only. (F) Lysate of* E. coli *K-12 (exconjugant no. 5, obtained from a cross between* S. dysenteriae *type 1 [Mexico] and* E. coli *K-12 W1485-1 Nx^r), resistant to ampicillin and containing the 5.5-megadalton plasmid. (G) Lysate of* E. coli *K-12 (exconjugant no. 25, obtained from a cross between* S. dysenteriae *type 1 and* E. coli *K-12 W1485-1 Nx^r), resistant to ampicillin and containing both 80- and 5.5-megadalton plasmids. Reprinted from Crosa et al. (5).*

TABLE 2. *Molecular characterization of plasmids coding for TEM beta-lactamase*

Plasmid	Origin	Molecular mass (daltons)	Mol fraction guanine + cytosine	Relative homology[a] with:		Replication in chloramphenicol	Polymerase I requirement
				RSF1030	pMR0360		
RSF1030	*Salmonella panama*	5.5×10^6	49	100	40	Yes	Yes
RJHC8	*Shigella dysenteriae*	5.5×10^6	ND[b]	87	ND	Yes	Yes
RJHC2	*Citrobacter freundii*	5.5×10^6	ND	98	ND	Yes	Yes
pMR0360	*Neisseria gonorrhoeae*	4.4×10^6	41	40	100	ND	ND
RSF0885	*Haemophilus parainfluenzae*	4.1×10^6	40	40	95	ND	ND
RSF007	*H. influenzae*	30.1×10^6	40	57	40	ND	ND

[a] Heteroduplex molecules prepared between RSF1030 and pMR0360 and examined in an electron microscope reveal that homology is restricted to Tn*A* sequences exclusively. RSF1030 and RSF007 contain a complete Tn*A* segment; pMR0360 and RSF0885 contain only a portion of the Tn*A* sequence.
[b] ND, Not determined.

to an R plasmid the next. Yet, the epidemiology of R-plasmid resistance in gram-negative organisms in the hospital environment, in epidemic enteric pathogens, and in nonenteric pathogens causing community-acquired infections discloses that this is not necessarily the case. Rather, it seems that we are witnessing epidemics of a single plasmid carried within a single bacterial clone.

The occurrence of similar organisms around the world could represent the evolution of similar plasmids arising from numerous analogous genetic events, a kind of convergent evolution. It seems more likely, however, that in most instances there has actually been a widespread clonal dissemination of a strain carrying a particular R plasmid. Implicit in this hypothesis is the concept that these "special" plasmid-carrying strains have a selective advantage, not necessarily related to antibiotic resistance, which allows them to proliferate. The nature of the plasmid-mediated properties which might provide such an advantage is entirely speculative at this time.

ACKNOWLEDGMENT

This work was supported by The Charles E. Merrill Trust.

REFERENCES

1. **Acar, J. F., D. H. Bouanchaud, and Y. A. Chabbert.** 1977. Evolutionary aspects of plasmid mediated resistance in a hospital environment, p. 21. *In* J. Drews and G. Hogenauer (ed.), R-factors: their properties and possible control. Springer-Verlag, New York.
2. **Anderson, E. S.** 1977. The geographical predominance of resistance transfer systems of various compatibility groups of Salmonellae, p. 25–47. *In* J. Drews and G. Hogenauer (ed.), R-factors: their properties and possible control. Springer-Verlag, New York.
3. **Barth, P. T., N. Datta, R. W. Hedges, and N. J. Grinter.** 1976. Transposition of a deoxyribonucleic acid sequence encoding trimethoprim and streptomycin resistances from R483 to other replicons. J. Bacteriol. **125:**800–810.
4. **Berg, D. E., J. Davis, B. Allet, and J. D. Rochaix.** 1975. Transposition of R-factor genes to lambda. Proc. Natl. Acad. Sci. U.S.A. **72:**3628–3632.
5. **Crosa, J. H., J. Olarte, L. J. Mata, L. K. Luttropp, and M. E. Penaranda.** 1977. Characterization of an R plasmid associated with ampicillin resistance in *Shigella dysenteriae* type 1 isolated from epidemics. Antimicrob. Agents Chemother. **11:**553–558.
6. **Dang Van, A., G. Beith, and D. H. Bouanchaud.** 1975. Resistance plasmidique a la tetracycline chez *H. influenzae.* C. R. Acad. Sci. **280:**1321–1323.
7. **Dang Van, A., F. Goldstern, S. F. Acar, and D. H. Bouanchaud.** 1975. A transferable kanamycin resistance plasmid from *H. influenzae.* Ann. Microbiol. (Paris) **126:**397–399.
8. **Falkow, S., P. Guerry, R. W. Hedges, and N. Datta.** 1974. Polynucleotide sequence relationships among plasmids of the I compatibility complex. J. Gen. Microbiol. **85:**65–76.
9. **Gottesman, M. M., and J. L. Rosner.** 1975. Acquisition of a determinant for chloramphenicol resistance by lambda. Proc. Natl. Acad. Sci. U.S.A. **72:**5041–5045.

10. **Hedges, R. W., and A. Jacob.** 1974. Transposition of ampicillin resistance from RP4 to other replicons. Mol. Gen. Genet. **132:**31–40.
11. **Heffron, F., C. Rubens, and S. Falkow.** 1975. Translocation of a plasmid DNA sequence which mediates ampicillin resistance: molecular nature and specificity of insertion. Proc. Natl. Acad. Sci. U.S.A. **72:** 3623–3627.
12. **Heffron, R., R. Sublett, R. W. Hedges, A. Jacob, and S. Falkow.** 1975. Origin of the TEM beta-lactamase gene found on plasmids. J. Bacteriol. **122:**250–256.
13. **Ingram, L. C., M. H. Richmond, and R. B. Sykes.** 1973. Molecular characterization of the R factor implicated in the carbenicillin resistance of a sequence of *Pseudomonas aeruginosa* strains isolated from burns. Antimicrob. Agents Chemother. **3:**279–288.
14. **Kleckner, N., R. K. Chan, B. K. Tye, and D. Botstein.** 1975. Mutagenesis by insertion of a drug resistance element carrying an inverted repeat. J. Mol. Biol. **97:**561–575.
15. **Mata, L. J., E. J. Gangarosa, A. Caceres, D. R. Perera, and M. C. Mejicanos.** 1970. Epidemic *Shiga bacillus* dysentery in Central America. I. Etiologic investigation in Guatemala in 1969. J. Infect. Dis. **122:**170–180.
16. **Olarte, J., L. Filloy, and E. Galindo.** 1976. Resistance of *Shigella dysenteriae* type 1 to ampicillin and other antimicrobial agents: strains isolated during a dysentery outbreak in a hospital in Mexico City. J. Infect. Dis. **133:**572–575.
17. **Petrocheilou, V., J. Grinsted, and M. H. Richmond.** 1976. R-plasmid transfer in vivo in the absence of antibiotic selection pressure. Antimicrob. Agents Chemother. **10:**753–761.
18. **Richmond, M. H.** 1975. R-factors in man and his environment, p. 27–35. *In* D. Schlessinger (ed.), Microbiology—1974. American Society for Microbiology, Washington, D.C.
19. **Richmond, M. H.** 1977. The survival of R plasmids in the absence of antibiotic selection pressure, p. 61–78. *In* J. Drews and G. Hogenauer (ed.), R-factors: their properties and possible control. Springer-Verlag, New York.
20. **Roberts, M., L. P. Elwell, and S. Falkow.** 1977. Molecular characterization of two beta-lactamase-specifying plasmids isolated from *Neisseria gonorrhoeae.* J. Bacteriol. **131:**557–563.
21. **Roberts, N. J., and R. G. Douglas, Jr.** 1978. Gentamicin use and *Pseudomonas* and *Serratia* resistance: effect of a surgical prophylaxis regimen. Antimicrob. Agents Chemother. **13:**214–220.
22. **Rubens, C., F. Heffron, and S. Falkow.** 1976. Transposition of a plasmid deoxyribonucleic acid sequence that mediates ampicillin resistance: independence from host *rec* functions and orientation of insertion. J. Bacteriol. **128:**425–434.
23. **Schaberg, D. R., A. K. Highsmith, and I. K. Wachsmuth.** 1977. Resistance plasmid transfer by *Serratia marcescens* in urine. Antimicrob. Agents Chemother. **11:**449–451.
24. **Stanisich, V. A., P. M. Bennett, and M. H. Richmond.** 1977. Characterization of a translocation unit encoding resistance to mercuric ions that occurs on a nonconjugative plasmid in *Pseudomonas aeruginosa.* J. Bacteriol. **129:**1227–1233.
25. **Vega, R., M. L. Sadoff, and M. J. Patterson.** 1976. Mechanism of ampicillin resistance in *Haemophilus influenzae* type B. Antimicrob. Agents Chemother. **9:**164–168.

Recombination Studies with *Cephalosporium acremonium*

P. F. HAMLYN AND C. BALL

Glaxo Laboratories Ltd., Ulverston, Cumbria, England

The strains of *Cephalosporium acremonium* used in the industrial production of cephalosporin C are derived from the Brotzu strain. In the absence of any detailed reports of genetic analysis in this species, the aim of this work was to investigate the possibility of attempting such analysis. We have used not only the conventional methods of crossing (5) but also the new technique of fungal protoplast fusion. The latter method enables parental nuclei to be brought into close proximity. In *C. acremonium* this is a particular advantage because hyphal cells are mainly uninucleate, and juxtapositioning of nuclei in such cells is infrequent (5).

MATERIALS AND METHODS

Media. Complete medium (CM) contained (per liter): maltose (British Drug Houses), 40 g; peptone (Oxoid), 10 g; malt extract (Oxoid), 24 g; and Oxoid No. 3 agar, 20 g; adjusted to pH 7.5 and then autoclaved at 121°C for 20 min. Minimal medium (MM) contained (per liter): sucrose, 30 g; $NaNO_3$, 2 g; KH_2PO_4, 1 g; $MgSO_4 \cdot 7H_2O$, 0.5 g; KCl, 0.5 g; $FeSO_4 \cdot 7H_2O$, 0.01 g; and Oxoid No. 3 agar, 20 g; adjusted to pH 6.8 and autoclaved as above. All media onto which protoplasts were plated for regeneration were osmotically stabilized by 0.8 M NaCl plus 2% sucrose (3).

Strains and type of cross. Three groups, A, B, and C, of genealogically related strains were used. Group C consisted of strains derived from M8650, a low-titer cephalosporin C producer closely related to the Brotzu strain. The other two groups were derived from group C by many mutation and selection steps.

The strains within the groups differed by only a few mutation steps, mainly those involved in introducing auxotrophic and morphological markers, with UV light used to give 1% spore/cell survival. The following markers are referred to, the group of origin being noted in parentheses:

Requirement for arginine	*arg-1* (C) *arg-2* (C) *arg-3* (B)
Requirement for proline	*pro-1* (A)
Requirement for methionine	*met-1* (C) *met-2* (C) *met-3* (A)
Requirement for aneurin	*ane-1* (C) *ane-2* (A) *ane-3* (B)
Requirement for nicotinamide	*nic-1* (A)
Requirement for leucine	*leu-1* (C)
Requirement for thiosulfate	*thi-1* (B)
Red mycelial pigment	*red-1* (C) *red-2* (A) *red-3* (A)

Some markers, e.g., *arg-1*, were leaky; others, e.g., *arg-2*, were not. Phenotypes were indicated by capitalizing the first letter, e.g., Arg for arginine-requiring phenotypes and Arg^+ for arginine-independent phenotypes. Three general types of crosses were carried out: sister-strain crosses of type A/A and C/C, divergent-strain crosses of type A/B, and ancestral crosses of type A/C.

Crossing techniques. All incubations were carried out at 25.5°C. The first method of crossing was the conventional one of Nüesch et al. (5), and prototrophs were selected on MM. The second method of crossing by protoplast fusion also involved selecting prototrophs on MM.

Protoplasts were prepared by the method of Fawcett et al. (3), with L1 enzyme used to remove the cell wall. Protoplasts were fused by the method of Anné and Peberdy (1), with polyethylene glycol as the fusogen. For both the conventional and the protoplast fusion methods of crossing, the parent strains each carried at least one marker that was nonleaky on MM and did not back-mutate spontaneously at high frequency. Prototrophs recovered from crosses were classified by plating spores, cells, and hyphal fragments onto CM and MM at different plating densities. Good agreement between the CM and MM plates indicated that the strain was a stable prototroph. Poor agreement indicated that the strain was an unstable prototroph. Certain types of unstable prototrophs were found to be heterokaryons; others were classified as heterozygous for markers that segregated on CM.

Also, various prototrophs so defined were tested to see if they segregated auxotrophs after treatment with the recombinogens *p*-fluorophenylalanine as described by Nüesch et al. (5) or gamma rays as described by Käfer (4).

Cephalosporin C titer testing. A method based on that described by Nüesch et al. (5) was used to determine cephalosporin C titers.

Electron microscopy. Protoplasts were fixed in 4% glutaraldehyde in 0.8 M mannitol for 1.5 h at room temperature and postfixed in 2% aqueous osmium tetroxide for 1 h. After being washed, the fixed cells were dehydrated and embedded in Spurr's resin. Thin sections were stained with uranyl acetate and lead citrate and were examined in an electron microscope at 80 kV.

RESULTS

Comparison of conventional and protoplast fusion crosses. More than 40 different conventional crosses were carried out. These included 8 sister-strain, 26 divergent-strain, and 6 ancestral crosses. Stable prototrophs that did not segregate after treatment with *p*-fluorophenylalanine and gamma rays were recovered in 10 of these crosses. Unstable prototrophs, other than heterokaryons, were detected in only 2 crosses, namely, *arg-1 leu-1/arg-2 met-1*, a sister-strain cross of type C/C, and *arg-2/pro-1 red-2*, an ancestral cross of type A/C. The unstable prototrophs recovered from the first cross spontaneously produced arginine-requiring segregants and the unstable prototrophs from the second cross segregated arginine-requiring and arginine-independent segregants as well as segregating for the *red-2* morphological marker. Phenotype frequencies were as follows: Red^+ Arg, 25; Red^+ Arg^+, 8; Red Arg, 22; Red Arg^+, 1. None of these heterozygotes showed any change in the type of segregant they produced when treated with the recombinogens detailed previously. These results from conventional breeding suggested that fusion between nuclei of different strains does not readily occur.

Protoplast fusion techniques were then adopted in the hope of improving the chances of such fusion. Electron microscopy indicated that up to 1% of the treated protoplasts exhibit nuclear fusion immediately after protoplast fusion (Fig. 1). Nuclear fusion was not observed in protoplasts that had not been fused.

Table 1 shows the results from eight protoplast fusion crosses. In contrast to the results obtained by conventional crossing, stable prototrophs were always recovered together with certain types of recombinant auxotrophs. Other recombinant auxotrophs were produced as spontaneous segregants from rare heterozygotes.

In the sister-strain cross, the heterozygotes recovered were unstable and were identical to those discovered by conventional crossing. In the divergent-line crosses the single heterozygote detected in cross *ane-2 nic-1 red-3/ane-3 arg-3* was unstable and spontaneously segregated all the markers of the cross except *arg-3*. The phenotypes were as follows: Red^+ Ane Nic, 4; Red^+ Ane^+ Nic, 6; Red^+ Ane Nic^+, 5; Red Ane Nic^+, 1; Red^+ Ane^+ Nic^+, 9; Red Ane Nic, 8; Red Ane^+ Nic^+, 1; Red Ane^+ Nic, 3. There was no change in the types of segregant produced when the heterozygote was treated with recombinogens. The stable prototrophs shown in Table 1 did not segregate after treatment with recombinogens. In addition, no individual recombinants with auxotrophic requirements from each parent were recovered by any route, but the consistent recovery of certain types of recombinant led us to use protoplast fusion in detailed genetic analysis and empirical breeding, both of which we shall describe.

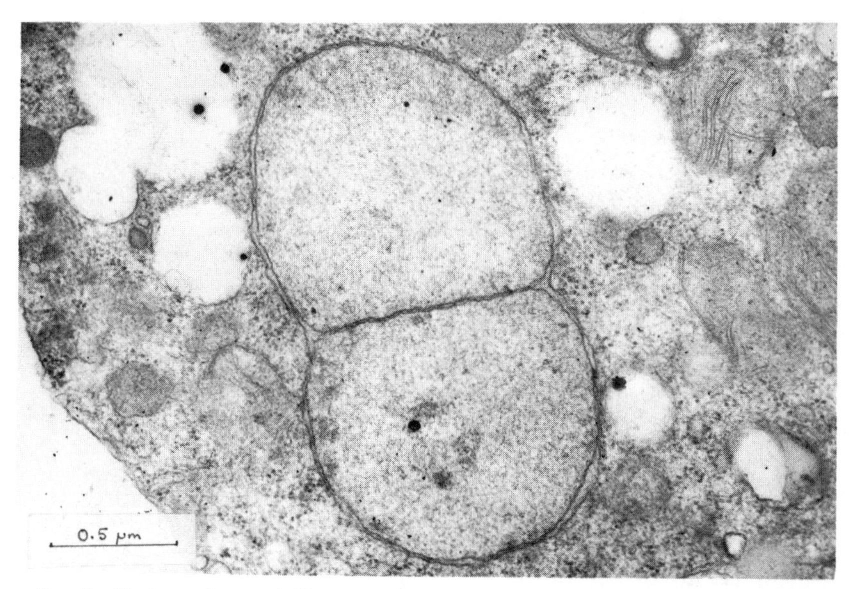

FIG. 1. *Electron micrograph illustrating putative nuclear fusion after protoplast fusion. After fusion, a small (<1%) proportion of the protoplasts contained nuclei apparently fusing, whereas no nuclear fusion was observed in unfused protoplast preparations.*

TABLE 1. *Detection[a] of prototrophs and recombinant auxotrophs after protoplast fusion*

Type of cross	Heterozygotes	Stable prototrophs	Recombinant auxotrophs[b]
Sister strains			
1. *arg-1 leu-1 red-1/arg-2 met-1* (C/C)	+	+	+
2. Repeat of 1	+	+	+
3. *arg-1 leu-1 red-1/ane-1 met-2* (C/C)	−	+	+
4. *ane-2 nic-1 red-3/met-3* (A/A)	−	+	+
Divergent strains			
1. *ane-2 nic-1 red-3/ane-3 arg-3* (A/B)	−	+	+
2. Repeat of 1	+	+	+
3. *ane-2 nic-1 red-3/arg-3* (A/B)	−	+	+
4. *ane-2 nic-1 red-3/thi-1* (A/B)	−	+	+

[a] +, Detected; −, not detected. The frequency of prototrophs was between 10^{-4} and 10^{-6} protoplasts in the fusion mixture.

[b] Recombinant auxotrophs were detected on CM after platings from either heterozygotes or heterokaryons.

Genetic analysis. The detection of recombinant auxotrophs and stable prototrophs from crosses in which heterozygotes were not recovered suggested that the diploid condition in *C. acremonium* is transient. Therefore, we plated out fused protoplasts from the sister cross *arg-1 leu-1 red-1/arg-2 met-1* onto a range of selective media on which the parent strains could not grow. Random samples of colonies growing on such media were then purified on CM and classified. In many cases the original colonies were confirmed as pure auxotrophs. The total classification results are shown in Table 2. The main features to be noted are the recovery of all the parental alleles and most of the possible recombinant phenotypes, including the double auxotrophic recombinant phenotype Leu Met.

TABLE 2. *Classification*[a] *of purified colonies from selective regeneration medium following the protoplast fusion cross* arg-1 leu-1 red-1/arg-2 met-1

Strain phenotypes	Selective regeneration medium					
	MM	MM + arginine	MM + methionine	MM + leucine	MM + methionine + leucine	Total
Red Met^+ Leu Arg	1	1	1	8	7	18
Red^+ Met Leu^+ Arg		11	8		13	32
Red Met Leu Arg^+					2	2
Red^+ Met Leu Arg^+					3	3
Red Met Leu^+ Arg					2	2
Red^+ Met^+ Leu Arg				1	4	5
Red Met^+ Leu^+ Arg	3	4	2	1		10
Red^+ Met^+ Leu^+ Arg	1	7	13	2	3	26
Red Met^+ Leu Arg^+				5	3	8
Red^+ Met^+ Leu Arg^+				3	11	14
Red Met Leu^+ Arg^+		1			1	2
Red^+ Met Leu^+ Arg^+		1	1	1	4	7
Red Met^+ Leu^+ Arg^+	3	1	1	1	2	8
Red^+ Met^+ Leu^+ Arg^+	18	6	12	8	15	59
Red Met Leu Arg						
Red^+ Met Leu Arg				1	1	2
Ratios						
Arg^+ to Arg	21:5	9:23	14:24	18:13	41:30	
Met^+ to Met	26:0	19:13	19:9	29:2	45:26	
Leu^+ to Leu	25:1	31:1	37:1	13:18	40:31	
Red^+ to Red	19:7	25:7	34:4	16:15	54:17	

[a] Colonies from each regeneration medium were purified on CM and then classified. Phenotypes involving Arg appear in all columns as a result of breakdown of heterozygotes (see Table 3). Other anomalous phenotypes in various columns arose as a result of unavoidable transfer of mixed colonies (e.g., heterokaryons) from the regeneration media to CM purification medium.

TABLE 3. *Heterozygotes*[a] *recovered on selective regeneration medium*[b] *from protoplast fusion cross* arg-1 leu-1 red-1/arg-2 met-1

Phenotypes		Possible aneuploid genotype of center			
Center	Sector	I[c]	II	III	IV
Red^+ Met^+ Leu^+ Arg^+	Red^+ Met^+ Leu^+ Arg	+	+	+	*arg-2* +
Red Met^+ Leu^+ Arg^+	Red Met^+ Leu^+ Arg	+	+	*red-1*	*arg-2* +
Red^+ Met Leu^+ Arg^+	Red^+ Met Leu^+ Arg	+	*met-1*	+	*arg-2* +
Red Met Leu^+ Arg^+	Red Met Leu^+ Arg	+	*met-1*	*red-1*	*arg-2* +
Red^+ Met^+ Leu Arg^+	Red^+ Met^+ Leu Arg	*leu-1*	+	+	*arg-2* +
Red Met^+ Leu Arg^+	Red Met^+ Leu Arg	*leu-1*	+	red-1	arg-2 +

[a] The Arg sectors were all identified as nonleaky like *arg-2*. The absence of Arg^+ sectors can be explained by invoking a translocation breakpoint involved in the determination of *arg-2*.

[b] See Table 2.

[c] Linkage groups.

The bottom of Table 2 shows the ratios of the various marker phenotypes. Column 1 shows a bias in favor of Arg^+ as prototrophs were favored by MM. In column 2 the 3:1 ratio of Arg to Arg^+ is as expected if the alleles *arg-1* and *arg-2* are freely recombining. In other columns the ratio of Arg to Arg^+ is not as biased in favor of Arg^+ as in column 1, probably because *arg-2* heterozygotes with extra requirements were allowed to grow on various selective media. Such heterozygotes break down on CM to give Arg phenotypes as shown in Table 3.

In column 4, the ratio of Leu^+ to Leu is as expected, and there is no selection for or against the allele *leu-1*. However, the ratios in other columns suggest that the *met-1* and *red-1* alleles have been selected against, possibly because strains carrying the *met-1* marker are slow-growing on MM supplemented only with methionine, and in the case of *red-1* because the allele might be linked to *leu-1*. Allowance can be made for this selection when estimating possible linkage between certain markers recovered from each regeneration medium. When this is done, the total data are compatible

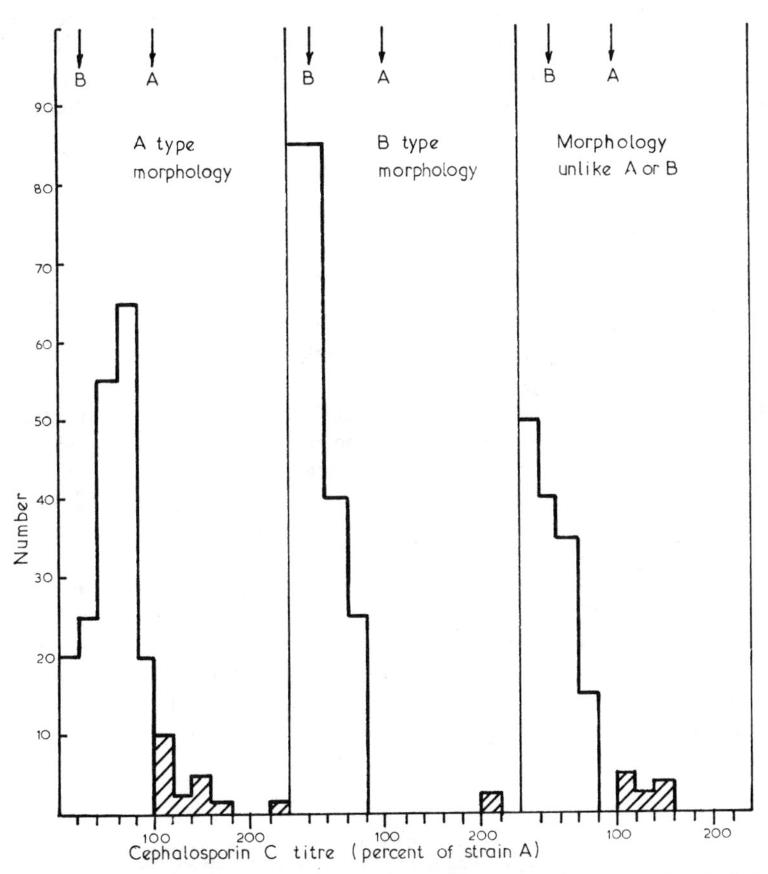

FIG. 2. *Cephalosporin C production by recombinants from protoplast fusion of marked derivatives of strains A and B. Cross* ane-2 nic-1 red-3/ane-3 arg-3 *(A/B) (Table 1). The marking of strain A caused a 30% reduction in titer. The recombinants were prototrophs or carried the* nic-1 *marker. Arrows indicate titers of prototrophic parents A and B. Shaded areas indicate improved titers.*

with free recombination between most of the alleles. This would be expected if segregation involves a high frequency of interchromosomal recombination (haploidization) and/or intrachromosomal recombination (mitotic crossing-over).

Further analysis might be aided by plating fused protoplasts onto CM before plating onto selective media since purification of colonies from these media would then be unnecessary.

Improvement of cephalosporin C titer and other characteristics by means of hybridization by protoplast fusion. The asporulating and slow-growing strain A could synthesize cephalosporin C from sulfate and had three times the cephalosporin C titer of the sporulating strain B, whereas strain B had approximately four times the growth rate of strain A. An extensive breeding program using protoplast fusion was carried out to recombine the merits of both strains (see Table 1, crosses 1 and 2).

Most of the tests for these characteristics were made with colonies arising on CM which had been used in stability testing of prototrophs. From one prototroph a wide range of growth rates and morphological types were seen.

Approximately 600 recombinants were titer-tested (Fig. 2). One isolate consistently showed more than a 40% improvement in cephalosporin C titer compared with the prototrophic parent A and also had a faster growth rate and better sporulation than strain A when cultured on CM. It could also synthesize cephalosporin C from sulfate.

CONCLUSION

In these studies, heterozygotes were rarely recovered, and multiple heterozygosity was detected only in crosses between strains separated by a large number of mutation steps. All heterozygotes were very unstable on CM.

Overall, the results suggest that in *C. acremonium* the parasexual cycle could resemble a meiotic cycle in that diploidization due to nuclear fusion is transient and is often followed by rapid chromosome segregation and possibly efficient intrachromosomal recombination, when one assumes that most of the segregants shown in Table 2 are haploid. Although this process may be typical only of hybridization by protoplast fusion, the generality of the phenomenon is supported by our results from conventional crosses. The theory of a high frequency of parasexual recombination in *C. acremonium* is further supported by work with the related *Emericellopsis* sp. (2) and *C. mycophyllum* (6). With the latter organisms, genetic recombination in supposed diploids occurs at high frequency.

Finally, the results we obtained after protoplast fusion indicate that fundamental genetic analysis of *C. acremonium* may be possible by use of selective media and the rationales of multifactor analysis. Furthermore, after protoplast fusion we have succeeded in breeding a strain of *C. acremonium* showing a 40% improvement in cephalosporin C titer and also improved growth rate and sporulation when compared with its highest-titer parent.

ACKNOWLEDGMENTS

We acknowledge valuable discussion with our colleagues R. T. Rowlands, J. A. Birkett, and M. P. McGonagle, and also J. F. Peberdy (Nottingham University), and technical assistance from M. Woods and A. Gray. We also thank P. J. Mason for carrying out the electron microscopy.

REFERENCES

1. **Anné, J., and J. F. Peberdy.** 1976. Induced fusion of fungal protoplasts following treatment with polyethylene glycol. J. Gen. Microbiol. **92:**413–417.

2. **Fantini, A. A.** 1962. Genetics and antibiotic production in *Emericellopsis* species. Genetics **47:**161–177.
3. **Fawcett, P. A., P. B. Loder, M. J. Duncan, T. J. Beesley, and E. P. Abraham.** 1973. Formation and properties of protoplasts from antibiotic-producing strains of *Penicillium chrysogenum* and *Cephalosporium acremonium.* J. Gen. Microbiol. **79:**293–309.
4. **Käfer, E.** 1963. Radiation effects and mitotic recombination in diploids of *Asperigillus nidulans.* Genetics **48:**27–45.
5. **Nüesch, J., H. J. Treichler, and M. Liersch.** 1973. The biosynthesis of cephalosporin C, p. 309–334. *In* Z. Vaněk, Z. Hošťálek, and J. Cudlín (ed.), Genetics of industrial microorganisms, vol. 2. Elsevier Publishing Co., Amsterdam.
6. **Tuveson, R. W., and D. O. Coy.** 1961. Heterokaryosis and somatic recombination in *Cephalosporium mycophilum.* Mycologia **53:**224–253.

New Approaches to Gene Transfer in Fungi

JOHN F. PEBERDY

Department of Botany, University of Nottingham, Nottingham, England

Genetic manipulations with industrially important fungi aimed at improving performance have invariably involved programs of mutation and selection. This approach has clearly been highly successful, but it is pertinent to question how long it can be continued in view of the finite nature of the genome of the organism being used. The lack of sexuality and consequently of a utilizable breeding system has been one mitigating factor in this approach. However, several studies reported in earlier symposia in this series have demonstrated that the parasexual and sexual cycles can be exploited fairly successfully (4, 6, 9, 20).

In the past decade, several novel ideas for the genetic modification of a variety of organisms have been proposed. In this arena the biotechnology involving cell and protoplast fusion has figured prominently. Fungal protoplasts have two virtues of value in genetic manipulations: (i) the relative ease with which they can be prepared and (ii) their ability to undergo reversion to the normal cellular form. Studies on cell fusion in animal cells which became firmly established in the 1960s (17) provided the impetus for similar studies with plant protoplasts (14) and, more recently, with filamentous fungi (1–3, 8, 10, 12, 18, 22) and yeasts (13, 21, 25–27). This paper reviews part of this published work with *Aspergillus* (18) and *Penicillium* (22) and discusses the significance of protoplast fusion and other manipulations involving protoplasts in applied fungal genetics.

METHODOLOGY FOR PROTOPLAST ISOLATION AND FUSION

The fungal strains used for protoplast fusion experiments are auxotrophic, allowing for selection of fusion products on the basis of nutritional complementation. The methods used are described in detail in the references listed above. For brevity, the experimental procedure is summarized in Fig. 1.

GENETIC ANALYSIS OF FUSION PRODUCTS

After fusion treatment and reversion of the protoplasts, the isolates were investigated by conventional procedures of parasexual analysis (23).

INTRASPECIFIC HYBRIDIZATION BY PROTOPLAST FUSION

Techniques for heterokaryon formation based on hyphal anastomoses have been practiced for many years (5, 23). In some species more elaborate adaptations have proved necessary (19). Once problems in protoplast isolation have been overcome, the fusion techniques provide a simple and rapid procedure for heterokaryon production. Anné and Peberdy (3) showed that the technique was readily applied to several fungal species, including some of industrial importance.

Recently published experiments on the hybridization of *Penicillium chrysogenum* and *P. cyaneo-fulvum* (22) have to be reinterpreted as an example of intraspecific breeding since the publication of a revision of the *P. chrysogenum* series (24).

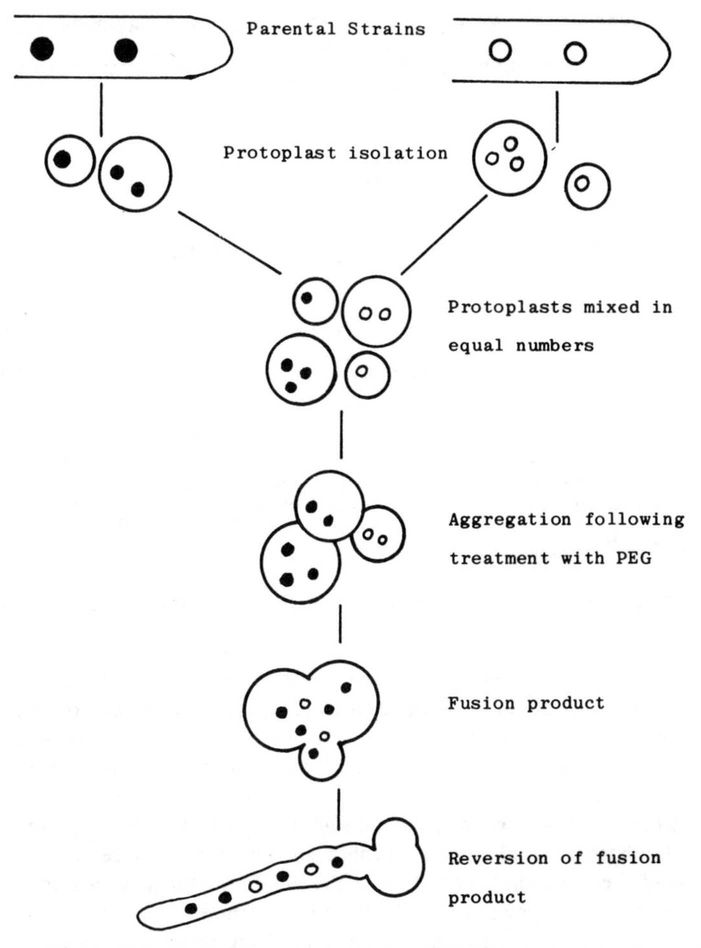

FIG. 1. *Hybridization by protoplast fusion. PEG, Polyethylene glycol.*

However, attempts to produce heterokaryons from the two strains by "conventional" techniques (19) have to date proved unsuccessful. This may mean that they belong to distinct incompatibility groups or that classification based on limited morphological criteria is too limited.

The heterokaryons obtained as fusion products from *P. chrysogenum* and *P. cyaneo-fulvum* were unstable when cultured on complete medium (CM), but yielded more stable, vigorously growing colonies at a relative high (>60%) frequency on minimal medium (B. Smith and J. F. Peberdy, unpublished data). The behavior of these isolates suggested that their development originated as a result of nuclear fusion, and they may therefore be regarded as diploids. The colonies from some of the crosses were very stable on CM, but supplementation with *p*-fluorophenylalanine or benomyl resulted in the development of sectors characterized by different spore colors. Many of these were prototrophic and gave rise to second-order segregants (Table 1).

TABLE 1. *First- and second-order segregants isolated from the cross* P. chrysogenum y, leu, paba × P. cyaneo-fulvum lys[a]

Segregants	Nutritional status	Spore color	Frequency (%)[b]
First-order			
Colony type 1	Prototrophic	Green	37
Colony type 2	Prototrophic	Yellow	15.6
Colony type 3	Prototrophic	White	3
Colony type 4	*lys*	Yellow	12.5
Colony type 5	*leu*	Yellow	12.5
Colony type 6	*leu, paba*	Yellow	19.4
Second-order			
Colony type 1[c]	*lys*	Green	6.2
Colony type 2	*lys*	Yellow	9.3
Colony type 3	Prototrophic	Yellow	6.2
Colony type 3	*lys*	Yellow	31.2

[a] First-order segregants were obtained after *p*-fluorophenylalanine treatment, and second-order segregants were obtained after growth on benomyl (22).

[b] For the second-order segregants 32 subcultures of each prototrophic isolate were grown on complete medium plus benomyl. Frequency values represent the score of sectors of the various types.

[c] Origin (first-order type).

INTERSPECIFIC HYBRIDIZATION BY PROTOPLAST FUSION

Successful recombination between two closely related species, *Asperigillus nidulans* and *A. rugulosus*, was obtained after protoplast fusion (18). As described above, a proportion of the heterokaryons isolated after fusion gave rise to sectors characterized by a more normal colony appearance and vigorous growth. These sectors were described as hybrids and could also be obtained by the reversion of protoplasts obtained from heterokaryons.

Hybrid colonies cultured on CM were stable, at least for 5 to 7 days, producing a few conidiophores with abnormal morphology. The conidia were all hybrid. After more extended cultivation (10 to 14 days), the stability of the hybrids started to break down, with the development of small clusters of normal conidiophores. Analysis of spores produced by hybrids at this stage showed the presence of hybrid, parental, and recombinant progeny (Table 2). Segregation could be induced much earlier if the hybrids were grown in the presence of benomyl, again yielding both parental and recombinant types. Many of the segregants showed only the parental nutritional markers, but the colony morphologies were greatly varied, suggesting an interaction between the greater unknown components of the two genomes. Evidence indicating that the hybrids arose as a result of nuclear fusion and were interspecific diploids was obtained from estimations of DNA content of protoplast nuclei (Table 3). The nuclei of hybrid protoplasts contained twice the amount of DNA found in parental protoplasts.

Further experiments using the more extensively marked *A. nidulans* master strains have been carried out (F. Kevei et al., unpublished data). Analysis of recombinant progeny, which were obtained at a much higher frequency, showed that there was no apparent selectivity for any particular *A. nidulans* chromosomes.

UPTAKE OF DNA BY PROTOPLASTS

The absence of the cell wall makes protoplasts potentially interesting structures for genetic manipulations involving the incorporation of foreign DNA. Several reports

TABLE 2. *Frequency of colony types among single-spore isolates of the hybrid from* A. nidulans paba lys y × A. rugulosus met (18)

Colony type	Spores derived from hybrid grown on:	
	MM[a]	CM
Hybrid	87.8 ± 7.8	36.2 ± 2.5
A. nidulans	0	12.8 ± 4.6
A. rugulosus	0	2.7 ± 0.7
Other segregants and recombinants	12.2 ± 1.8	48.1 ± 6.3

[a] MM, Minimal medium; CM, complete medium.

TABLE 3. *DNA content of nuclei (18)*

Strain	Nuclei per protoplast	DNA content per nucleus (μg)	Ratio of DNA content, hybrid/parent
A. nidulans	1.50 ± 0.2	4.7×10^{-6}	2.29
A. rugulosus	2.16 ± 0.3	5.43×10^{-6}	1.98
Hybrid	2.79 ± 0.3	10.78×10^{-6}	

on the uptake of foreign DNA by plant protoplasts have been published. However, the question regarding the degradation of DNA taken up and de novo synthesis of DNA in the protoplasts using the fragments still remains to be answered (7). The potential for protoplasts, both plant and fungal, in this area is very apparent.

Fungal protoplasts are capable of uptake of larger bodies, as shown in the case of chloroplasts (28) and bacteria (15). Whether such protoplasts can be subjected to reversion and the sustained activity of the incorporated organelles or bacteria requires further investigation.

Mitochondrial and plasmid DNAs are potentially useful vectors in the transfer of genetic material by use of protoplast technology. Ferenczy and Maraz (11) showed that mitochondrial transfer by protoplast fusion can be achieved in yeast. The isolation and genetic characterization of the 2-μm plasmid in yeast (16) also adds a new dimension. Extrapolation of studies on higher plants suggests that uptake of the plasmid by protoplasts should be attainable, and protoplast fusion provides a possible vehicle for the transfer of this DNA vector.

CONCLUSIONS

Protoplasts used in fusion and hybridization have rapidly become acceptable tools for the fungal geneticist. It has been suggested that these methods will probably be very useful in conventional genetic studies, and an undoubted value lies in the potential for interspecific hybridization. The extent to which recombination can be attained between fungal species is an unknown quantity, but the fusion technique at least provides the opportunity to overcome any preconjugation incompatibility barriers. The applications for interspecific genetics and breeding are potentially very wide. On the one hand, the technique provides the opportunity for a broadening of the gene pool resulting from general mixing and segregation of the genetic material. Alternatively, attempts at more controlled hybridizations are equally possible with a view to modification of existing products, synthesis of new products, or simply changes in the physiology of the producer organism which are ultimately reflected in improved product yields or utilization of new substrates.

The application of protoplast technology related to gene vectors is more long-term and speculative. Clearly, the possibility of cloning the genes required for the biosynthesis of a particular metabolite onto a plasmid which could then be transferred by protoplast fusion provides an exciting prospect for the future.

REFERENCES

1. **Anné, J., H. Eyssen, and P. de Somer.** 1976. Somatic hybridization of *Penicillium roqueforti* with *P. chrysogenum* after protoplast fusion. Nature (London) **262:**719–721.
2. **Anné, J., and J. F. Peberdy.** 1975. Conditions for induced fusion of fungal protoplasts in polyethylene glycol solutions. Arch. Microbiol. **105:**201–205.
3. **Anné, J., and J. F. Peberdy.** 1976. Induced fusion of fungal protoplasts following treatment with polyethylene glycol. J. Gen. Microbiol. **92:**413–417.
4. **Ball, C., and J. L. Azevedo.** 1976. Genetic instability in parasexual fungi, p. 243–252. *In* K. D. Macdonald (ed.), Second international symposium on the genetics of industrial microorganisms. Academic Press, London.
5. **Barron, G. L., and B. H. MacNeill.** 1962. A simplified procedure for demonstrating the parasexual cycle in Aspergillus. Can. J. Bot. **40:**1321–1327.
6. **Calam, C. T., L. B. Daglish, and E. P. McCann.** 1976. Penicillin: tactics in strain improvement, p. 273–290. *In* K. D. Macdonald (ed.), Second international symposium on the genetics of industrial microorganisms. Academic Press, London.
7. **Cocking, E. C.** 1977. Uptake of foreign genetic material by plant protoplasts. Int. Rev. Cytol. **48:**323–343.
8. **Dales, R. B. G., and J. H. Croft.** 1977. Protoplast fusion and the isolation of heterokaryons and diploids from vegetatively incompatible strains of *Aspergillus nidulans.* FEMS Microbiol. Lett. **1:**201–204.
9. **Elander, R. P., M. A. Espenshade, G. G. Pathak, and C. H. Pan.** 1973. The use of parasexual genetics in an industrial strain improvement programme with *Penicillium chrysogenum,* p. 239–253. *In* Z. Vaněk, Z. Hošťálek and J. Cudlín (ed.), Genetics of industrial microorganisms, vol. 2. Elsevier Publishing Co., Amsterdam.
10. **Ferenczy, L., F. Kevei, and M. Szegedi.** 1975. High frequency fusion of fungal protoplasts. Experientia **31:**1028–1030.
11. **Ferenczy, L., and A. Maraz.** 1977. Transfer of mitochondria by protoplast fusion in *Saccharomyces cerevisiae.* Nature (London) **268:**524–525.
12. **Ferenczy, L., M. Szegedi, and F. Kevei.** 1977. Interspecific protoplast fusion and complementation in aspergilli. Experientia **33:**184.
13. **Fournier, P., A. Provost, C. Bourguignon, and H. Heslot.** 1977. Recombination after protoplast fusion in the yeast *Candida tropicalis.* Arch. Microbiol. **115:**143–149.
14. **Galun, E., D. Aviv, D. Raveh, A. Vardi, and A. Zecler.** 1977. Protoplasts in studies of cell genetics and morphogenesis, p. 302–312. *In* W. Barz, E. Reinhard, and M. H. Zenk (ed.), Plant tissue culture and its biotechnological application. Springer-Verlag, Berlin.
15. **Giles, K. L., and H. Whitehead.** 1974. Uptake and continued metabolic activity and *Azotobacter* within fungal protoplasts. Science **193:**1125–1126.
16. **Guerineau, M., C. Granchamp, and P. Slominski.** 1976. Structure and genetics of the 2 μm circular DNA in yeast, p. 557–564. *In* T. Bucher (ed.), Genetics and biogenesis of chloroplasts and mitochondria. Elsevier/North Holland Biomedical Press, Amsterdam.
17. **Handmaker, S. D.** 1973. Hybridization of eukaryotic cells. Annu. Rev. Microbiol. **27:**189–204.
18. **Kevei, F., and J. F. Peberdy.** 1977. Interspecific hybridization between *Aspergillus nidulans* and *A. rugulosus* by fusion of somatic protoplasts. J. Gen. Microbiol. **102:**255–262.
19. **Macdonald, K. D., J. M. Hutchinson, and W. A. Gillett.** 1963. Heterokaryon studies and the genetic control of penicillin and chrysogenin production in *Penicillium chrysogenum.* J. Gen. Microbiol. **33:** 375–383.
20. **Merrick, M. J.** 1976. Hybridisation and selection for penicillin production in *Aspergillus nidulans*—a biometrical approach to strain improvement, p. 229–242. *In* K. D. Macdonald (ed.), Second international symposium on the genetics of industrial microorganisms. Academic Press, London.
21. **Morgan, A. J., J. Heritage, and P. A. Whittaker.** 1977. Protoplast fusion between petite and auxotrophic mutants of the petite-negative yeast, *Kluyveromyces lactis.* Microbios Lett. **4:**103–107.
22. **Peberdy, J. F., H. Eyssen, and J. Anné.** 1977. Interspecific hybridization between *Penicillium chrysogenum* and *P. cyaneo-fulvum* following protoplast fusion. Mol. Gen. Genet. **157:**281–284.
23. **Pontecorvo, G., J. A. Roper, D. W. Hemmons, K. D. Macdonald, and A. W. Bufton.** 1953. The genetics of *Aspergillus nidulans.* Adv. Genet. **5:**141–238.
24. **Samson, R. A., R. Hadlok, and A. C. Stolk.** 1977. A taxonomic study of the *Penicillium chrysogenum* series. Antonie van Leeuwenhoek J. Microbiol. Serol. **43:**169–175.
25. **Sipiczki, M., and L. Ferenczy.** 1977. Fusion of *Rhodosporidium (Rhodotorula)* protoplasts. FEMS Microbiol. Lett. **2:**203–205.
26. **Sipieczki, M., and L. Ferenczy.** 1977. Protoplast fusion of *Schizosaccharomyces pombe* auxotrophic mutants of identical mating type. Mol. Gen. Genet. **151:**77–81.
27. **van Solingen, P., and J. B. van der Plaat.** 1977. Fusion of yeast spheroplasts. J. Bacteriol. **130:**946–947.
28. **Vasil, I. K., and G. L. Giles.** 1975. Induced transfer of higher plant chloroplasts into fungal protoplasts. Science **190:**680.

Novel Methods of Genetic Analysis in Fungi

A. UPSHALL, B. GIDDINGS, S. C. TEOW, AND I. D. MORTIMORE

Department of Biological Sciences, University of Lancaster, Lancaster LA1 4YQ, United Kingdom

Mutation followed by selection has been successful in the commercial exploitation of microorganisms. However, as Hopwood (12) pointed out, success has to be balanced against the effort expended, since there is no comparison as to what could have been achieved by recombination. Reports on the recombination approach to yield improvement (e.g., 3, 6, 10) have recorded a limited success, primarily because the recombination approach has been used late in the industrial history of a particular fungus. Two observations about the mutation-selection approach can be made. (i) Mutagenic treatments damage genome structure. Interstrain chromosome homology is thus limited, with the consequent lack of recombination in heterozygous diploids (e.g., 16). Ball (4) overcame some problems by using closely related sister strains, especially when the diploids were treated with "recombinogens" (Morrison and Ball, cited in 5). (ii) A structured breeding program requires a knowledge of the basic genetic architecture i.e., ploidy and chromosome number, and a collection of mutant strains for the selective analytical procedures. Routinely, mapping follows the induction of gene mutations by a convenient mutagen, but these coincidentally induce chromosome aberrations which, if undetected, cause complications in subsequent crosses.

We will discuss our study of the feasibility of satisfying the basic requirements using only spontaneous mutant strains, under the assumption that they will be unlikely to carry multiple aberrations. *Aspergillus nidulans* was used as a model since considerable genetic information exists on this fungus (7) and a meiotic method for detecting translocations is available (17). *A. terreus* was also used as an example of an industrial fungus for which no genetic information has been available. For parasexual analysis, selective markers are generally few and are inappropriately located on the chromosome. In *A. nidulans,* mitotic linkage group V carries no selective marker and is not amenable to conventional parasexual analysis. In this paper we expand our previous observations upon recombination in disomic isolates which do not require selective markers and indicate the potential value of this alternative parasexual recombination system. The final part of this report discusses problems associated with the disomic analysis and outlines approaches aimed at overcoming the more serious problems.

SPONTANEOUS MUTANTS IN *A. NIDULANS* AND *A. TERREUS*

Figure 1 indicates the arrangement of the gene loci identified in *A. nidulans*, and Table 1 lists the details of the mutant types recovered for *A. terreus*. Although all mutations serve as genetic markers, in both species specific types were required. For diploid recognition, spore color mutants were isolated from within the lawn generated from a dense plating of spores. To enable the selection of a heterokaryon and diploid, clean auxotrophs were isolated via the filtration enrichment technique (1), occasionally modified by a final selection with nystatin (9). Thirteen experiments with *A. nidulans* and eight with *A. terreus*, each screening approximately 2×10^9 spores, have

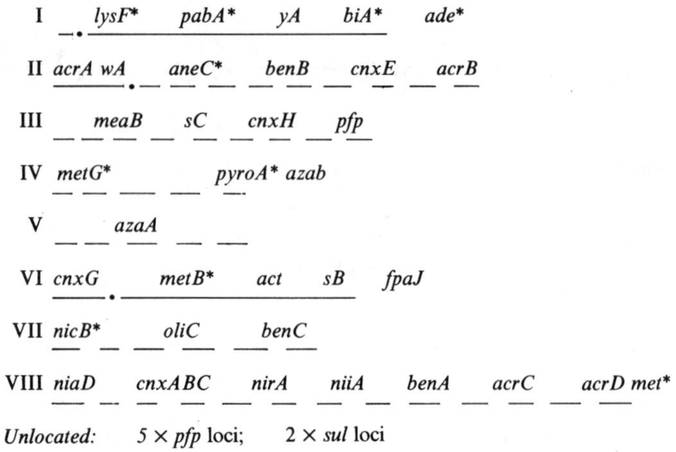

FIG. 1. *Linkage relationships (and orientation where appropriate) of spontaneously isolated mutations in* A. nidulans. *Solid lines: orientation determined. Broken lines: orientation undetermined. Asterisks indicate mutations obtained after filtration enrichment. For details of abbreviation of gene loci, see reference 11.*

TABLE 1. *Mutations spontaneously isolated within* A. terreus, *their linkage group, and their characteristics*[a]

Designation	Symbol	Locus	Linkage group	Comment
Morphologicals				
Buff conidiospores	*buf*	A	VI	
White conidiospores	*whi*	A	VII	Epistatic to buff
Compact colony	*com*	A	—	
Drug resistance				
Acriflavine	*acr*	A	I	Semidominant
Azaguanine	*aza*	A	—	
Benlate	*ben*	A	—	
p-Fluorophenylalanine	*fpa*	A	—	
Sulfanilamide	*sul*	A	—	
		B	—	
Auxotrophs				
Adenine	*ade*	A	I	All obtained via nystatin or filtration enrichment
Biotin	*bio*	A	III	
Methionine	*met*	A	II	
		B	—	
Nitrogen	*nia*	D	IV	Obtained as resisters to potassium chlorate
	cnx	A	V	
		B	—	
		C	—	
Sulfur	*s*	B	—	Obtained as resisters to sodium selenate
		C	—	

[a] For details of abbreviation of gene loci, see reference 11.

yielded the auxotrophs detailed. The overall recovery was low, but predictable, and the majority were of the type required. The spore color mutants allow the detection of the products of somatic recombination. Strains which are resistant to various toxins are also convenient for analyzing the products of recombination. Particularly useful are those resistant to potassium chlorate and sodium selenate since these are simultaneously auxotrophic for nitrogen (8) and sulfur (2), respectively. Under some circumstances, e.g., where haploid parents grow slowly (leaky), complementing auxotrophic mutants will select faster-growing heterozygous diploids from heterokaryons (C. Ball, personal communication). Crossover and non-disjunctional progeny can be recovered as vigorous sectors emerging from the poor growth of a heterozygous diploid on toxic medium (14; B. Giddings, unpublished data; A. Upshall and N. McKenzie, unpublished data). The resistance to acriflavine (*A* locus), azaguanine, *p*-fluorophenylalanine, cycloheximide, and oligomycin are all similarly selectable. In *A. nidulans*, loci were identified by allelism tests and were assigned to mitotic linkage groups by conventional parasexual and disomic (15) analyses.

From over 500 mutant strains of *A. nidulans*, 42 gene loci were identified; 30 of these were located to a mitotic linkage group, labeling all eight chromosomes in the process. Four of the gene loci were previously unknown, and the position of the centromere of linkage group VI has been determined. No translocation has been detected. In *A. terreus*, we have identified at least seven chromosomes and have undertaken parasexual analyses.

RECOMBINATION IN MITOTIC LINKAGE GROUP I AND V DISOMICS OF *A. NIDULANS*

Disomics of *A. nidulans* have well-known properties. Each extra chromosome state has a specific phenotype (15), and all are somatically unstable following loss of the extra chromosome. A heterozygous disomic will therefore show a genetic segregation among the haploid sectors (13). In our experiments, from each daughter disomic derived by plating spores from a single purified disomic of known genotype, a single haploid sector was sampled and its genotype was determined.

Linkage group I. The results of the analyses are summarized in Fig. 2. The two control disomics differed in the number of heterozygous gene loci. The difference in the frequency of recombinant sectors appears anomalous but is a consequence of the *trpB* mutation. This mutation in a haploid confers a requirement for a high level of tryptophan, and it became apparent that our level of supplementation was insufficient. Thus, "control 2" disomic was examined in a regimen of partial selection against the *trpB* mutation during sector production. It remains a valid control for the experiment designed to specifically select for the products of recombination since these media were supplemented with the same tryptophan concentration. Two features stand out. (i) There are two "high frequency of recombination" regions, namely, the *proA-lysA* centromere interval and the *camC-trpB* (*galD*) interval. The latter shows a greater frequency. (ii) There is a high incidence of multiply recombined sectors, with 82% of "control 1" and all of "control 2" sectors in this category. In all cases, either or both of the two "high-frequency" regions were involved.

The value of this recombination system as a routine procedure is limited by the low frequency of recombinant sectors and the low incidence of crossing-over in some intervals. However, "control disomic 2" indicates that environmental selective pressures can increase the former. Disomics with the genotype

camC	*trpB*	+	+	*adeG*	*proA*	:	+	+	+	+
+	+	*galD*	*ribA*	+	+	:	*lysF*	*hisB*	*pabA*	*yA*

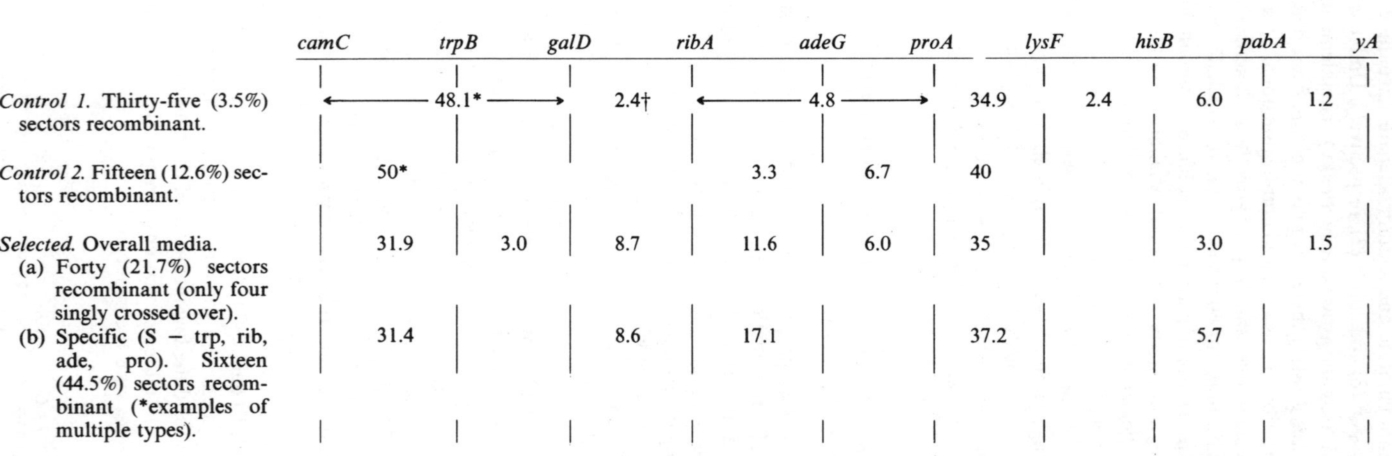

	camC–trpB	trpB–galD	galD–ribA	ribA–adeG	adeG–proA	proA–lysF	lysF–hisB	hisB–pabA	pabA–yA
Control 1. Thirty-five (3.5%) sectors recombinant.	← 48.1* → (camC–galD)		2.4†	← 4.8 → (ribA–proA)		34.9	2.4	6.0	1.2
Control 2. Fifteen (12.6%) sectors recombinant.	50*			3.3	6.7	40			
Selected. Overall media. (a) Forty (21.7%) sectors recombinant (only four singly crossed over).	31.9	3.0	8.7	11.6	6.0	35		3.0	1.5
(b) Specific (S − trp, rib, ade, pro). Sixteen (44.5%) sectors recombinant (*examples of multiple types).	31.4		8.6	17.1		37.2		5.7	

Control 1: * 2.5% doubly recombinant with *lysF-hisB*; 67.5% doubly recombinant with *proA-lysF*; 10% doubly recombinant with *ribA-proA*
† all doubly recombinant with *proA-lysF*

Control 2: * all doubly recombinant in one or another of the recombining regions

Selected: *(a) *camC-trpB, ribA-adeG, proA-lysF*
(b) *camC-trpB, ribA-adeG, proA-lysF, hisB-pabA*

FIG. 2. *Summary of recombination in linkage group I disomics (frequency per region expressed as a percentage of total recombination events). For details of abbreviation of gene loci, see reference 11.*

were plated on (in addition to the general tryptophan deficiency) a variety of media differentially supplemented by the omission of one or more of the growth factors (e.g., S − trp, rib, ade, pro). The results show that, on the average, selection increases the frequency of recombinant sectors from 12.6% to 21.7%. The highest frequency recorded was 44.5%, and this occurred when there was simultaneous selection against four mutant alleles. A 100% recovery of recombinant sectors was not achieved, as many proved to be heterokaryons between disomic and nonrecombinant haploids. The two "high-frequency" regions still predominate among the recombinant haploids, but significant increases in recovery were observed for crossovers in the other regions, notably, the *ribA-adeG* interval. Only four of the total recombinant haploids were the consequence of a single crossover event. Indeed, triply and quadruply recombined haploids were recovered.

Linkage group V. Figure 3 summarizes the results. In the controls (mean data of five analyses), the frequency of recombinant sectors was 2.7%, with the vast majority of crossovers between the *lysA* and *pA* loci, the centromere region. Subsequently the mutation *mnrA* (where sensitivity at 42°C is relieved by incorporating mannose in the medium [18]), which maps between these two loci, became available. In a small analysis, all crossovers were restricted to between the *lysA* and *mnrA* intervals. Only one sector was doubly recombinant, that in the *camD-nicA* and *lysA-pA* intervals. With a disomic of the genotype

camD	*nicA*	*lysA*	:	+	*pA*	*ribD*
+	+	+		*mnrA*	+	+

attempts were made to increase the frequency of recombinant haploids and to modify the position of crossovers by growth on differentially supplemented media, coupled with incubation at 42°C (without mannose). Although the frequency of recombinant sectors increased to 18.2%, no significant change in the distribution of crossover events was observed.

In many organisms, chromosome aberrations are known to interfere with the pairing and recombination of homologs. The frequency of recombination was therefore examined in "typical" heterozygous disomics (17) obtained from among the progeny of crosses involving T1 (V; VI) (see 7) and a newly induced T1 (V; VII) to a translocation-free strain. In the former, the breakpoint is between the *mnrA* and *pA* loci, and the disomic was heterozygous for *nicA*, *lysA*, *mnrA*, *pA*, and *facA*, with

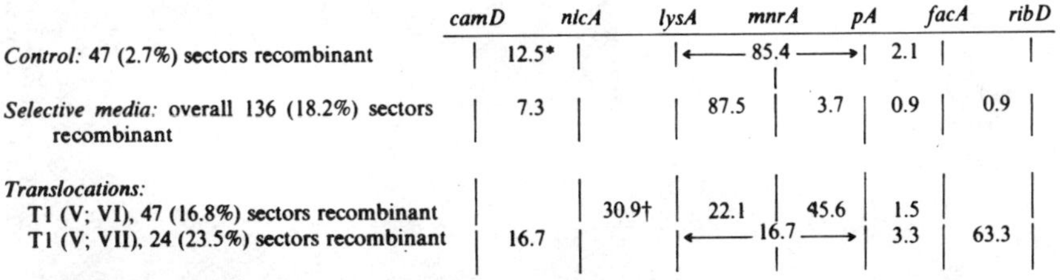

	camD–*nicA*	*nicA*–*lysA*	*lysA*–*mnrA*	*mnrA*–*pA*	*pA*–*facA*	*facA*–*ribD*
Control: 47 (2.7%) sectors recombinant	12.5*		←85.4→ (*lysA*–*pA*)		2.1	
Selective media: overall 136 (18.2%) sectors recombinant	7.3		87.5	3.7	0.9	0.9
Translocations:						
T1 (V; VI), 47 (16.8%) sectors recombinant		30.9†	22.1	45.6	1.5	
T1 (V; VII), 24 (23.5%) sectors recombinant	16.7		←16.7→ (*lysA*–*pA*)		3.3	63.3

* 2.1% doubly recombinant.
† all doubly recombinant in either *lysA-mnrA* or *mnrA-pA* regions.

FIG. 3. *Summary of recombination in linkage group V disomics (frequency per region expressed as a percentage of total crossover events). For details of abbreviation of gene loci, see reference 11.*

the *pA*$^-$ and *facA*$^+$ alleles on the segment attached to the centromere of linkage group VI. In the latter, the breakpoint is not defined. The disomic was heterozygous for the *camD, nicA, lysA, pA, facA,* and *ribD* loci. In both cases the frequency of recombinant sectors was increased when compared to the control. For T1 (V; VI), a significant increase in crossovers was observed in the *nicA-lysA* and *mnrA-pA* intervals, as well as a reduced frequency in the centromere region. Multiple recombination events increased, with all recombinant sectors in the *nicA-lysA* region also being recombinant in the *lysA-mnrA* or *mnrA-pA* intervals. Interestingly, we recovered a sector which was recombined between the two loci on the segment attached to the VI centromere. For the T1 (V; VII) translocation, we believe that the considerable increase in crossovers in the *facA-ribD* region, coupled with the concomitant decrease in the *lysA-pA* interval, indicates that the breakpoint is in the vicinity of the *ribD* locus.

RECOVERY AND STABILIZATION OF DISOMICS

The results in the preceding section indicate the value of disomic analysis as a tool for the generation of strains known to be recombinants in chromosomes previously unavailable for direct analysis. Additionally, haploids can be recovered which are the consequence of multiple recombination events. Both environmental and genetic modifications can enhance exchange in specific chromosome regions. However, this method of analysis could be made more valuable if two of the major problems of disomics were overcome, namely, the recovery of a specific disomic type and their rapid instability. Heterozygous disomics can most simply be recovered by treating conidiospores of heterozygous diploids with a haploidizing agent. The restriction is that, for example in *A. nidulans,* with the exception of linkage group V which can be specifically and routinely isolated from p-fp medium (15), the complete spectrum of hyperhaploid and hyperdiploid aneuploids is generated, from which one has to select the required type. Stabilization, based upon balanced lethals, has been proposed by

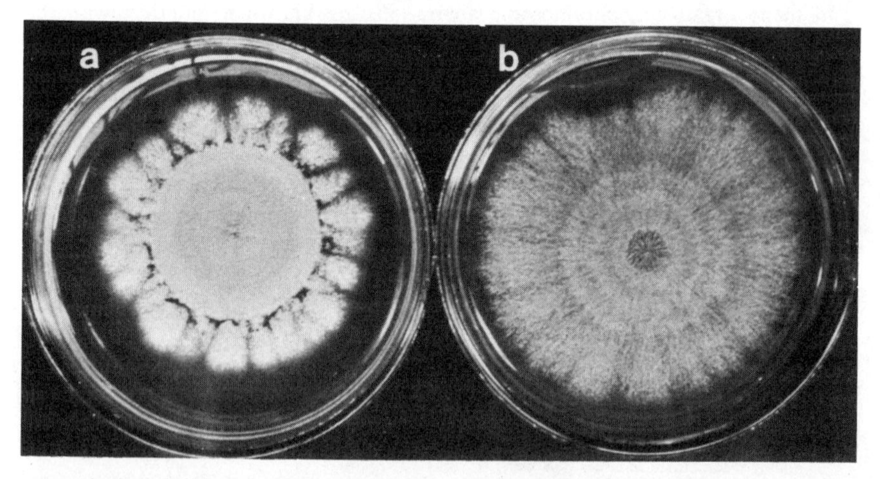

FIG. 4. *Growth behavior of a strain carrying a mutation specifically yielding and stabilizing disomy for mitotic linkage group III of* A. nidulans. *Mutation designated* sod^{III} A1, *(a) at 37°C (6 days) and (b) at 37°C (5 days) followed by 30°C (3 days).*

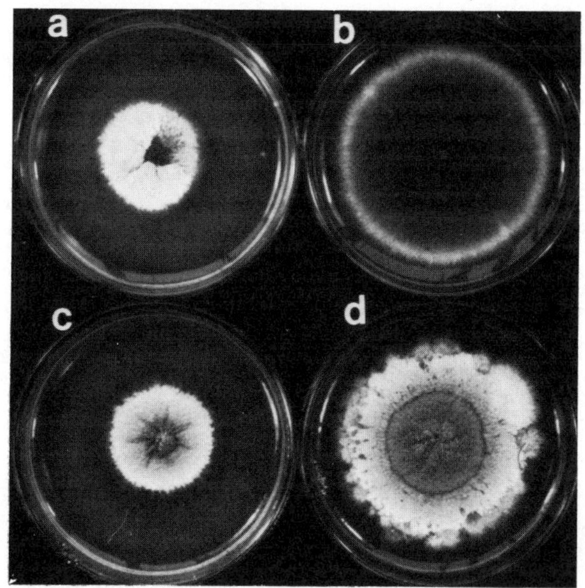

FIG. 5. *Growth behavior of a strain carrying a mutation specifically yielding and stabilizing disomy for mitotic linkage group VI of* A. nidulans. *Mutation designed* sod ^{VI} A1, *previously* tsB *(11) at (a) 37°C (6 days) and (b) 30°C (6 days). (c) Sector off (a) at 37°C (5 days); (d) sector off (a) at 37°C (5 days) followed by 30°C (2 days).*

Ball (4). However, to ensure complete stability, these lethals need to be either allelic or in chromosome regions where recombination is rare.

In a program designed to obtain mutants of *A. nidulans* defective in anaphase chromosome segregation, we have identified four mutants which yield specific aneuploid colonies during vegetative growth. All four fail to grow at 42°C, but on prolonged incubation at 37°C they yield colonies disomic for a specific chromosome. During subculturing at 37°C, these colonies remain in the stable state. A transfer of the stable aneuploid to 30°C allows reversion to haploidy (Fig. 4 and 5). These mutations have been designated *sod* for stabilization of disomics. Three colonies yield and stabilize the n + VI state, and all three gene loci are located on mitotic linkage group VI. The fourth yields and stabilizes the n + III state, the gene locus being situated on mitotic linkage group VIII. Disomy has been proved by observing a genetic segregation in disomics obtained from diploids homozygous for a *sod* mutation but heterozygous for other gene loci of the relevant linkage group.

REFERENCES

1. **Armitt, S., W. McCullough, and C. F. Roberts.** 1976. Analysis of acetate non-utilising (acu) mutants in *Aspergillus nidulans.* J. Gen. Microbiol. **92:**413–417.
2. **Arst, H. N.** 1968. Genetic analysis of the first steps of sulphate metabolism in *Aspergillus nidulans.* Nature (London) **219:**268–270.
3. **Ball, C.** 1973. Improvement of penicillin productivity in *Penicillium chrysogenum* by recombination, p. 227–237. *In* Z. Vaněk, Z. Hošťálek, and J. Cudlin (ed.), Genetics of industrial microorganisms. Academia, Prague.
4. **Ball, C.** 1973. The genetics of *Penicillium chrysogenum*. Prog. Ind. Microbiol. **12:**47–72.
5. **Ball, C., and J. L. Azevedo.** 1976. Genetic instability in parasexual fungi, p. 243–251. *In* K. D. Macdonald

(ed.), Second international symposium on the genetics of industrial microorganisms. Academic Press, London.

6. **Calam, C. T., L. B. Daglish, and E. P. McCann.** 1976. Penicillin: tactics in strain improvement, p. 273–287. *In* K. D. Macdonald (ed.), Second international symposium on the genetics of industrial microorganisms. Academic Press, London.
7. **Clutterbuck, A. J.** 1974. *Aspergillus nidulans,* p. 447–510. *In* R. C. King (ed.), Handbook of genetics, vol. 1. Plenum Press, New York.
8. **Cove, D. J.** 1976. Chlorate toxicity in *Aspergillus nidulans*: the selection and characterisation of chlorate resistant mutants. Heredity **36:**191–203.
9. **Ditchburn, P., and K. D. McDonald.** 1971. The differential effect of nystatin on growth of auxotrophic and prototrophic strains of *Aspergillus nidulans.* J. Gen. Microbiol. **67:**229–306.
10. **Elander, R. P., M. A. Espenshade, S. G. Pathak, and C. H. Pan.** 1973. The use of parasexual genetics in an industrial strain improvement programme with *Penicillium chrysogenum,* p. 239–254. *In* Z. Vaněk, Z. Hošťálek, and J. Cudlin (ed.), Genetics of industrial microorganisms. Academia, Prague.
11. **Giddings, B., and A. Upshall.** 1975. Spontaneous mapping of *Aspergillus terreus* and *Aspergillus nidulans. Aspergillus* Newsl. **13:**14–17.
12. **Hopwood, D. A.** 1977. Genetic recombination and strain improvement. Dev. Ind. Microbiol. **18:**9–21.
13. **Käfer, E.** 1961. The process of spontaneous recombination in vegetative nuclei of *Aspergillus nidulans.* Genetics **46:**1581–1609.
14. **Käfer, E.** 1977. Meiotic and mitotic recombination in Aspergillus and its chromosomal aberrations. Adv. Genet. **19:**33–124.
15. **Käfer, E., and A. Upshall.** 1973. The phenotypes of the eight disomics and trisomics of *Aspergillus nidulans.* J. Hered. **64:**35–38.
16. **Macdonald, K. D.** 1968. The persistence of parental genome segregation in *Penicillium chrysogenum* after nitrogen mustard treatment. Mutat. Res. **5:**302–305.
17. **Upshall, A., and E. Käfer.** 1974. Detection and identification of translocations by increased specific non-disjunction in *Aspergillus nidulans.* Genetics **76:**19–31.
18. **Valentine, B. P., and B. W. Bainbridge.** 1975. Properties and chromosomal locations of two mannose mutants. *Aspergillus* Newsl. **12:**31.

Industrial Microorganisms Tailor-Made by Removal of Regulatory Mechanisms

JUAN F. MARTÍN, JOSÉ A. GIL, GERMÁN NAHARRO, PALOMA LIRAS, AND JULIO R. VILLANUEVA

Instituto de Microbiología Bioquímica, Consejo Superior de Investigaciones Científicas, and Departamento de Microbiología, Facultad de Ciencias, Universidad de Salamanca, Salamanca, Spain

REGULATION OF CANDICIDIN SYNTHESIS BY AROMATIC AMINO ACIDS

The biosynthesis of candicidin represents an adequate model for the study of the regulatory mechanisms of antibiotic synthesis. This antibiotic is a polyene macrolide (16) produced by *Streptomyces griseus.* Its structure consists of a macrolide ring which contains a seven-double-bond chromophore. It has an amino sugar, mycosamine, attached by a glycosidic bond to the macrolide ring and a *p*-aminoacetophenone moiety (16). Candicidin is synthesized by a head-to-tail condensation of C_2 and C_3 units, in the forms of malonylcoenzyme A and methylmalonylcoenzyme A, on a *p*-aminobenzoylcoenzyme A primer unit (9). The mycosamine moiety seems to be attached during the secretion process.

The *p*-aminoacetophenone moiety of candicidin is synthesized from chorismic acid via *p*-aminobenzoic acid (PABA) (9) (Fig. 1). Aromatic amino acids inhibiting candicidin synthesis have been described (6). In our experiments, candicidin synthesis was measured either by the increase of antibiotic activity or by the incorporation of labeled precursors ([^{14}C]PABA or [^{14}C]propionate) and was found to be strongly inhibited by tryptophan and anthranilic acid. On the other hand, tyrosine and phenylalanine had low inhibitory effects (Fig. 2). The inhibitory effect of anthranilic acid was not reverted by PABA (J. A. Gil and J. F. Martín, submitted for publication), which excludes the possibility of competitive inhibition of incorporation of PABA into candicidin by anthranilic acid.

Both effectors inhibited the incorporation of either [^{14}C]propionate or [^{14}C]PABA into candicidin, which suggests that the regulatory effect is exerted directly at the candicidin synthetase level or that they interfere with the biosynthesis of the amino sugar moiety of candicidin. An inhibitory effect at the PABA synthetase level (dotted arrow, Fig. 1), which would deprive the cell of intracellular PABA, seems less likely, since the inhibition is not reverted upon addition of exogenous PABA, which is easily taken up by the cells (15; Gil and Martín, submitted for publication).

The addition of either tryptophan or anthranilic acid to candicidin-producing resting cells results in considerable stimulation of in vivo protein synthesis. It appears that tryptophan is limiting for protein synthesis in these cells, probably because the flow of intermediates is largely directed toward candicidin (Fig. 1). The inhibition of candicidin synthesis by anthranilate is not due to a stimulation of protein synthesis, since the regulatory effect is also exerted in the absence of protein synthesis (Gil and Martín, submitted for publication).

The results suggest that *S. griseus* has developed a cross-pathway feedback regulation that prevents candicidin synthesis from taking place when the intracellular tryptophan pool is high, and therefore protein synthesis is still possible. If tryptophan

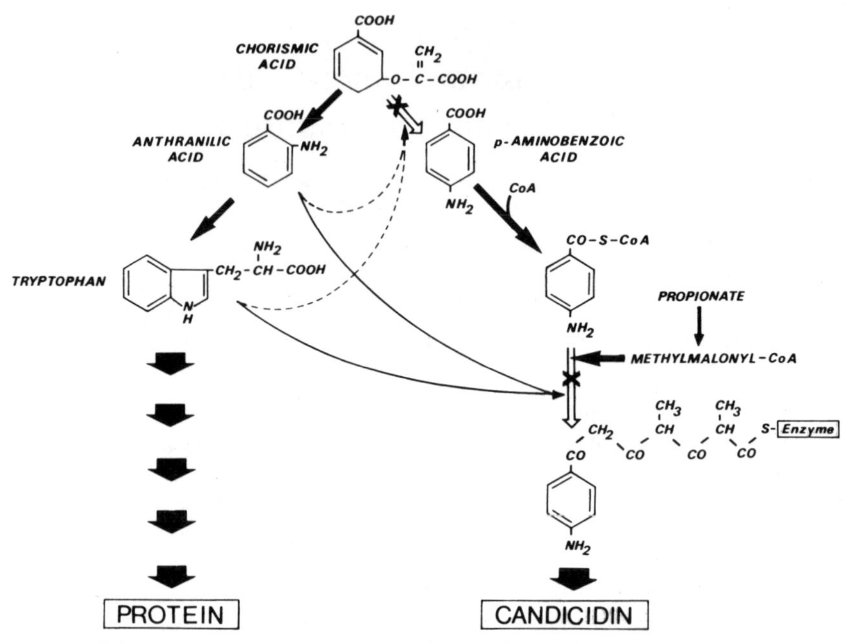

FIG. 1. *Biosynthetic pathway of the aromatic moiety of candicidin and crossed regulation by tryptophan and anthranilic acid. CoA, Coenzyme A.*

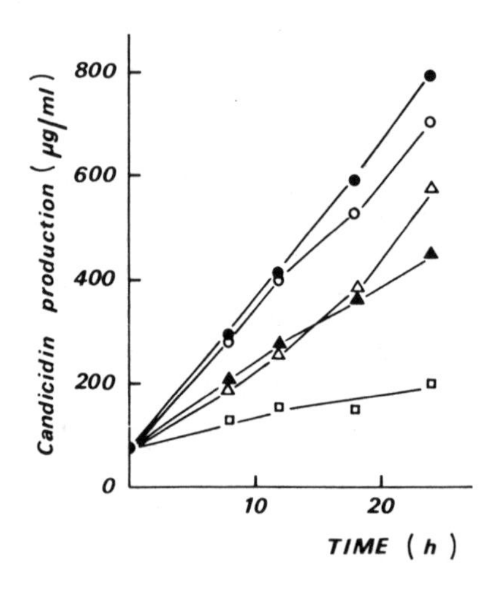

FIG. 2. *Cumulative inhibition of candicidin synthesis by aromatic amino acids in resting cells of* S. griseus. *Symbols:* ●, *control;* ○, *10 mM tyrosine;* △, *10 mM phenylalanine;* ▲, *10 mM tryptophan;* □, *10 mM phenylalanine, tyrosine, and tryptophan.*

is not available, protein synthesis is not possible and the aromatic intermediates are channeled toward candicidin. When tryptophan is added again, it rapidly inhibits candicidin synthesis to save precursors for protein synthesis.

CONTROL OF ANTIBIOTIC SYNTHESIS BY P_i

The synthesis of a number of antibiotics belonging to different biosynthetic groups is controlled by the P_i in the medium (see review by Martín [10]). P_i strongly inhibits

the biosynthesis of polyketide-derived antibiotics (tetracyclines a
all aminoglycoside and some peptide antibiotics. Their industria
be carried out under growth-limiting P_i concentrations.

The molecular mechanism underlying such control is comp
controls the central pathways of intermediary metabolism as
specific for special metabolites, and the onset of the synthesis of sp
is the result of a shiftdown in primary metabolism (10). Phosphate depletion appears to be the major nutritional change involved in slowing down primary metabolism and initiating the formation of antibiotic. It also causes a decrease in RNA synthesis and triggers the onset of candicidin synthesis (5). The late appearance of candicidin synthetase is due to de novo protein synthesis, as demonstrated by the inhibition of its formation with chloramphenicol. Similar results have been described for the phosphate control of tetracycline, vancomycin, and levorin (10). Nutrient limitation makes balanced growth impossible and brings on biochemical differentiation. Different degrees of nutrient limitation (i.e., different low specific growth rates) are required for the derepression of the synthesis of different antibiotics.

The phosphate regulation may be explained by assuming that high phosphate concentrations inhibit the derepression of secondary metabolism (Fig. 3). This is a negative control mechanism in which the concentration of a repressor decreases at the end of the rapid growth phase. However, inducers of the synthesis of some special metabolites have been described (1, 11). The biosynthesis of the ergot alkaloids is induced by tryptophan or tryptophan analogs (4, 17). This may suggest a positive regulatory model of the onset of secondary metabolism at least in some cases.

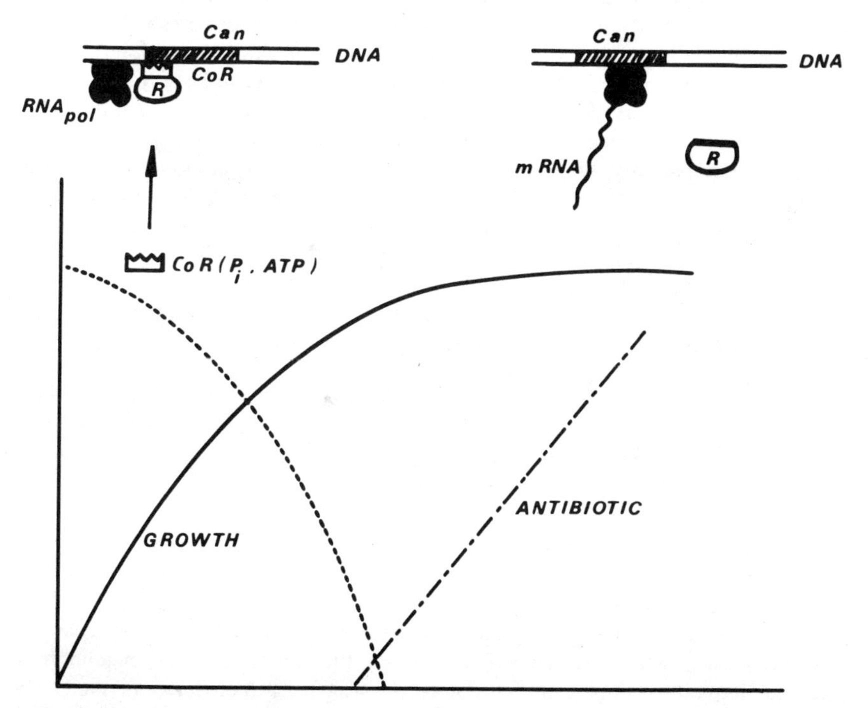

FIG. 3. *Proposed model of control by phosphate or ATP of the onset of candicidin (Can) synthesis at the transcription level (see text). Abbreviations: RNA_{pol}, RNA polymerase; CoR, corepressor R; R, repressor.*

he inhibition of candicidin synthesis by resting cells is dependent on phosphate oncentration and independent of the reactivation of the protein or RNA synthesis (8, 12). Such inhibition is also produced in the absence of protein synthesis in the cells already committed to the antibiotic production (13). This suggests a second effect of phosphate, namely, the inhibition of the activity of candicidin synthetases. Several phosphatases which are involved in the formation of streptomycin are inhibited by phosphate (19–22). The alkaline phosphatase of neomycin-producing *S. fradiae* is inhibited and repressed by P_i (7).

The ultimate intracellular effector controlling antibiotic synthesis during phosphate regulation may be phosphate itself or, more likely, an intracellular nucleotide (such as ATP) or the adenylate energy charge. Our recent results indicate that after phosphate addition to the candicidin-producing resting cell system, the intracellular ATP pool increases rapidly two- to threefold before the inhibition of antibiotic synthesis is observed (14). Such an ATP increase is reflected in a small increase of the adenylate energy charge: [ATP + (½ADP)]/(ATP + ADP + AMP). The relative importance of ATP, cyclic nucleotides, polyphosphates, and highly phosphorylated nucleotides as effectors controlling gene expression in relation to antibiotic synthesis has been reviewed elsewhere (10).

DIRECTED REMOVAL OF REGULATORY MECHANISMS

Removal of regulatory mechanisms is a procedure to obtain tailor-made overproducer strains. It is possible to design ways to select deregulated mutants altered in those regulatory mechanisms which are better known. Removal of feedback regulation of antibiotic synthesis has been described in a number of cases (11). Several techniques, including resistance to analogs of the end product, supression of auxotrophic mutations, and reversion of nonproducer mutants, have been described for directed removal of control mechanisms (2). We have recently isolated mutants of *S. griseus* resistant to tryptophan analogs which produce more candicidin than the parent strain, apparently due to the removal of tryptophan regulation of candicidin synthesis. We are also studying the removal of phosphate regulation of candicidin synthesis in *S. griseus.* Two different approaches have been followed: isolation of overproducer mutants in media containing an excess of P_i and isolation of phosphate permeability mutants (Martín et al., submitted for publication).

Uptake of phosphate in bacteria is mediated by a phosphate-binding protein of the cell membrane and by at least four phosphate permeases (18, 23). Obviously, a phosphate auxotroph would be a conditionally lethal mutant. Partial phosphate auxotrophs of *S. griseus* (leaky auxotrophs or bradytrophs) with a decreased phosphate uptake have been obtained. Such mutants are resistant to up to 200 mM arsenate (a toxic phosphate analog) and are partially insensitive to phosphate control of candicidin synthesis. They are useful for the study of the phosphate regulatory mechanism at the molecular level.

A detailed knowledge of the regulatory mechanisms controlling antibiotic synthesis is essential to understand and to remove those mechanisms that exert a control in antibiotic synthesis. The above-described procedures may be used for the selection of mutants altered in feedback regulatory mechanisms of antibiotic synthesis. In those cases when there is evidence that the biosynthesis of a particular metabolite is under an inducible control, mutants may be designed which have an increased synthesis of the inducer. Inducers of secondary metabolism are frequently primary metabolites (3).

REFERENCES

1. **Demain, A. L.** 1972. Cellular and environmental factors affecting the synthesis and excretion of metabolites. J. Appl. Chem. Biotechnol. **22:**345–362.
2. **Demain, A. L.** 1973. Mutation and the production of secondary metabolites. Adv. Appl. Microbiol. **16:** 177–202.
3. **Drew, S. W., and A. L. Demain.** 1977. Effect of primary metabolites on secondary metabolism. Annu. Rev. Microbiol. **31:**343–356.
4. **Krupinski, V. M., J. E. Robbers, and H. G. Floss.** 1976. Physiological study of ergot: induction of alkaloid synthesis by tryptophan at the enzymatic level. J. Bacteriol. **125:**158–165.
5. **Liras, P. L., J. R. Villanueva, and J. F. Martín.** 1977. Sequential expression of macromolecule biosynthesis and candicidin formation in *Streptomyces griseus.* J. Gen. Microbiol. **102:**269–277.
6. **Liu, C. M., L. E. McDaniel, and C. P. Schaffner.** 1975. Factors affecting the production of candicidin. Antimicrob. Agents Chemother. **7:**196–202.
7. **Majumdar, M. J., and S. K. Majumdar.** 1971. Relationship between alkaline phosphatase and neomycin formation in *Streptomyces fradiae.* Biochem. J. **122:**397–404.
8. **Martín, J. F.** 1976. Phosphate regulation of gene expression in candicidin biosynthesis, p. 548–552. *In* D. Schlessinger (ed.), Microbiology—1976. American Society for Microbiology, Washington, D.C.
9. **Martín, J. F.** 1977. Biosynthesis of polyene macrolide antibiotics. Annu. Rev. Microbiol. **31:**13–38.
10. **Martín, J. F.** 1977. Control of antibiotic synthesis by phosphate. Adv. Biochem. Eng. **6:**105–127.
11. **Martín, J. F.** 1978. Manipulation of gene expression in the development of antibiotic production. *In* R. Hütter, T. Leisinger, J. Nüesch, and W. Wehrli (ed.), Antibiotics and other secondary metabolites. Biosynthesis and production. Academic Press, London, in press.
12. **Martín, J. F., and A. L. Demain.** 1976. Control by phosphate of candicidin production. Biochem. Biophys. Res. Commun. **71:**1103–1109.
13. **Martín, J. F., P. Liras, and A. L. Demain.** 1977. Inhibition by phosphate of the activity of candicidin synthases. FEMS Microbiol. Lett. **2:**173–176.
14. **Martín, J. F., P. Liras, and A. L. Demain.** 1978. ATP and adenylate energy charge during phosphate-mediated control of antibiotic synthesis. Biochem. Biophys. Res. Commun. **83:**822–828.
15. **Martín, J. F., and L. E. McDaniel.** 1975. Specific inhibition of candicidin biosynthesis by the lipogenic inhibitor cerulenin. Biochim. Biophys. Acta **411:**186–194.
16. **Martín, J. F., and L. E. McDaniel.** 1977. Production of polyene macrolide antibiotics. Adv. Appl. Microbiol. **21:**2–52.
17. **Robbers, J. E., L. W. Robertson, K. M. Hornemann, A. Jindra, and H. G. Floss.** 1972. Physiological studies on ergot: further studies on the induction of alkaloid synthesis by tryptophan and its inhibition by phosphate. J. Bacteriol. **112:**791–796.
18. **Sprague, G. F., R. M. Bell, and J. E. Cronan.** 1975. A mutant of *E. coli* auxotrophic for organic phosphate: evidence for two defects in inorganic phosphate transport. Mol. Gen. Genet. **143:**71–77.
19. **Walker, J. B., and M. Skorvaga.** 1973. Phosphorylation of streptomycin and dihydro-streptomycin by *Streptomyces.* Enzymatic synthesis of different diphosphorylated derivatives. J. Biol. Chem. **248:** 2435–2440.
20. **Walker, J. B., and M. Skorvaga.** 1973. Streptomycin biosynthesis and metabolism. Phosphate transport from dihydrostreptomycin-6-phosphate to inosamine streptamine and 2-deoxystreptamine. J. Biol. Chem. **248:**2441–2446.
21. **Walker, M. S., and J. B. Walker.** 1970. Streptomycin biosynthesis and metabolism. Enzymatic phosphorylation of dihydrostreptobiosamine moieties of dihydrostreptomycin- (streptidino) phosphate and dihydrostreptomycin by *Streptomyces* extracts. J. Biol. Chem. **245:**6683–6689.
22. **Walker, M. S., and J. B. Walker.** 1971. Streptomycin biosynthesis. Separation and substrate specificities of phosphatases acting on guanidino-deoxy-scyllo-inositol phosphate and streptomycin- (streptidino) phosphate. J. Biol. Chem. **246:**7034–7041.
23. **Yagil, E., N. Silberstein, and R. G. Gerdes.** 1976. Co-regulation of the phosphate-binding protein and alkaline phosphatase synthesis in *Escherichia coli.* J. Bacteriol. **127:**656–659.

Regulatory Interrelationships of Nitrogen Metabolism and Cephalosporin Biosynthesis

YAIR AHARONOWITZ

Faculty of Life Sciences, Department of Microbiology, Tel Aviv University, Tel Aviv, Israel

Highly developed mechanisms for the regulation of metabolic processes enable microorganisms to respond efficiently to nutritional changes in their environment. The presence of an excess of a required metabolite generally results in the repression or inhibition of enzymes which catalyze its biosynthesis and reduces the concentration of intermediates of such a metabolite. Moreover, exposure of cells which can utilize different compounds as sole sources of energy, carbon, or nitrogen to such compounds frequently leads to the synthesis of new enzymes and the disappearance of others. One feature of such catabolic regulation is that a single factor can activate the transcription of different operons that have the same physiological function. For example, cyclic AMP together with a specific cyclic AMP-binding protein activates transcription of certain catabolic operons (13).

A similar effect was described in the case of glutamine synthetase, which activates several catabolic pathways (operons) and generates ammonia and glutamate (see Table 1). Such a regulatory process operates at the level of transcription independently of the specific operon control system and has been termed a "post-classical mode of regulation" (13). Glutamine synthetase has also been implicated in sporulation (6). In eucaryotic cells such as *Saccharomyces* and *Aspergillus*, the synthesis of arginase (4), amidases, and extracellular proteases (11) is regulated in a similar manner by glutamate dehydrogenase. In actinomycetes, uricase (22) and xanthine dehydrogenase (15) were shown to be regulated by the kind of nitrogen available.

NITROGEN REGULATION IN ANTIBIOTIC BIOSYNTHESIS

The effect of primary nitrogen metabolism and its possible regulatory consequences for the production of secondary metabolites (antibiotics) have not been studied extensively. The reason for this may stem from the current inability to study the regulation of antibiotic production at the level of the specific gene expression. The steps which may be the targets for such regulation are not well understood. However, many antibiotic molecules are derived from precursors which are themselves end products or intermediates of primary metabolism. Thus, the mechanisms that control the supply of the required substrates for antibiotic production may affect the ability of the cells to produce antibiotics. This paper will consider the possible interrelationship of primary nitrogen metabolism and the biosynthesis of cephalosporin.

Several findings suggest that antibiotic production is regulated by the kind of nitrogen in the culture medium. For example, proline was found to be a better nitrogen source than ammonia for the production of streptomycin (5). In addition, when *Streptomyces niveus* was grown in a chemically defined medium which contained proline and ammonium sulfate as nitrogen sources, ammonia was utilized prior to the amino acid, and its consumption preceded the production of novobiocin (12). By increasing the concentration of ammonium sulfate from 10 mM to 40 mM, novobiocin production was inhibited by 90%. Of interest was the finding that the replacement of ammonium sulfate in the medium by a high concentration of proline resulted in higher yields of novobiocin. A correlation was also found between erythromycin biosynthesis and alanine dehydrogenase levels in *S. erythreus* (19). The addition of alanine to the medium resulted in higher levels of both erythromycin and alanine dehydrogenase. *o*-Aminobenzoic acid was found to alter activities of glutamine synthetase and of glutamate dehydrogenase in *S. noursei* (10). It was suggested that the production of nourseothricin could be improved by altering the nitrogen metabolism and particularly by the enhancement of glutamate dehydrogenase.

These and other findings dealing with the involvement of nitrogen catabolic regulation in sporulation (6, 21) may be due in part to the activation of antibiotic precursors which are normal constituents of vegetative cells, and they suggest targets which may be controlled by primary nitrogen catabolism.

NITROGEN REGULATION AND CEPHALOSPORIN BIOSYNTHESIS IN STREPTOMYCETES

For a study of the relationship that might exist between nitrogen metabolism and antibiotic production, we chose the system in which cephalosporin is produced by bacteria.

A chemically defined medium in which the pH remains constant during fermentation (1) enabled us to study individual growth parameters and their effect on cephalosporin production. When *S. clavuligerus* was grown under nutritionally defined conditions with asparagine as the sole source of nitrogen, growth preceded cephalosporin production. The addition of NH_4Cl to the asparagine-containing culture did not alter the growth kinetics but reduced the antibiotic production by 80%. As can be seen in Fig. 1, the effect of NH_4Cl in inhibiting yields and specific production of cephalosporin was observed only under the conditions in which NH_4Cl was no longer growth-limiting. The same phenomenon occurred in cells grown with NH_4Cl as the sole N source. Thus, when the culture was grown in increasing concentrations of NH_4Cl, a lower cephalosporin yield was observed only after nitrogen ceased to be limiting.

The inhibitory effect of NH_4Cl could be explained by the inhibition either of the cephalosporin biosynthetic enzyme(s) or of other steps required for synthesis of the antibiotic molecule. To distinguish between these alternatives, cells were grown in

TABLE 1. *Pathways regulated by glutamine synthetase*

Pathway	Organism	Reference
Degradative		
Histidine	*Klebsiella aerogenes*	16
Proline	*K. aerogenes*	17
Tryptophan	*K. aerogenes*	14
Urea	*K. aerogenes*	7
Asparagine	*K. aerogenes*	18
Arginine	*K. aerogenes*	8
γ-Aminobutyrate	*Escherichia coli*	23
Others		
Nitrogen fixation	*K. pneumoniae*	20
Nitrogen fixation	*Spirillum lipoferum*	9
Glutamine transport	*Salmonella typhimurium*	3

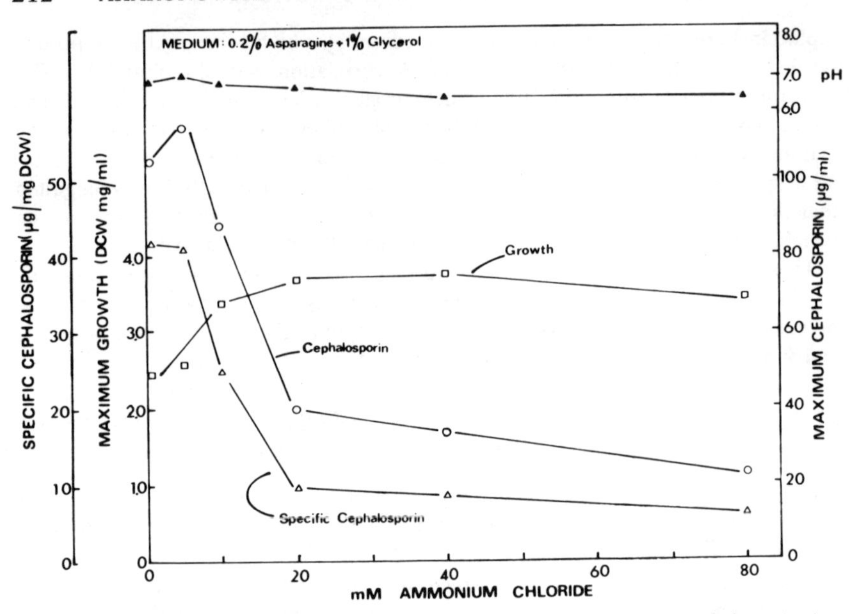

FIG. 1. *Effect of NH_4Cl concentration on maximum values of biomass, cephalosporin titer, specific cephalosporin production, and minimum pH in the presence of 0.2% L-asparagine and 1% glycerol. Maximum specific cephalosporin titer is the maximum cephalosporin titer per milliliter (in micrograms) divided by the maximum biomass per milliliter (in milligrams). DCW, Dry cell weight.*

glycerol-asparagine medium to which excess amounts of NH_4Cl were added at different times. The data in Fig. 2 show that the specific rate and extent of cephalosporin production were severely inhibited when NH_4Cl was added at time zero. The addition of NH_4Cl after 24 h caused a partial decrease in the specific rate of production and the yields. In contrast, no effect was found when the addition was made any time after 48 h. It appears from this experiment that the inhibitory action of NH_4Cl is not exerted on the antibiotic biosynthetic enzymes themselves, but rather on earlier steps preceding the biosynthesis of these enzymes. The NH_4Cl effect was most pronounced when the cells were in their growth phase, during which they utilize nitrogen-assimilatory systems which have been shown in other microorganisms to be regulated by the availability of the nitrogen source. Of various ammonia-assimilatory enzymes of *S. clavuligerus* which may be related to the production of cephalosporin, glutamine synthetase, glutamate synthase, and reductive amination of α-keto acids were detected. In contrast to the enteric bacteria, in *S. clavuligerus* glutamic dehydrogenase activity was very low, but high levels of alanine dehydrogenase were found. Table 2 shows the effect of different nitrogen sources on the ammonia-assimilatory enzymes. The specific activity of glutamine synthetase was highest when cells were grown with asparagine or alanine as the sole nitrogen source, but the addition of NH_4Cl depressed the glutamine synthetase activity. Glutamate synthase, on the other hand, was largely unaffected, and alanine dehydrogenase levels were enhanced by growth on alanine and in the presence of NH_4Cl. The mycelial extracts always contained some glutamine synthetase, even when grown on NH_4Cl alone. However, when grown on asparagine or when the ammonia supply in the medium

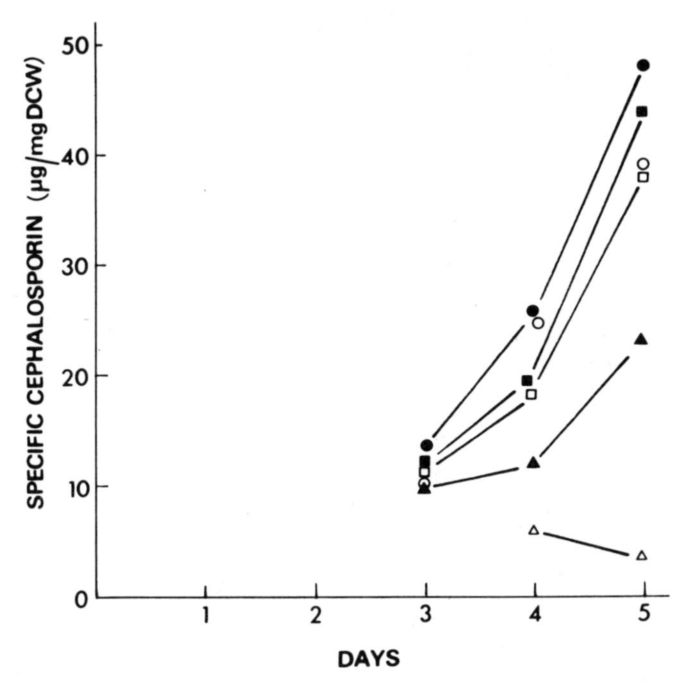

FIG. 2. *Effect of time of addition of NH_4Cl on cephalosporin production. Cells were grown in 0.2% asparagine—1% glycerol medium and divided into fermentation flasks at time zero. Each flask received 40 mM NH_4Cl at the time indicated:* ●, *none added;* △, *time zero;* ▲, *24 h;* ○, *48 h;* □, *72 h;* ■, *96 h.*

TABLE 2. *Effect of nitrogen nutrition on ammonia-assimilatory enzymes in* S. clavuligerus[a]

Nitrogen source	Specific activity (U/mg of protein)		
	Glutamine synthetase	Glutamate synthase	Alanine dehydrogenase
L-Asparagine	1.2	0.013	0.27
L-Asparagine + NH_4Cl	0.24	0.023	0.86
NH_4Cl	0.15	0.019	1.64
L-Alanine	1.0	0.022	1.43
L-Alanine + NH_4Cl	0.23	0.020	1.95

[a] Cells were grown in minimal media supplemented with 0.2% L-asparagine or L-alanine and 40 mM NH_4Cl. At 72 h, cells were harvested and enzyme activities were measured in cell-free extracts. Glutamine synthetase was assayed by the γ-glutamyltransferase procedure (2). Glutamate synthase and alanine dehydrogenase were assayed spectrophotometrically by measuring the rate of oxidation of reduced nicotinamide adenine dinucleotide at 340 nm.

had been exhausted, the glutamine synthetase activity was induced to its highest value.

The interrelationship of glutamine synthetase activity, cephalosporin yields, and specific production of cephalosporin is illustrated in Table 3. It can be seen that the highest specific production was obtained under growth conditions which favored high specific activities of glutamine synthetase. When cells were grown either on asparagine plus NH_4Cl or on high concentrations of NH_4Cl alone, both the glutamine synthetase activity and specific cephalosporin production decreased. Thus, the ability

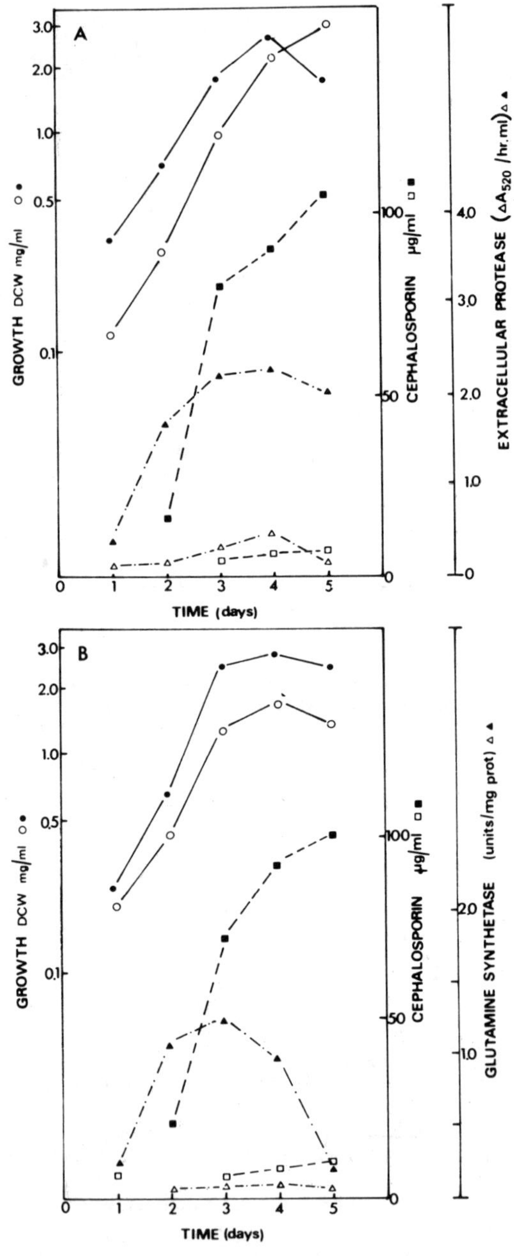

FIG. 3. *Effect of nitrogen on growth of* S. calvuligerus, *cephalosporin production, glutamine synthetase, and extracellular proteases. (A) Extracellular proteolytic activity in relation to cephalosporin production; (B) specific activity of glutamine synthetase in cell-free extracts in relation to cephalosporin production. Asparagine was used at 0.2% and NH_4Cl was used at 40 mM. Curves show growth in asparagine (●) and in NH_4Cl (○), enzymatic activities of asparagine-grown cultures (▲) and of NH_4Cl-grown cultures (△), and cephalosporin production by asparagine-grown cultures (■) and by NH_4Cl-grown cultures (□).*

of the cells to produce high levels of antibiotics is not simply a function of the biomass, since even under growth-limiting conditions specific production was elevated. A similar correlation was found when alanine was substituted for asparagine.

S. clavuligerus produces extracellular proteases which are repressed by ammonia.

TABLE 3. *Interrelationship of glutamine synthetase activity, cephalosporin yields, and specific cephalosporin production*[a]

Nitrogen source		GS (U/mg of protein)	CEPH (μg/ml)	DCW (mg/ml)	CEPH/DCW (μg/mg)
L-Asparagine (%)	NH_4Cl (mM)				
0.2	0	2.2	100	2.8	38
0.2	10	0.91	120	2.4	50
0.2	20	0.48	40	2.4	16
0.2	40	0.28	20	2.8	7
0	10	1.2	20	0.72	27
0	20	0.55	20	1.20	16
0	40	0.34	7	1.6	4

[a] GS, Glutamine synthetase; CEPH, cephalosporin; DCW, dry cellular weight. *S. clavuligerus* cells were grown in minimal media supplemented with asparagine and/or NH_4Cl as sole nitrogen sources. Glutamine synthetase activity, biomass, and antibiotic activity represent the highest levels achieved during growth cycle.

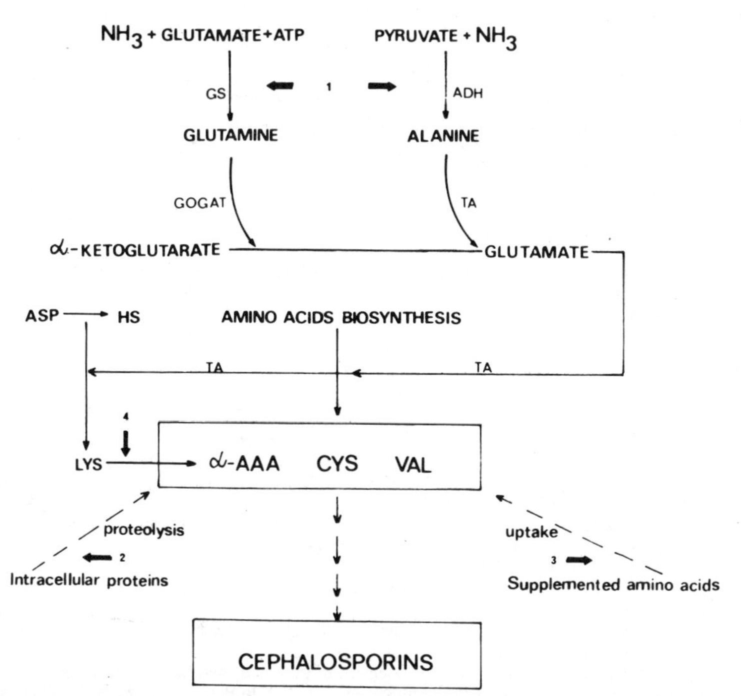

FIG. 4. *Possible targets for interaction of cephalosporin biosynthesis with primary nitrogen metabolism. GS, Glutamine synthetase; ADH, alanine dehydrogenase; GOGAT, glutamate synthase; TA, transamination; ASP, aspartate; HS, homoserine; LYS, lysine; α-AAA, α-aminoadipate; CYS, cysteine; VAL, valine. Numbered arrows point to the possible targets for interaction of primary nitrogen metabolism with cephalosporin biosynthesis.*

Of interest is the similarity between the kinetics of appearance of glutamine synthetase, extracellular proteases, and cephalosporin production (Fig. 3). Again, it can be seen that growth in the presence of NH_4Cl caused repression of glutamine synthetase, with a concomitant decrease of protease and cephalosporins.

CONCLUDING REMARKS

We have demonstrated a correlation between the capacity of *S. clavuligerus* to synthesize cephalosporin and its physiological state with respect to nitrogen assimilation. Although these two phenomena are probably genetically independent, control mechanisms which operate at the level of nitrogen assimilation may affect the ability of the antibiotic biosynthetic machinery of the cells to operate under more favored conditions. Several possible targets for such regulation are suggested in Fig. 4:

1. The provision of glutamate which is required for the formation of cephalosporin precursors.
2. The catabolism of other nitrogenous compounds which may provide cells with these precursors. Such a target may be proteolytic activity.
3. Active uptake of precursors from the environment.
4. The catabolism of lysine to α-aminoadipate which provides the cephalosporin side chain.

In addition, a direct regulatory effect may be exerted on the cephalosporin-synthesizing enzymes themselves. The regulation at any of the targets might be mediated by a nonspecific effector which senses the state of nitrogen availability for the cell. An understanding of the correlation between the primary nitrogen regulation and cephalosporin production may lead to the development of a rational basis for mutant selection and strain improvement. Among such mutants may be those which are released from the inhibition or repression by ammonia.

ACKNOWLEDGMENTS

I thank Arnold L. Demain for his interest and support during the course of this work and C. G. Friedrich and B. Friedrich for their help in the initiation of this project. I am also grateful to J. Piret, N. Shuster, J. Bahat, and V. Kuper for carrying out biochemical assays and to D. Gutnick for help in the preparation of the manuscript.

REFERENCES

1. **Aharonowitz, Y., and A. L. Demain.** 1977. Influence of inorganic phosphate and organic buffers on cephalosporin production by *Streptomyces clavuligerus.* Arch. Microbiol. **115:**169–173.
2. **Bender, R. A., K. A. Janssen, A. D. Resnick, M. Blumenberg, F. Foor, and B. Magasanik.** 1977. Biochemical parameters of glutamine synthetase from *Klebsiella aerogenes.* J. Bacteriol. **129:**1001–1009.
3. **Betteridge, P. R., and P. D. Ayling.** 1976. The regulation of glutamine transport and glutamine synthetase in *Salmonella typhimurium.* J. Gen. Microbiol. **95:**324–334.
4. **Dubois, E., M. Grenson, and J. M. Wiam.** 1974. The participation of the anabolic glutamate dehydrogenase in the nitrogen catabolic repression of arginase in *Saccharomyces cerevisiae.* Eur. J. Biochem. **48:**603–616.
5. **Dulaney, E. L.** 1948. Observations of Streptomyces griseus. II. Nitrogen sources for growth and streptomycin production. J. Bacteriol. **56:**305–313.
6. **Elmerich, C., and J.-P. Aubert.** 1975. Involvement of glutamine synthetase and the purine nucleotide pathway in repression of bacterial sporulation, p. 385–390. *In* P. Gerhardt, R. N. Costilow, and H. L. Sadoff (ed.), Spores VI. American Society for Microbiology. Washington, D.C.
7. **Friedrich, B., and B. Magasanik.** 1977. Urease of *Klebsiella aerogenes:* control of its synthesis by glutamine synthetase. J. Bacteriol. **131:** 446–452.
8. **Friedrich, B., and B. Magasanik.** 1978. Utilization of arginine by *Klebsiella aerogenes.* J. Bacteriol. **133:** 680–685.
9. **Gauthier, D., and C. Elmerich.** 1977. Relationship between glutamine synthetase and nitrogenase in

Spirillum lipoferum. FEMS Microbiol. Lett. **2:**101–104.

10. **Grafe, U., H. Bocker, and H. Thrum.** 1977. Regulative influence of o-aminobenzoic acid on the biosynthesis of nourseothricin in cultures of *Streptomyces noursei* JA3890b. Z. Allg. Mikrobiol. **17:** 201–209.
11. **Hynes, M. J.** 1974. Effects of ammonium, L-glutamate, and L-glutamine on nitrogen catabolism in *Aspergillus nidulans.* J. Bacteriol. **120:**1116–1123.
12. **Kominek, L. A.** 1972. Biosynthesis of novobiocin by *Streptomyces niveus.* Antimicrob. Agents Chemother. **1:**123–134.
13. **Magasanik, B.** 1976. Classical and postclassical modes of regulation of the synthesis of degradative bacterial enzymes. Prog. Nucleic Acid Res. Mol. Biol. **17:**99–115.
14. **Magasanik, B., M. J. Prival, J. E. Brenchley, B. M. Tyler, A. D. Deleo, S. L. Streicher, R. A. Bender, and C. G. Paris.** 1974. Glutamine synthetase as a regulator enzyme synthesis. Curr. Top. Cell. Regul. **8:**119–138.
15. **Ohe, T., and Y. Watanabe.** 1977. Effect of glucose and ammonium on the formation of xanthine dehydrogenase of *Streptomyces sp.* Agric. Biol. Chem. **41:**1161–1170.
16. **Prival, M., J. Brenchley, and B. Magasanik.** 1973. Glutamine synthetase and the regulation of histidase formation in *Klebsiella aerogenes.* J. Biol. Chem. **248:**4334–4344.
17. **Prival, M., and B. Magasanik.** 1971. Resistance to catabolite repression of histidase and proline oxidase during nitrogen limited growth of *Klebsiella aerogenes.* J. Biol. Chem. **246:**6288–6296.
18. **Resnick, A. D., and B. Magasanik.** 1976. L-Asparaginase of *Klebsiella aerogenes.* Activation of its synthesis by glutamine synthetase. J. Biol. Chem. **251:**2722–2728.
19. **Roszkowski, J., A. Rafalski, and K. Raczynska-Bojanowska.** 1969. Alanine and alanine dehydrogenase in *Streptomyces erythreus.* Acta Microbiol. Pol. **1**(18):59–68.
20. **Streicher, S. L., K. T. Shanmugam, F. Ausubel, C. Morandi, and R. B. Goldberg.** 1974. Regulation of nitrogen fixation in *Klebsiella pneumoniae*: evidence for a role of glutamine synthetase as a regulator for nitrogenase synthesis. J. Bacteriol. **120:**815–821.
21. **Vitković, L., and H. L. Sadoff.** 1977. In vitro production of bacitracin by proteolysis of vegetative *Bacillus licheniformis* cell protein. J. Bacteriol. **131:**897–905.
22. **Watanabe, Y., T. Ohe, and M. Morita.** 1976. Control of the formation of uricase in *Streptomyces sp.* by nitrogen and carbon sources. Agric. Biol. Chem. **40:**131–139.
23. **Zaboura, M., and Y. S. Halpern.** 1978. Regulation of γ-aminobutyric acid degradation in *Escherichia coli* by nitrogen metabolism enzymes. J. Bacteriol. **133:**447–451.

Regulation of Aerial Mycelium Formation in Streptomycetes

BURTON M. POGELL

Department of Microbiology, St. Louis University School of Medicine, St. Louis, Missouri 63104

SIMULTANEOUS LOSS OF MULTIPLE DIFFERENTIATED FUNCTIONS IN STREPTOMYCETES

In addition to the phenotypic control by catabolite repression (7), there is specific regulation of aerial mycelium formation and expression of other specialized functions in streptomycetes at the genetic level, most probably involving extrachromosomal DNA. Thus, when spores of several species of streptomycetes were allowed to germinate and grow in the presence of intercalating dyes, the occurrence of a high frequency (2 to 20%) of stable mutants unable to form aerial mycelium (Amy$^-$) was observed. In addition, a few spontaneous Amy$^-$ colonies were obtained by spore germination in the absence of dye. The morphological characteristics of some typical isolates grown on Hickey-Tresner agar are illustrated in Fig. 1. All of these isolates were still capable of sporulating, although to a much lower extent than the Amy$^+$ wild types. Detailed studies of these strains revealed a simultaneous loss of other differentiated functions including their ability to produce the characteristic "earthy" odor (geosmin formation) and wild-type pigments. In addition, this work has led to the isolation of specific factors from *Streptomyces alboniger* which both stimulate and inhibit formation of aerial mycelium.

Our observations on the loss of differentiated functions in a large number of *S. alboniger* isolates are summarized in Table 1. The *S. alboniger* Amy$^-$ colonies no longer formed the characteristic black pigment excreted by the organism or the specific hexane-soluble stimulator of aerial mycelium formation, and they still remained Amy$^-$ when grown in the presence of added stimulator. Significant amounts of aerial mycelium inhibitor activity were detected in most Amy$^-$ extracts, and all of the Amy$^-$ isolates still produced antibiotic activity (presumably puromycin). All of the Amy$^-$ isolates of *S. alboniger* and a high percentage of those of *S. scabies* (27%) and *S. violaceus-ruber* (39%) were arginine auxotrophs. The remaining *S. scabies* and *S. violaceus-ruber* Amy$^-$ strains were prototrophs. The specific absence of argininosuccinate synthetase in these auxotrophs as well as in two *S. coelicolor* Amy$^-$ Arg$^-$ isolates was established by growth studies with arginine precursors and direct enzyme measurements. Thus, selected colonies grew on minimal media containing argininosuccinate or arginine but not on citrulline or ornithine. Extracts of these isolates contained normal levels of argininosuccinase but no detectable synthetase activity.

The occurrence of a genetic deletion in these pleiotropic mutants appeared probable since no revertants to Amy$^+$ or Arg$^+$ were detected in any strain. With all three streptomycetes, large numbers of cells from different Amy$^-$ isolates have been tested for reversion but not a single revertant has been obtained (see Table 1, vii). In addition, no Amy$^+$ revertants were seen after treatment of *S. alboniger* Amy$^-$ cells with *N*-methyl-*N'*-nitro-*N*-nitrosoguanidine, ethidium bromide, or acriflavine. There was no evidence for inhibition of Amy$^+$ colony formation by the presence of excess Amy$^-$ cells.

I propose the molecular model shown in Fig. 2 as a plausible explanation for these

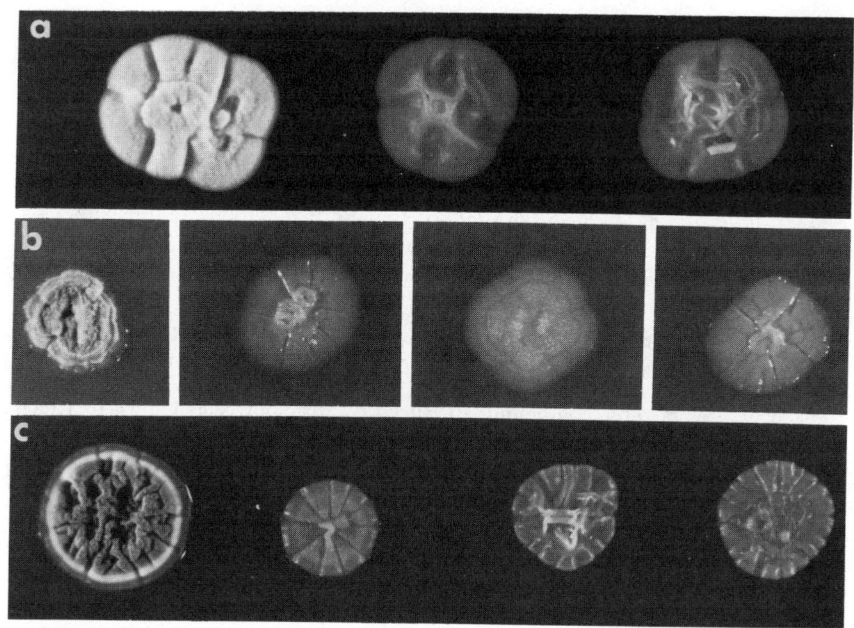

FIG. 1. *Morphology of Amy$^+$ and Amy$^-$ colonies of* Streptomyces. *(a)* S. alboniger *ATCC 12461, grown for 4 days at 28°C: (left to right) Amy$^+$, spontaneous Amy$^-$ Arg$^-$, ethidium bromide-"cured" Amy$^-$ Arg$^-$. (b)* S. scabies *PA10, grown for 3 days at 37°C: (left to right) Amy$^+$, spontaneous Amy$^-$ prototroph, acriflavine-"cured" Amy$^-$ prototroph, acriflavine-"cured" Amy$^-$ Arg$^-$. (c)* S. violaceus-ruber *ATCC 3355, grown for 5 days at 37°C: (left to right) Amy$^+$, spontaneous Amy$^-$ prototroph, acriflavine-"cured" Amy$^-$ prototroph, acriflavine-"cured" Amy$^-$ Arg$^-$.*

TABLE 1. *Properties of* S. alboniger *Amy$^-$ isolates*

Function	Frequency
i. Cannot form aerial mycelium, but can sporulate	84/84
ii. Do not have characteristic "earthy" odor	84/84
iii. Do not produce "niger" pigment	48/48
iv. a. No measurable formation of aerial mycelium-stimulating factor	15/15
b. Remain Amy$^-$ in presence of stimulating factor	10/10
c. Produce inhibitors of aerial mycelium formation	10/12
v. Continue to excrete antibiotic activity against *S. lutea*	84/84
vi. Require arginine for growth on minimal medium and remain Amy$^-$	84/84
vii. No detectable reversion to Amy$^+$ or Arg$^+$ (six isolates, 2×10^6 to 8×10^6 organisms plated; four isolates, 1.4×10^8 to 1.8×10^9 organisms plated)	10/10

results. Many reports in the literature suggest the possible involvement of extrachromosomal DNA in controlling the expression of differentiated functions in streptomycetes. I suggest that specific functions required for aerial mycelium formation, as well as other aspects of streptomycete differentiation, are coded for by one or more extrachromosomal elements. This plasmid DNA can be free or alternatively can exist at some stage of the streptomycete growth cycle in episomal integration in the chromosome at or near the site coding for argininosuccinate synthetase. Movement of the episome in and out of the chromosome could occur by translocation mechanisms (3). "Curing" of the free plasmid could result in the invariable (in *S. alboniger*) or frequent (in *S. scabies* or *S. violaceus-ruber*) excision of part of this *arg* gene in a manner analogous to the occasional loss of *gal* genes upon the induction of λ bacteriophage in *Escherichia coli* (5). A loss of a specific plasmid, or portion thereof, involved in the regulation of expression of specialized gene function could also explain our findings. Preliminary experiments supporting such a mechanism for aureothricin production by streptomycetes were presented by M. Okanishi (this volume). Intercalating dyes can also induce frameshift mutations and deletions, so the possibility that the dyes have acted as mutagens, or have caused a large deletion or inversion of chromosomal DNA, cannot be excluded.

ISOLATION, PROPERTIES, AND PRELIMINARY CHARACTERIZATION OF AERIAL MYCELIUM STIMULATOR

The structures of three compounds produced by streptomycetes known to affect aerial mycelium and antibiotic production are shown in Fig. 3. A-factor, isolated and characterized from *S. bikiniensis* and *S. griseus* by Khokhlov and co-workers, restored aerial mycelium formation as well as streptomycin production in mutants which had lost these functions (1, 4). Methylenomycin A inhibited aerial mycelium formation in *S. coelicolor* SCP1$^-$ strains, and Kirby et al. (2) showed that both the synthesis of and resistance to this antibiotic were determined by the SCP1 plasmid. Lincomycin, an inhibitor of protein synthesis produced by *S. lincolnensis*, was shown in our laboratory (6) to exert concentration-dependent effects on development in *S. alboniger*. Low levels (0.002 to 1 μg) caused a marked enhancement of aerial mycelium formation, manifested as an increase in white powder formation. In contrast, higher levels (2 to 10 μg) completely repressed aerial mycelium formation, and above 10 μg, vegetative growth began to be inhibited.

These observations led to a search for endogenous differentiation effectors in *S. alboniger*. Several compounds have been isolated which differentially stimulate or inhibit aerial mycelium formation in this organism. One specific stimulator of aerial mycelium formation, pamamycin, has been purified to homogeneity and partially characterized.

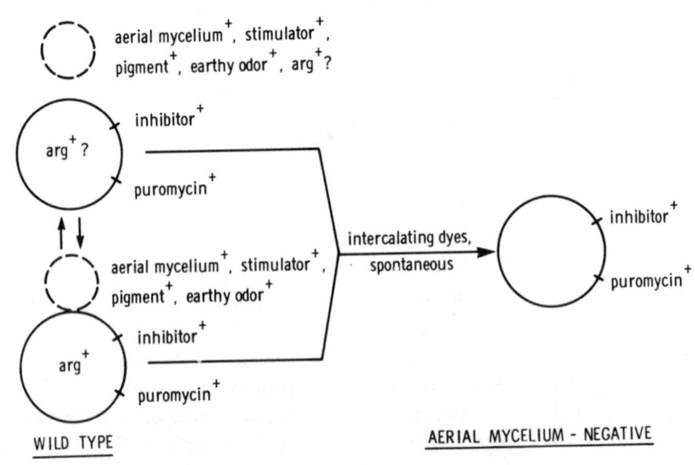

FIG. 2. *Proposed model of the development of mutants unable to form aerial mycelium.*

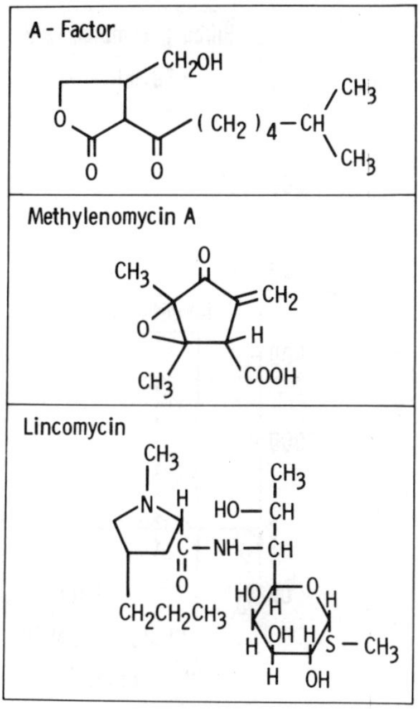

A - Factor

Methylenomycin A

Lincomycin

FIG. 3. *Structures of the differentiation factors produced by streptomycetes.*

The stimulatory activity of pamamycin was quantitated by a disk assay on a slightly modified Hickey-Tresner agar, which was balanced to give more reproducible zones of aerial mycelium stimulation. Disks containing different samples were placed on plates inoculated uniformly with *S. alboniger* hyphae, and the plates were incubated at 37°C. After 24 to 48 h, the zones of aerial mycelium stimulation were clearly visible as circular areas of intense white powder formation. A plot of zone diameter versus log of stimulator concentration gave a linear response curve. In addition to its role as a streptomycete differentiation effector, pamamycin has very high antimicrobial activities. Routine measurements of pamamycin during purification were carried out by disk diffusion assays against *Sarcina lutea*. A unit was defined as the amount of material producing a zone of 17-mm diameter. Activities at various stages of purification were expressed as units per milligram of dry weight.

The highest yields of pamamycin were obtained from *S. alboniger* mycelium grown on Hickey-Tresner agar. The dried cells were extracted with methanol in a Soxhlet apparatus, and the methanol extracts were concentrated by rotary evaporation. The residue was then triturated into a nonpolar solvent (either hexanes or benzene) to obtain a substantially purified extract.

Silicic acid column chromatography with stepwise elution was then used to separate pamamycin from two inhibitors of aerial mycelium formation. Typical results are illustrated in Fig. 4. Aerial mycelium inhibitor activity was also measured by disk assay against *S. alboniger* on Hickey-Tresner medium without $CoCl_2$. This deletion enhanced the contrast between the zones of inhibited aerial mycelium and white powdery background. Chloroform eluted the bulk of inactive material, chloroform-acetone (9:1) and chloroform-ethanol (95:5) eluted one major inhibitor component, and chloroform-methanol (95:5) eluted greatly purified pamamycin (329 U/mg). A second distinct fraction containing inhibitor activity was eluted with methanol.

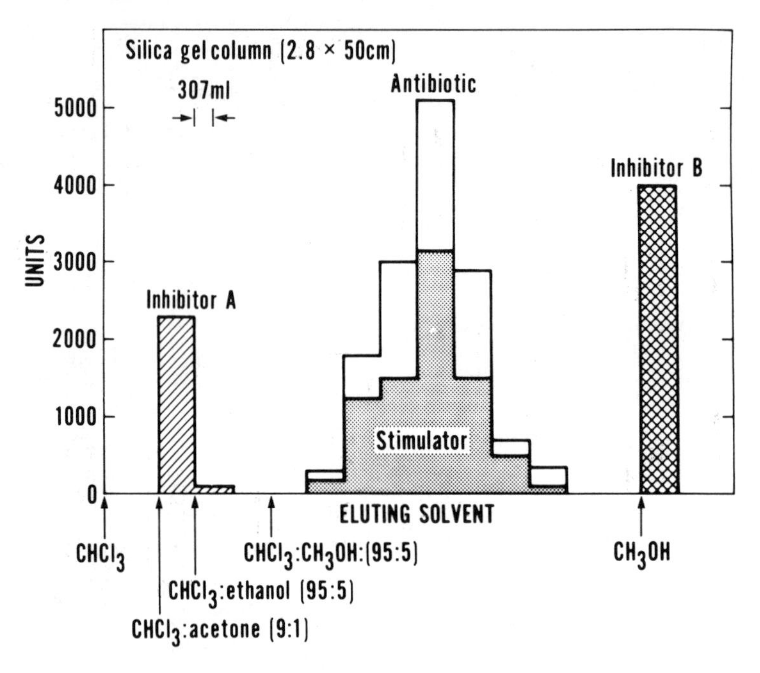

FIG. 4. *Separation of pamamycin from the inhibitors of aerial mycelium formation. This elution profile was obtained with 400 mg of hexane-soluble material (extracted from 15 g of cells).*

Although there was considerable variation in starting specific activities, the yields of pamamycin obtained after the silica gel column separation were relatively constant per unit weight of dry cells (0.64 to 1.1 U/mg of dried cells). The large increase in yields (150 to 2,000%) after the silica gel columns was accounted for by the observation that both of the aerial mycelium inhibitor fractions competitively inhibited pamamycin in the *S. lutea* assay (see Fig. 5).

Pamamycin was further purified by column chromatography on neutral alumina and purified to homogeneity by thin-layer chromatography on alumina glass plates in methanol-ethyl acetate (R_f, ca. 0.5). The peak fractions had a specific activity of 1,000 to 1,200 U/mg.

Pamamycin is a hydrophobic, neutral compound, insoluble in water but soluble in a wide range of organic solvents including hexanes, ether, benzene, chloroform, methanol, and dimethyl sulfoxide. The molecular weight of purified pamamycin was determined to be 621 by field desorption mass spectroscopy. The elemental composition of this mass ion, $C_{36}H_{63}NO_7$, and those of other major fragments listed in Table 2 were determined by peak matching and by computer analysis of the high resolution numbers. The infrared and UV spectra indicated that pamamycin was highly aliphatic and showed the absence of aromatic, -OH, -NH, and amide groups. Therefore, the nitrogen is probably in a tertiary linkage. A peak at 1,725 cm^{-1} indicated the presence of a carbonyl group, but, since this peak was not as intense as the hydrocarbon band at 2,880 to 2,960 cm^{-1}, there are probably only one or two carbonyl groups. A literature search (*Chemical Abstracts* and other natural product indexes) and antibiotic screening at Eli Lilly & Co. indicate that there is no previously

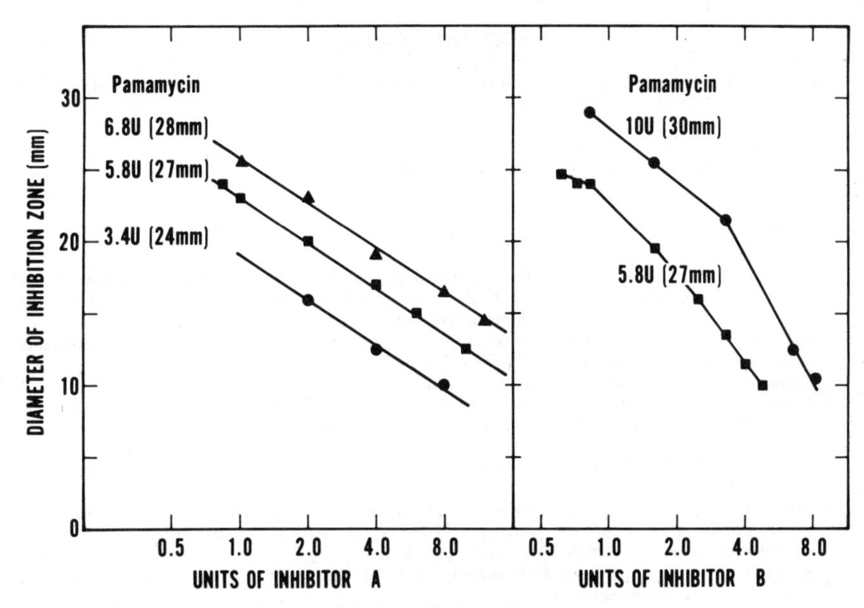

FIG. 5. *Reversal of pamamycin inhibition of* S. lutea *by aerial mycelium inhibitors. A unit of each inhibitor fraction was arbitrarily defined as the amount of extract decreasing the diameter of a pamamycin inhibition zone from 27 to 23 mm.*

TABLE 2. *Pamamycin fragments obtained by mass spectroscopy*

Fragment mass	Molecular formula
621	$C_{36}H_{63}NO_7$
578	$C_{33}H_{56}NO_7$
508	$C_{30}H_{54}NO_5$
352a	$C_{22}H_{42}NO_2$
352b	$C_{20}H_{34}NO_4$
254	$C_{16}H_{32}NO$
227	$C_{13}H_{23}O_3$
184	$C_{11}H_{22}NO$
143	$C_7H_{11}O_3$
100	$C_6H_{14}N$

described compound with this molecular composition, chromatographic behavior, and profile of antibiotic activity.

The relative ratios of activities of pamamycin as a stimulator of aerial mycelium formation and as an inhibitor of *S. lutea* remained constant during purification. Pamamycin also had high antimicrobial activity against *Bacillus subtilis, Staphylococcus aureus, Mycobacterium phlei, M. smegmatis,* and *Neurospora crassa,* but not against *E. coli, Proteus mirabilis,* and *P. morganii.*

Very little is known about the nature of the inhibitors of aerial mycelium, but the two fractions have the unusual property of competitively reversing the inhibition of *S. lutea* growth by pamamycin (Fig. 5). Preliminary in vivo results with *S. aureus* indicate that pamamycin is a potent inhibitor of both DNA and RNA synthesis. These results are compatible with the regulation of development in streptomycetes by differential effects on transcription (8), which would be dependent on the relative concentrations of stimulator and inhibitors during growth.

ACKNOWLEDGMENTS

Research described in this presentation was carried out by Pamela A. McCann, Peggy A. Redshaw, and Wen-Gang Chou. Thanks are due to the many scientists at Eli Lilly & Co. for their invaluable assistance in the characterization of pamamycin.

This work was supported by grants from the National Institutes of Health (CA-12080), Eli Lilly & Co. (7043), and the National Science Foundation (PCM77 16482).

REFERENCES

1. **Khokhlov, A. S., L. N. Anisova, I. I. Tovarova, E. M. Kleiner, I. V. Kovalenko, O. I. Krasilnikova, E. Ya. Kornitskaya, and S. A. Pliner.** 1973. Effect of A-factor on the growth of asporogenous mutants of *Streptomyces griseus*, not producing this factor. Z. Allg. Mikrobiol. **13**:647–655.
2. **Kirby, R., L. F. Wright, and D. A. Hopwood.** 1975. Plasmid-determined antibiotic synthesis and resistance in *Streptomyces coelicolor*. Nature (London) **254**:265–267.
3. **Kleckner, N.** 1977. Translocatable elements in prokaryotes. Cell **11**:11–23.
4. **Kleiner, E. M., S. A. Pliner, V. S. Soifer, V. V. Onoprienko, T. A. Balashova, B. V. Rozynov, and A. S. Khokhlov.** 1976. Structure of the A-factor, a bioregulator from *Streptomyces griseus*. Bioorg. Khim. **2**: 1142–1147.
5. **Morse, M. L., E. M. Lederberg, and J. Lederberg.** 1956. Transduction in *Escherichia coli* K-12. Genetics **41**:142–156.
6. **Pogell, B. M., L. Sankaran, P. A. Redshaw, and P. A. McCann.** 1976. Regulation of antibiotic biosynthesis and differentiation in streptomycetes, p. 543–547. *In* D. Schlessinger (ed.), Microbiology—1976. American Society for Microbiology, Washington, D.C.
7. **Redshaw, P. A., P. A. McCann, L. Sankaran, and B. M. Pogell.** 1976. Control of differentiation in streptomycetes: involvement of extrachromosomal deoxyribonucleic acid and glucose repression in aerial mycelia development. J. Bacteriol. **125**:698–705.
8. **Sankaran, L., and B. M. Pogell.** 1973. Differential inhibition of catabolite-sensitive enzyme induction by intercalating dyes. Nature (London) New Biol. **245**:257–260.

Specific Primary Pathways Supplying Secondary Biosynthesis

Z. HOŠŤÁLEK, V. BĚHAL, EVA ČURDOVÁ, AND VENDULKA JECHOVÁ

Czechoslovak Academy of Sciences, Institute of Microbiology, Prague, Czechoslovakia

The enzymes participating in the formation of secondary metabolites can be divided into two groups: those of the metabolic pathways yielding the precursors or building blocks and those of the secondary matabolism itself which accomplish the condensation of primary precursors or the transformation of biosynthetic intermediates into final secondary products. Of importance in this context is another group that includes anaplerotic enzymes which provide energy, oxidation or reduction equivalents, or donors of various functional groups substituting biosynthetic intermediates (9).

The activity of enzymes involved in the formation of precursors for secondary biosynthesis was studied in the chlortetracycline-producing actinomycete *Streptomyces aureofaciens.* To obtain information on the nature of metabolic changes associated with the production of the antibiotic, we compared the enzyme activity of a low-production strain of the standard type (strain RIA 57) with that of a production variant (strain 8425) obtained from a series of mutagenic treatments.

TRICARBOXYLIC ACID CYCLE AND LIPOGENESIS

Metabolic pathways which could compete with the utilization of acetate units in the biosynthesis of the tetracycline nucleus include the oxidation of acetate in the tricarboxylic acid cycle or lipid synthesis. It was found that the production variant exhibited a significantly lowered activity of enzymes of the tricarboxylic acid cycle (6). The study of incorporation of labeled acetate into fatty acids showed that lipogenesis was active only during the initial period of cultivation (Fig. 1) and did not interfere with the biosynthesis of chlortetracycline (2).

FORMATION OF MALONYL CoA

Acetyl coenzyme A (CoA) carboxylase, the enzyme that yields malonyl CoA which serves as a building block of the tetracycline molecule, exhibited an activity course identical with the rate of incorporation of labeled acetate into fatty acids (Fig. 1). Its activity was close to zero in the period of intensive chlortetracycline synthesis (4). The low activity of acetyl CoA carboxylase in *S. aureofaciens* 8425 during the production phase was paralleled by low activities of the preceding enzymes of the pathway, i.e., pyruvate kinase and the pyruvate dehydrogenase complex. This indicates that the conversion of phosphenolpyruvate to acetyl CoA is suppressed under the conditions of high antibiotic production. On the other hand, phosphoenolpyruvate carboxylase, the enzyme yielding oxalacetate, had a strikingly high activity during the intensive production period (Fig. 1). The assumption that oxalacetate is utilized for malonyl CoA synthesis was experimentally verified in a cell-free preparation by use of radio-gas-liquid chromatography. After incubation with [4-^{14}C]oxalacetate, the peak of the resulting malonate contained relatively high

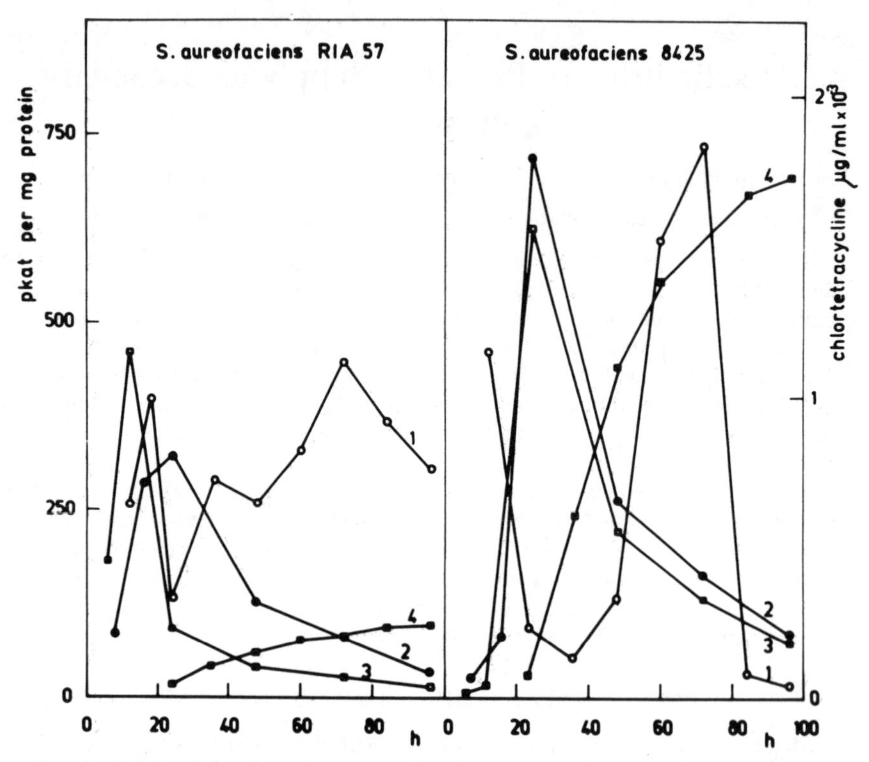

FIG. 1. *Activity of phosphoenolpyruvate carboxylase, activity of acetyl CoA carboxylase, and incorporation of labeled acetate into fatty acids during cultivation of low-producing (RIA 57) and production (8425) strains of* S. aureofaciens. *(1) Phosphoenolpyruvate carboxylase (picokatals per milligram of protein); (2) acetyl CoA carboxylase (picokatals per milligram of protein); (3) incorporation of* $CH_3{}^{14}COO^-$ *(microcuries per minute per milligram of dry weight* $\times 10^4$*); (4) chlortetracycline production (micrograms per milliliter* $\times 10^{-3}$*).*

radioactivity; the reaction required the presence of HSCoA and nicotinamide adenine dinucleotide.

The existence of the alternative pathway of malonyl CoA formation was confirmed in vivo by studying the incorporation of ^{14}C-labeled substrates. The radioactivity of the carbamoyl group of tetracycline, determined after Hofmann's degradation, was about 2.8% during the incorporation of [2-^{14}C]acetate, [3-^{14}C]pyruvate, and [2-^{14}C]-malonate. This points to a high randomization of acetate units since the carbamoyl group comes from the carboxyl group of malonate (acetate) and should not be active if malonyl CoA were formed in the classical pathway by acetyl CoA carboxylation (3).

PHOSPHORYLATION OF SUGARS

Another group of metabolic changes associated with increased antibiotic biosynthesis involves alterations of the character of energy metabolism. The low activity of the tricarboxylic acid cycle and the low ATP level in the production variant (10) led to the study of phosphorylation during cultivation.

Investigation of ATP glucokinase levels showed that they reach a maximum after about 12 h of cultivation and then gradually decline concomitantly with the decrease in the cellular level of ATP. Enzyme activity was minimal during the stationary growth phase. The study of glucose 6-phosphate formation showed the presence of polyphosphate glucokinase which provides building intermediates for secondary biosynthesis. The enzyme was present in the culture only after 12 h of cultivation, its activity rising sharply concomitantly with the decrease of ATP glucokinase activity (Fig. 2). The rise was steeper in the high-production variant (7).

POLYPHOSPHATE METABOLISM

All polyphosphate fractions in the cell reached a maximum only after the decrease of the ATP level. This fact is obviously connected with the shift from the adenylate phosphorylation mechanism to the polyphosphate system. The production variant possessed only about 10% of the polyphosphate level in strain RIA 57.

Determination of the activity of the two known enzymes of polyphosphate synthe-

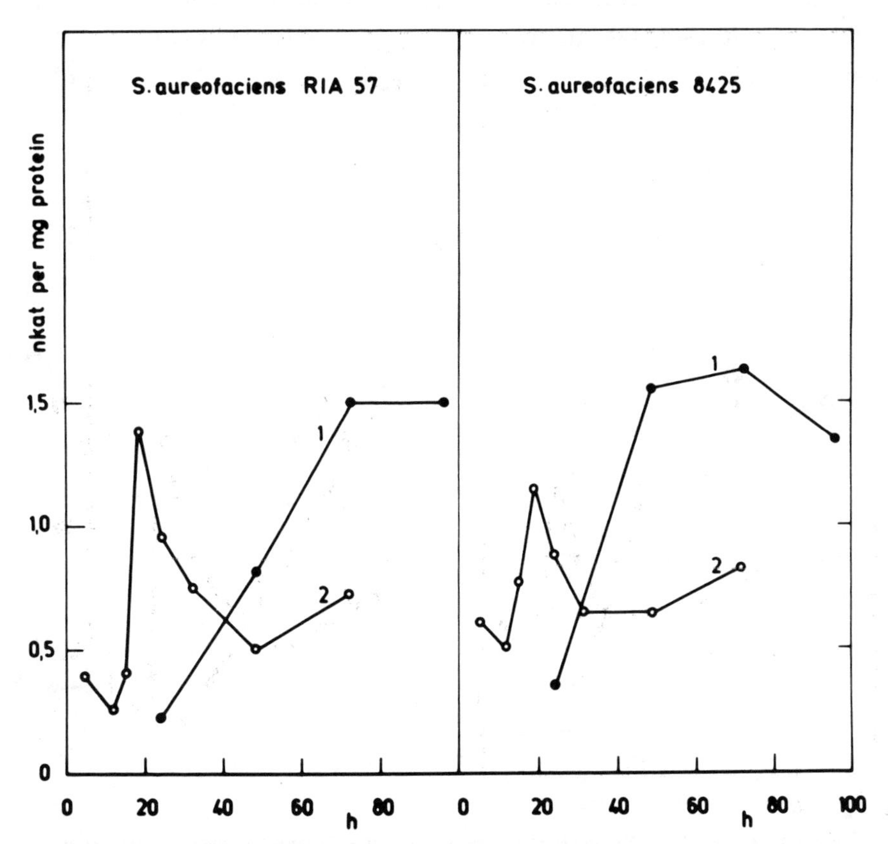

FIG. 2. *Activity of polyphosphate glucokinase and ATP glucokinase during cultivation of low-producing (RIA 57) and production (8525) strains of* S. aureofaciens. *(1) Polyphosphate glucokinase (nanokatals per milligram of protein); (2) ATP glucokinase (nanokatals per milligram of protein × 0.05).*

sis, ATP, polyphosphate phosphotransferase and 1,3-diphosphoglycerate: polyphosphate phosphotransferase, revealed appreciable differences in the level of the latter enzyme in the two strains (Fig. 3). During the ATP deficiency, the production variant compensates the increased demand for polyphosphates by an increased synthesis of 1,3-diphosphoglycerate:polyphosphate phosphotransferase in the stationary phase, whereas the wild strain retains the low level of the enzyme (8).

Study of ^{32}P incorporation into individual polyphosphate fractions showed that the highest turnover occurred in the high-polymer, acid-insoluble fraction. High-polymer polyphosphates are thus likely to participate in the phosphorylation processes associated with chlortetracycline biosynthesis (7).

ENERGY CHARGE

The biosynthesis of elevated amounts of chlortetracycline represents an energetically favorable system for the utilization of C3 compounds which accumulate during

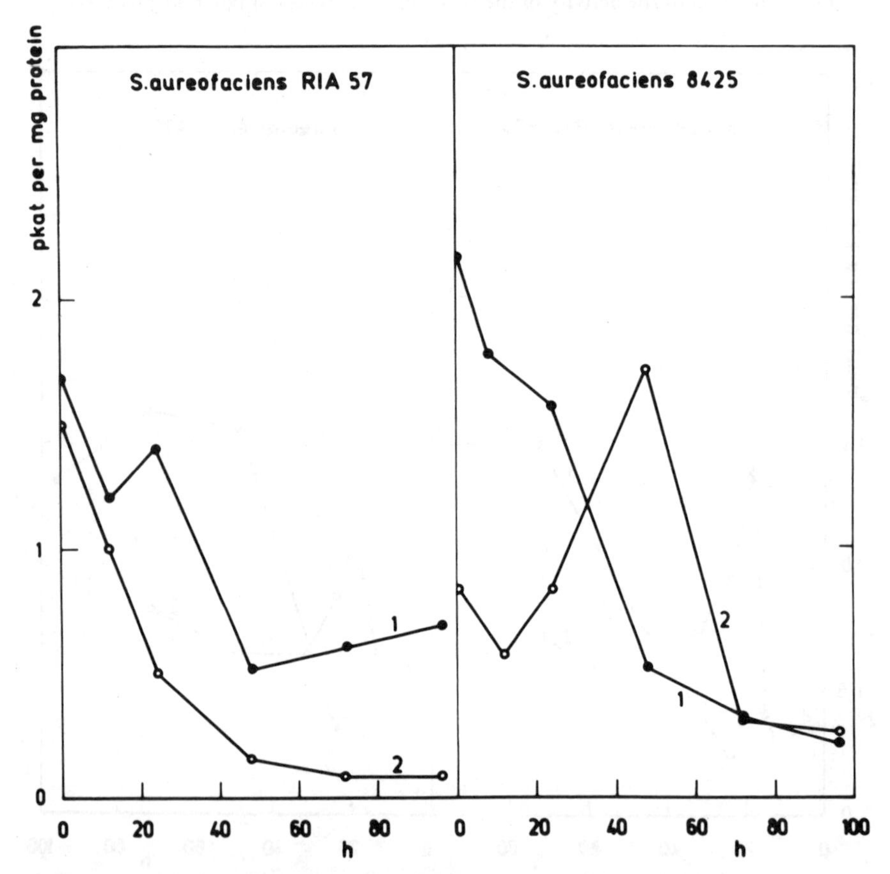

FIG. 3. *Activity of ATP:polyphosphate phosphotransferase and 1,3-diphosphoglycerate:polyphosphate phosphotransferase during cultivation of the low-producing (RIA 57) and production (8425) strains of* S. aureofaciens. *(1) ATP:polyphosphate phosphotransferase (picokatals per milligram of protein × 0.05); (2) 1,3-diphosphoglycerate:polyphosphate phosphotransferase (picokatals per milligram of protein).*

the initial high-rate sugar dissimilation. To evaluate the intensity of energy metabolism and its regulatory role, we determined the so-called energy charge during cultivation. According to Atkinson (1), this parameter, defined as [(ATP) + (½ADP)]/[(ATP) + (ADP) + (AMP)], reflects the energy state of the cell.

The intracellular level of the adenylates in strain RIA 57 was 10 times higher as compared with the production variant; thus, the latter strain is characterized by an overall suppression of adenylate synthesis. On the other hand, the values of the energy charge are practically the same in the two strains (Fig. 4). Their decrease in the stationary phase does not reflect a lowering in the level of the energy-rich ATP bonds but, more probably, an increase in ADP and AMP levels (5). The striking rise in the AMP level of the low-production strain may be caused by induction of the increased AMP synthesis, or by nonhomogeneity of the population. A regulation

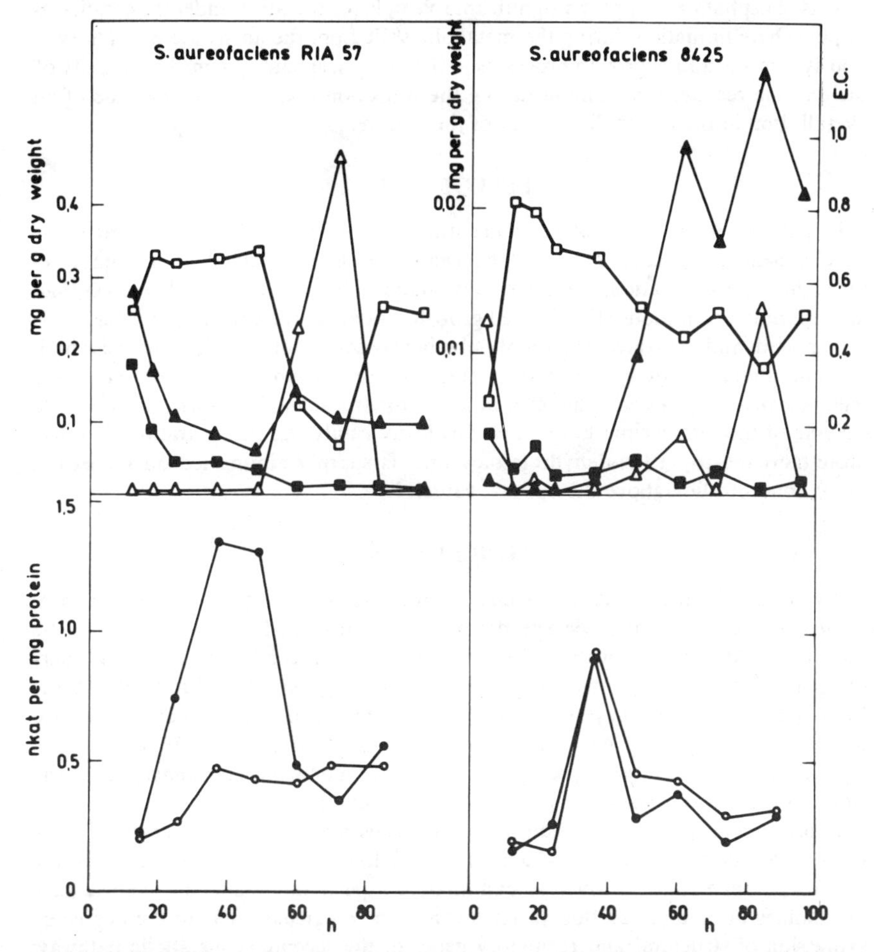

FIG. 4. *Intracellular concentration (milligrams per gram of dry weight) of ATP (■), ADP (▲), and AMP (△), energy charge (E. C., □), and phosphatase activity (nanokatals per milligram of protein) of membrane (●) and supernatant (○) fractions during cultivation of low-producing (RIA 57) and production (8425) strains of* S. aureofaciens.

mechanism analogous to the adenylate kinase reaction, which would stabilize the relative proportions of individual adenylates, is probably absent. Under these circumstances, the so-called energy charge is insufficient for the characterization of energy relations in the cell.

PHOSPHATASE ACTIVITY

The other possible explanation of the rise in the AMP level is the splitting of diphosphate bonds in ATP and ADP. The phosphatase activity in a culture of *S. aureofaciens* increased sharply after 24 h of cultivation and attained a maximum after 36 h, i.e., during the decrease of the ATP level (Fig. 4). The enzyme is a nonspecific phosphatase that cleaves diphospate bonds and is repressed in the presence of inorganic phosphate in the medium.

This phosphatase may play a significant role in P_i regeneration under the conditions of phosphate limitation during the metabolic shift from the adenylate phosphorylation system sustaining growth processes to the polyphosphate system. The activity of the enzyme reaches a maximum during the transition phase, i.e., at the time of an overall drop in the metabolic activity of the culture.

EFFECT OF P_i

In cultures with an increased concentration of P_i the customary character of development disappears, the ATP concentration is maintained at a high level throughout the cultivation, and the energy charge values are considerably lower than under limiting conditions (Fig. 5). Cell growth is favored, the content of nucleic acids is increased, and the concentration of polyphosphates doubles. Simultaneously, both the phosphatase activity and the synthesis of the enzyme system of secondary metabolism are repressed. The level of anhydrotetracycline hydratase, an inducible enzyme of the tetracycline biosynthetic pathway catalyzing the hydration of anhydrotetracycline to 5a,11a-dehydrotetracycline, is sharply diminished on increasing the P_i concentration above the optimal value.

CONCLUSION

The results indicate that the primary precursors of secondary metabolism in *S. aureofaciens* are generated via specific pathways different from those yielding the same intermediates for growth. A rise in the intensity of secondary metabolism occurs under conditions unfavorable for growth, e.g., during phosphate limitation. Apart from the synthesis of enzymes of secondary metabolism, these conditions induce other metabolic changes, e.g., the activation of regulation mechanisms governing P_i regeneration. The process represents an overall economizing rearrangement of the culture metabolism immediately after growth termination.

Improvement of the production microorganisms may result in the acquisition of hereditary metabolic changes which, in the wild strain, represent merely a purely phenotypic feature. However, the quantitative aspects of production of secondary metabolites in high-production variants cannot be restricted solely to the degree of expression of structural and regulatory genes of the secondary metabolic pathway; this facet represents only one of the details of the regulation changes which are to take place. The term "deregulation," used with high-production strains, denotes the rearrangement and a harmonious balancing of the large amount of regulatory

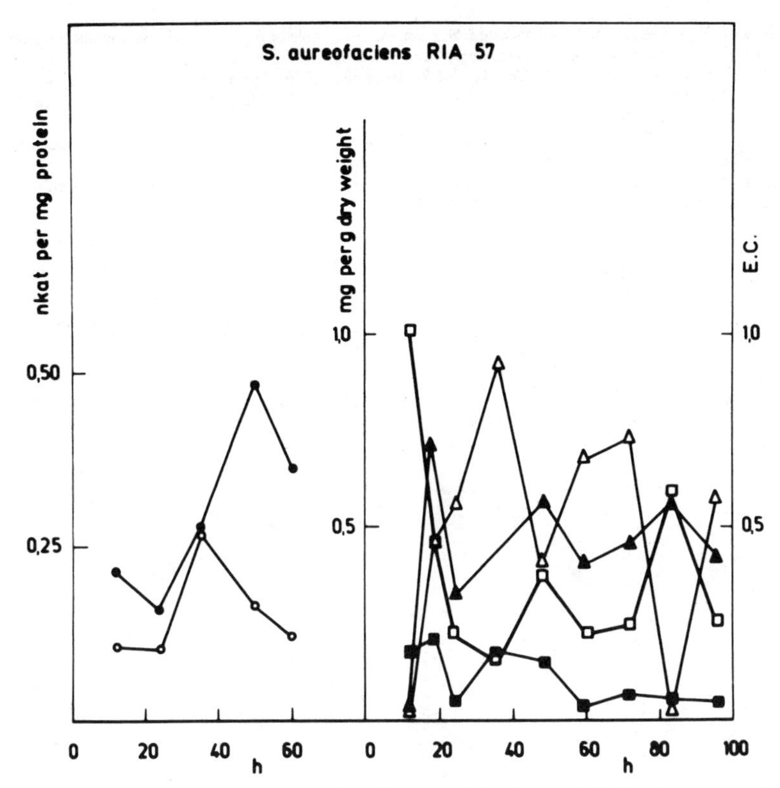

FIG. 5. *Intracellular concentrations of ATP* (■), *ADP* (▲), *and AMP* (△), *energy charge (E. C.,* □), *and phosphatase activity of membrane* (●) *and supernatant* (○) *fractions during cultivation of* S. aureofaciens *RIA 57 in a medium with increased* P_i *concentration (1.46 μmol/ml).*

mechanisms in the cell achieved by genetic manipulation. These alterations include chiefly changes in regulatory mechanisms at the primary metabolic level, a suppression of competing pathways and promotion of anaplerotic ones, and an establishment of an intact function of the metabolic systems providing basic biosynthetic intermediates.

REFERENCES

1. **Atkinson, D. E., and G. M. Walton.** 1967. Adenosine triphosphate conservation in matabolic regulation. Rat liver citrate cleavage enzyme. J. Biol. Chem. **242:**3239–3241.
2. **Běhal, V., J. Cudlín, and Z. Vaněk.** 1969. Regulation of biosynthesis of secondary metabolites. III. Incorporation of 1-^{14}C-acetic acid into fatty acids and chlortetracycline in *Streptomyces aureofaciens.* Folia Microbiol. (Prague) **14:**117–120.
3. **Běhal, V., V. Jechová, Z. Vaněk, and Z. Hošťálek.** 1977. Alternate pathways of malonylCoA formation in *Streptomyces aureofaciens.* Phytochemistry **16:**347–350.
4. **Běhal, V., and Z. Vaněk.** 1970. Regulation of biosynthesis of secondary metabolites. XII. AcetylCoA carboxylase in *Streptomyces aureofaciens.* Folia Microbiol. (Prague) **15:**354–357.
5. **Čurdová, E., A. Křemen, Z. Vaněk, and Z. Hošťálek.** 1976. Regulation of biosynthesis of secondary metabolites. XVIII. Adenylate level and chlortetracycline production in *Streptomyces aureofaciens.* Folia Microbiol. (Prague) **21:**481–487.
6. **Hošťálek, Z., M. Tintěrová, V. Jechová, M. Blumauerová, J. Suchý, and Z. Vaněk.** 1969. Regulation of biosynthesis of secondary metabolites. I. Biosynthesis of chlortetracycline and tricarboxylic acid cycle activity. Biotechnol. Bioeng. **11:**539–548.

7. **Hošťálek, Z., I. Tobek, M. A. Bobyk, and I. S. Kulayev.** 1976. Role of ATP-glucokinase and polyphosphate glucokinase in *Streptomyces aureofaciens.* Folia Microbiol. (Prague) **21:**131–138.
8. **Kulayev, I. S., M. A. Bobyk, I. Tobek, and Z. Hošťálek.** 1976. The possible role of high-molecular polyphosphates in the biosynthesis of chlortetracycline in *Streptomyces aureofaciens* (in Russian). Biokhimiya **41:**343–348.
9. **Vaněk, Z., J. Cudlín, M. Blumauerová, and Z. Hošťálek.** 1971. How many genes are required for the synthesis of chlortetracycline? Folia Microbiol. (Prague) **15:**225–240.
10. **Vaněk, Z., and Z. Hošťálek.** 1972. Some aspects of the genetic control of the biosynthesis of chlortetracycline. Postepy Hig. Med. Dosw. **26:**445–467.

Background and Legislative Activities Related to Recombinant DNA Research During the 95th Congress

JAMES M. McCULLOUGH

Congressional Research Service, Library of Congress, Washington, D.C. 20540

For those who work in science policy, the legislative activities generated during the 95th Congress in response to debates about the recombinant DNA technique have been most interesting. It has been possible to observe firsthand the dynamics of an activity which is viewed as a classic case history of the decision-making process in science and public policy. This issue illustrates the difficulty confronting the Congress in determining how public policy decisions should be made when technologies are viewed as potentially dangerous. This is one of the reasons that the National Science Foundation funded a project with the Department of History at the Massachusetts Institute of Technology to construct a real time national depository of information on this issue to permit scholarly study of the processes involved. Since Federal legislative actions are not complete, and there are some indications that such actions may not be completed by this Congress, it is not possible to write a conclusion to the legislative positions. To gain an understanding of the congressional reactions, it does seem appropriate, in fact almost essential, to look at part of the data base from which the Congress began to examine the need for regulation of recombinant DNA research before summarizing the specific activities of the 95th Congress.

The following comments should not be interpreted as evidence of all of the earlier discussions in the Congress relevant to this issue. However, the examples selected are believed to be sufficient to make the point that the Congress and the public were not (or should not have been) entirely unprepared for the announcement by the scientific community calling for restrictions on certain types of recombinant DNA experiments following the 1973 Gordon Conference of Nucleic Acids. In fact, if the documents cited in the following commentary are carefully examined, and if the terminology is accepted in the broad terms in which it is generally perceived by lay persons, it is possible to see a real similarity to the current DNA issue. It is particularly interesting to note the time spans between these earlier discussions and the evolution of the legislative activity on the recombinant DNA issue—the forecasters of 1968 were not too far afield in identifying the proximity of breakthroughs in genetics although the precise methodology or areas of research in which these breakthroughs occurred were not specified.

On 7 March 1968, Fred Harris (then Chairman of the Subcommittee on Government Research, Committee on Government Operations, and Senator from Oklahoma) opened hearings on S. J. Res. 145, a joint resolution for the establishment of a National Commission on Health Science and Society (introduced by Walter Mondale, then Senator from Minnesota). In his statement, Harris included the comment (6), " . . . Investigations in areas such as genetics research promise to provide new tools for changing the character and quality of life." Mondale pointed out in his opening remarks that the public response to his proposal had identified certain areas, including "genetic engineering," requiring studies by the proposed commission. ("Genetic engineering" from the public perspective is thought of in the broadest possible terms; lay persons have been conditioned by the news media to view basic research in such techniques as recombinant DNA, transformation, transduction, cell fusion, cloning, etc., as all part of the scientific programs in "genetic engineering." Joshua Lederberg frequently has pointed out the communication problems associated with the use of this term and has attempted to gain acceptance of more precise terminology, but the public view is still quite general and broad in scope when this term is used.) In further discussion of his proposed legislation, Mondale said (6):

> These new developments [in biomedical technology; it should be kept in mind that the 1968 and 1971 hearings discussed herein had as a major focus of attention the public concern at that time with such issues as organ transplantation, kidney dialysis, and abuses in human experimentation] are as dramatic as the dawning of the nuclear age. And some of them, like genetic manipulation and behavioral control, are potentially as dangerous. Their potential benefits to human physical and mental health are tremendous of course. But our experience with the atom teaches us that we must look closely at the implications of what we do. [A paper presented by a Representative during the 1978 American Association for the Advancement of Science conference in Washington, D.C., drew a critical analogy between recombinant DNA research and applications in nuclear technology.]

One of the witnesses at these hearings, John S. Najarian (University of Minnesota), pointed out that although he wished to focus his attention during the discussion primarily on transplantation of organs, the other issue of interest to him was (6):

> ... the possibility of changing the DNA, the essential nucleic acid molecules of the cell, and what implications this is going to have on the genetic makeup of the future and of humans today. I think that the important consideration is that the DNA replication, or the change in DNA, or DNA engineering, as Senator Mondale pointed out is something that is in the future.

Arthur Kornberg (Stanford University) discussed his work on the synthesis of DNA copies during these hearings. He referred to his initial irritation at a reporter's question as whether his work could be cited as having created life in a test tube and noted:

> ... What's wrong with my saying "enzymatic synthesis of viral DNA"? I then realized that the reporter was simply asking a question in language that would be asked by the average citizen. Semantic problems flourish when there is a lack of understanding. ... We spoke several hours going over ... the work.

Other comments by Kornberg are as relevant to the recombinant DNA controversy today as they were when he made them in a different context about his own research 10 years ago. As he noted (6):

> We desperately need more information about the molecular anatomy and behavior of genes to understand and cope with *nature's genetic engineering* [emphasis added] ... One way [to prepare ourselves for developments in genetic engineering] is to educate all our citizens so that biologists and chemists are not the only people called upon to make the crucial judgments. ... What you do [he stated in discussion with the committee members] in congressional legislation has an enormous impact on our future. ... For my part, I see no alternative but to learn more and share my knowledge.

Joshua Lederberg was another witness at these earlier hearings, as he has been on other subjects related to genetics and controversial issues in biomedical research. He remarked (6) that the discussion held in public by committe members with Kornberg was "one of the most fascinating and important colloquies I have ever had the opportunity to attend—this kind of confrontation of political and scientific wisdom does not occur often enough."

While Lederberg was not in full agreement with all sections of the legislative proposals for technical reasons, he was in agreement with the need to insure a full consideration of the scientific work marked for concern before the legislative decisions were made. In his testimony following Kornberg, Lederberg was asked to discuss the status of the cloning of vertebrate animals (note here that then, as today, there was an overlap in the perception of the work on the molecular biology of the gene and the techniques of cloning).

During hearings in 1971 on a similar proposal to establish a commission to evaluate the ethical, social, and legal implications of advances in biomedical research (S. J. Res. 75, introduced again by Mondale), Edward Kennedy (Senator from Massachusetts) noted in his opening remarks (10):

> Advances in modern medical science have lengthened the span and changed the quality and very meaning of human life. At the same time, these advances have opened a Pandora's box of ethical, social, and legal issues. The hearings will focus on these problems in areas such as heart transplants, artificial kidneys, test-tube babies, genetic intervention, and experiments on humans.

In these same hearings, Mondale again noted his concern about issues in the general field of genetics by stating (10):

> With fertilization of human egg cells in the laboratory, with the possibility of successful implantation of such eggs in human beings, and with the potential for developing so-called duplicate people, we may be facing fundamental changes in the status of the individual, the role of the family, and in the very character of our society.

During his testimony on this resolution Merlin K. Duval, then Assistant Secretary for Health and Scientific Affairs, Department of Health, Education, and Welfare (DHEW), highlighted a topic a little closer to the recombinant DNA issue when he observed that (10):

> The extraordinary advances in genetics have illuminated to a remarkable degree the mechanisms through which biological information is transmitted from generation to generation. Parallel advances in virology have provided at least one lead to a method for replacing the genetic material of a cell with different information.
>
> Thus the very real possibility of "genetic engineering" is on the horizon. What, if any, ethical limits are appropriate in this area? What unique or special considerations come into play when the experimental species is man? How rigid should be the criteria for determining that an induced genetic alteration is not, in the long run, compatible with species survival? What attitudes should we take on the question of cloning? —possible asexual replication of individuals in the form of genetically identical products.

Hans Jonas (Department of Philosophy, The New School for Social Research) said (10):

> The more modest part [of genetic control] . . . not yet genetic engineering . . . aims at upholding rather than changing the norm of nature. They are essentially conservative. But there are more ambitious dreams abroad, summed up in the phrase that man will take charge of his own evolution, which means not watching over the integrity of the species, by modifying it by improved designs.
>
> . . . This is what is properly called biological engineering and it is innovative in intent.

Whether we are qualified for that creative role is the most serious question that can be posed to man finding himself suddenly in possession of such fateful powers.

Incidentally, Robert Sinsheimer, quoted so often today as urging caution because of his views about the potential for impact on evolutionary processes of modern recombinant DNA research, had already expressed this concern about "the potential for change in and control of the living world comparable to the mastery we have already achieved over our physical environment" during these same hearings (10).

In his letter to the committee which was submitted for inclusion in the hearing record (10), John T. Edsall (Harvard University) in reference to cloning of human beings asked, "Are we to ban certain types of experimentation as my colleague, James D. Watson has suggested might be desirable?"

To understand the relevance of these earlier discussions vis-à-vis our current recombinant DNA discussion, one must remember that subjects such as in vitro fertilization, the accomplishments by H. Gobind Khorana on gene synthesis, and correction of gene deficiencies by gene transfer were headline topics in newspapers and feature articles in weekly and other journals. (For example, *Time*, 19 April 1971, featured an article subtitled "The Promise and Peril of the New Genetics.")

James Watson, in a letter to the *Washington Post* following his discussion on cloning before the House Science and Technology Committee Science Panel in January 1971, indicated that, "The matter is much too important to be left in the hands of the scientists whose careers might be made by achieveing of a given experiment." At the House Science and Technology Panel Conference in 1971 (preceding the 1971 Senate hearings on S. J. Res. 75), Watson had chosen to speak on the potential for cloning and the need to examine this issue. Following his appearance, this same committee identified a number of issues in genetics as warranting a continuing examination and commissioned a series of studies on new developments in genetics research. These were published in 1972, 1974, and 1976, with the last study placing emphasis upon the developments in the use of recombinant DNA research.

Thus, an environment of concern was present in the Congress for reaction to the public announcement of scientists concerning the developments in recombinant DNA beginning in 1973. The fact that this announcement by the scientific community was one of concern served to increase the degree of this reaction within the several committees which had been sensitized to the potential for socially significant developments in genetics generally.

The first formal committee hearing directed specifically toward the recombinant DNA issue was held on 22 April 1975 following the Asilomar Conference (7). The Subcommittee on Health (Senate Committee on Labor and Public Welfare) asked a small group of witnesses to examine questions such as: What is the nature of this research which so disturbed the investigators that they felt compelled to stop it for a time? Is there a safety threat to the general population? What are the implications of the research for the society as a whole? Was it proper for scientists alone to decide to stop and then resume the research? How could nonscientists participate in the process, even if that were desirable; what should be done now in terms of public policy in this area? What are the potential dangers of Federal intervention? Naturally, a great deal of attention also was paid to the issue of potential hazard which might be associated with this research since this was the topic of the original concern discussed during the Asilomar Conference which had led to the establishment of the National Institutes of Health (NIH) Recombinant DNA Advisory Committee. Another topic of interest was the extent to which the public participation could occur in any potential regulation of scientific research. Obviously, congressional concern was not alleviated by what was learned during this hearing.

The next formal examination by a congressional committee was a joint hearing in September 1976 before the Subcommittee on Health and a subcommittee of the Senate Committee on the Judiciary. This time the topic was an oversight examination of the implementation of the NIH guidelines on recombinant DNA research (8). In these hearings, Edward Kennedy clearly indicated that his concern extended beyond the safety issues posed specifically by the recombinant technique. In his words (8):

> Recombinant DNA research presents a prototype of the problems our society will face over and over again as technology develops. The issues go far beyond safety questions—in many ways, those are the easiest to answer. The real problem is to understand the social consequences of what science can now enable us to do . . . It is my belief that at the very least, all Federal agencies should comply with the NIH guidelines.These guidelines need not be viewed as fixed or final. But they should be observed while appropriate modifications are made. I also believe that industrial research in this area should conform to NIH guidelines. . . . I would hope that industry would voluntarily comply with the guidelines. If they do not, we will have to consider legislative action.

Although not directed specifically toward legislation, the Subcommittee on Science, Research, and Technology (Chairman, Ray Thornton, Representative from Arkansas) of the House Committee on Science and Technology (Chairman, Olin Teague, Representative from Texas) planned and held 12 days of hearings on the science policy implications of recombinant DNA molecule research beginning in March and continuing into September 1977. Thus, these policy hearings were initiated during the period that legislative hearings were being held. The Science Committee hearings were preceded by 6 years of interest in issues related to genetic research, during which three reports on the subject were issued as noted earlier (3; see also 4). Fifty-six witnesses testified on subjects ranging from the significance for basic research policies of the issues generated by the high level of controversy over recombinant DNA research to an examination of the several legislative proposals introduced during the 95th Congress. On 3 May 1977, Teague requested sequential referral of whatever bill on the subject of recombinant DNA research might be reported from the Committee on Interstate and Foreign Commerce. He also requested that any bills on this subject which came to the House from the Senate be duly referred to the Science Committee.

The activities of this committee are mentioned before summarizing the legislative hearings for several reasons: (i) the hearings of the Science Committee accompanied the legislative hearings in other committees and included sequential referral of bills on the subject because of the implications for basic research of the legislative proposals for regulation, (ii) the issues were deemed of sufficient importance to science and public policy issues to warrant examination by the Science Committee, and (iii) the primarily nonlegislative orientation by the Science Committee provided a less controversial environment for examination of many of the peripheral science policy issues highlighted by the recombinant DNA issue without emphasizing the emotional and debative approach which frequently accompanies the construction of regulatory legislation. Thus, comments on the legislative hearings can be examined with the knowledge that nonlegislative hearings afforded opportunities for interested parties to examine these issues in a different type of committee environment. (The

Senate Subcommittee on Science, Research, and Space [Chairman, Adlai E. Stevenson] also initiated science policy hearings later in the stages of the legislative hearings.)

In January, February, and March 1977, a number of bills were introduced in the House which called for some combination of regulation and study of the recombinant DNA issue. These earlier bills would have resulted in rather stringent licensing, inspection, fines, administration, and similar monitoring and control measures if they had been enacted. On 6 April 1977, the administration's proposal was introduced as H.R. 6158 by Paul G. Rogers (Representative from Florida, Chairman of the Subcommittee on Health and Environment, Committee on Interstate and Foreign Commerce). Following subcommittee mark-up, a discussion draft bill, H.R. 7418, was introduced by Rogers on 24 May 1977. Following further mark-up sessions in June 1977, a "clean" bill, H.R. 7897, was introduced on 20 June 1977 by Rogers and a number of cosponsors. H.R. 7897 was considered in full committee mark-up sessions in September and October 1977 with several amendments being accepted. This mark-up was not completed during the first session, and therefore the bill was not reported by the committee.

On 19 January 1978, Harley O. Staggers (Representative from West Virginia, Chairman of the Committee on Interstate and Foreign Commerce), introduced H.R. 10453. This bill was never considered in committee. On 28 February 1978, Staggers and Rogers introduced H.R. 11192, a significantly revised bill in comparison with H.R. 7897. H.R. 11192 was subsequently considered by the full Committee on Interstate and Foreign Commerce on 15 March 1978, amended, and reported on 24 March 1978 (1). Dissenting views were included in the report.

H.R. 11192 was sequentially referred to the Committee on Science and Technology for a period ending not later than 21 April 1978. The Subcommittee on Science, Research, and Technology held hearings on this bill on 11 April and reported the bill on 13 April 1978. On 18 April 1978, the full Committee on Science and Technology considered and reported the bill as amended. Thus, at this time, H.R. 11192 is the bill most likely to be considered in the House (2). Again, dissenting views are included in the report.

Final House committee action is not completed, however, since the Interstate and Foreign Commerce Committee and the Science and Technology Committee versions of the same bill differ slightly as amended by the Science Committee. A ruling and decision on these differences, with the resolutions needed if a conference is required, may be necessary before the bill goes to the floor for a calendar date for debate and vote.

In the Senate, the status of legislation at the moment is even more tenuous. Hearings were held in April 1977 on three bills which had been referred to the Subcommittee on Health and Scientific Research, Committee on Human Resources (9).

S. 1217 as reported by the Committee on Human Resources on 22 July 1977 would provide for the establishment of a National Recombinant DNA Safety Regulation Commission. Duties of the Commission would include the promulgation of rules and regulations, monitoring of compliance by owners or operators of licensed facilities or by licensed persons, development of methods of safety monitoring, and the encouragement of risk assessment studies (5).

An early indication of dissatisfaction with S. 1217 was Amendment 754 by way of substitution for S. 1217 submitted to the Senate by Gaylord Nelson (Senator from Wisconsin) on 2 August 1977. In his remarks accompanying his submittal, Nelson noted that the amendment reflected the new information and views concerning the nature and extent of regulation necessary, definitions of recombinant DNA activities to be regulated, and a reduced emphasis on penalties and preemption. This proposal appeared to be given serious consideration in the revisions of legislation which later appeared in both the House and the Senate.

On 22 September 1977, in a floor statement about the recombinant DNA issue, Adlai E. Stevenson (Senator from Illinois), Chairman of the Subcommittee on Science, Technology, and Space (Senate Commerce, Science, and Transportation Committee), indicated his concern about the need for careful examination of the potential impact on such issues as freedom of inquiry before promulgation of restricted legislation on recombinant DNA research. At this time, he indicated that he would urge the adoption of interim legislation which would provide for flexibility in accommodating scientific evidence as it developed. He stated that he had several amendments to ensure such an approach. The Subcommittee on Science, Technology, and Space held their hearings in November 1977. During these hearings the issues identified by Stevenson in his September floor statement were examined (11).

On 1 November 1977, the day before the Senate Science Committee hearings, Harrison Schmitt (Senator from New Mexico) introduced S. 2267, a proposal which is broader in scope than any of the recombinant DNA legislation. As he remarked during the Senate Subcommittee on Science, Technology, and Space hearings, it is his hope that the proposal for the establishment of a National Science Policy Commission would permit the Congress to secure information on hazardous research and enable consistent policy guidelines to be developed so as to avoid unnecessary Federal interference with valid scientific endeavors. His proposal would essentially provide for the establishment of a special study commission and an examination of their recommendations by the Congress before legislation, such as that proposed for recombinant DNA, would be enacted.

On 1 March 1978, Edward Kennedy (Senator from Massachusetts) submitted to the Senate a proposed amendment, 1713, which in essence is a substitute for the text of S. 1217 as reported out by the Committee on Human Resources. This proposed amendment would significantly modify S. 1217 and represents a major change in the proposal first examined by Kennedy's Subcommittee on Health and Scientific Research. Amendent 1713, except for minor differences in wording and a major exclusion of a section on Federal preemption, is very similar to H.R. 11192 reported from the House Interstate and Foreign Commerce Committee and sequentially referred to the House Science and Technology Committee. Amendment 1713 would, if adopted by the Senate, completely replace S. 1217 which had been agreed to in committee.

An examination of the proposals in the Senate, primarily Amendment 1713, for regulating recombinant DNA had been scheduled before the Human Resources Committee on 4 and 5 May 1978, but this examination was postponed. It now appears that the Human Resources Committee is examining the possibility of requiring the Secretary of DHEW to use section 361 of the Public Health Service Act to extend the recombinant DNA guidelines to commercial and private research. This section of the Public Health Service Act authorizes the Surgeon General—Secretary DHEW—to make and enforce such regulations as in his judgment are necessary to prevent the introduction, transmission, or spread of communicable diseases from foreign countries into the States, or possessions, or from one State or possession into any other State or possession.

In summary, then, the current status of legislation in the House is a duly considered bill, H.R. 11192, which would provide the authority for the Secretary of DHEW to extend the recombinant DNA guidelines to non-Federally funded recombinant DNA research and to provide for review and inspection, for the maintenance of records, and for procedures for imposing fines in the event of violations requiring an action. Federal regulations would preempt local regulations unless local communities could demonstrate the necessity for deviation from the Federal guidelines. Provisions for exempting certain categories of research and for modification of the guidelines are included. Title II of the bill would provide for the establishment of a commission to study the problem and report back to the Congress as to the need for further legislation. The life of the legislation would be limited to 2 years. The Congress would then have to determine the need for further legislation. Indications are that the regulatory agency would be the Center for Disease Control, Atlanta, Ga., with the NIH Recombinant DNA Advisory Committee continuing to function as a technical source of recommendations for revisions of the guidelines.

In the Senate, S. 1217 retains the much more extensive regulation, with most of the regulatory and study responsibilities assigned to a national commission. This bill would permit local communities more opportunity to impose additional requirements found to be relevant and material to health and the environment. It now appears unlikely that this bill will be enacted in this form. If accepted, Amendment 1713 would establish requirements very similar to those of H.R. 11192, with a major difference in the area of Federal preemption, which is not addressed as specifically as it is within H.R. 11192. Nelson's position, as established by Amendment 754, could be considered a position containing aspects of both of these proposals. Although extension of authority to act under section 361 of the Public Health Service Act apparently would be precedent setting if used to regulate basic research, some opinions suggest that this is a possible alternative to specific legislation. Other opinions suggest that a resolution in court might be required as to the appropriateness of section 361.

The evidence at hand indicates that the Congress has devoted a tremendous amount of time to this issue relative to the actual funding levels involved for the research being carried out. This effort is a reflection not only of the awareness of the Congress of the long-term significance of this particular research technique and of the potential for some undefined or possibly only conjectural level of risk but also of the implications that the issue has for the regulation of basic research and the establishment of national science policies.

Legislative activity on the recombinant DNA issue has been interesting from a number of perspectives. The modification of proposed legislation in both the House and the Senate has resulted in a substantial move away from a strong regulatory approach to a stance which now can only be described as consisting of more lenient proposals to extend controls to industry, with reduced local intervention. This shift in position appears to have been the result of a number of factors at work, including the changes on DHEW committees to permit greater public participation in review of NIH guideline development, the steady and careful examination of the issues in public hearings by the several committees and in local communities, and the broadened examination of potential risk within the biomedical research community which has reduced scientific concern about risk. This change in perception of risk has been effectively communicated to the appropriate congressional committees by the scientific community.

REFERENCES

1. **U.S. Congress. House. Committee on Interstate and Foreign Commerce.** 1978. Report. Recombinant DNA Act. 95th Congress, 2nd session. Report no. 95-1005, part 1. 24 March 1978. U.S. Government Printing Office, Washington, D.C.
2. **U.S. Congress. House. Committee on Science and Technology.** 1978. Report. Recombinant DNA Act. To accompany H. R. 11192. Report no. 95-1005, part 2. 21 April 1978. U.S. Government Printing Office, Washington, D.C.
3. **U.S. Congress. House. Subcommittee on Science, Research, and Technology, Committee on Science and Technology.** 1977. Hearings. Science Policy Implications of DNA Recombinant Molecule Research. 95th Congress, 1st session. 29, 30, and 31 March, 27 and 28 April, 3, 4, 5, 25, and 26 March, and 7 and 8 September 1977 (no. 24). U.S. Government Printing Office, Washington,D.C.
4. **U.S. Congress. House. Subcommittee on Science, Research, and Technology, Committee on Science and Technology.** 1978. Report. Science Policy Implications of DNA Recombinant Molecule Research (Serial X). March 1978. U.S. Government Printing Office, Washington, D.C.
5. **U.S. Congress. Senate. Committee on Human Resources.** 1977. Recombinant DNA Safety Regulation Act. Report Together with Supplemental Views to Accompany S. 1217. 95th Congress, 1st Session. Report 95-359. U.S. Government Printing Office, Washington, D.C.
6. **U.S. Congress. Senate. Subcommittee on Government Research. Committee on Government Operations.** 1968. Hearings on S. J. Res. 145. National Commission on Health Science and Society. 90th Congress, 2nd session. 7, 8, 21, 22, 27, and 28 March; 2 April 1968. U.S. Government Printing Office, Washington, D.C.
7. **U.S. Congress. Senate. Subcommittee on Health, Committee on Labor and Public Welfare.** 1975. Hearings. Examination of the Relationship of a Free Society and Its Scientific Community. Genetic Engineering, 1975. 94th Congress, 1st session. U.S. Government Printing Office, Washington, D.C.
8. **U.S. Congress. Senate. Subcommittee on Health, Committee on Labor and Public Welfare. Subcommittee on Administrative Practice and Procedure, Committee on the Judiciary.** 1976. Oversight Hearing on Implementation of NIH Guidelines. 94th Congress, 2nd session. 22 September 1976. U.S. Government Printing Office, Washington, D.C.
9. **U.S. Congress. Senate. Subcommittee on Health and Scientific Research, Committee on Human Resources.** 1977. Hearings on S. 1217. Recombinant DNA Act, 1977. 95th Congress, 1st session. 6 April 1977. U.S. Government Printing Office, Washington, D.C.
10. **U.S. Congress. Senate. Subcommittee on Health and Subcommittee on National Science Foundation. Committee on Labor and Public Welfare.** 1972. Joint Hearings on S. J. Res. 75. National Advisory Commission on Health Science and Society. 92nd Congress, 1st session. 9 November 1971. U.S. Government Printing Office, Washington, D.C.
11. **U.S. Congress. Senate. Subcommittee on Science, Technology and Space, Committee on Commerce, Science and Transportation.** 1977. Hearings on the Regulation of Recombinant DNA Research (Serial 95-52). 95th Congress, 1st session. 2, 8, and 10 November 1977. U.S. Government Printing Office, Washington, D.C.

A Scientist's Perspective on Regulation of Recombinant DNA Activities

ROY CURTISS III

Department of Microbiology, Institute of Dental Research, Cancer Research and Training Center, University of Alabama in Birmingham, Birmingham, Alabama 35294

The ability to join together DNA molecules from disparate organisms in a test tube environment and reintroduce them into a living cell was first reported in 1973 (17). This, along with antecedent (34, 36) and subsequent (11, 16, 32, 39, 52, 57) discoveries, sparked the commencement of what is now referred to as the "recombinant DNA debate" (29). It was members of the scientific community, including the early practitioners of this art, who initially called attention to the potential risks of certain recombinant DNA experiments while lauding the benefits (6, 46). This led to a succession of meetings to consider the potential benefits and biohazards of recombinant DNA research (5, 7, 23, 40, 45). In this country, these and other considerations ultimately led to the formulation of a set of guidelines by the National Institutes of Health (NIH) for the conduct of recombinant DNA research (41); similar sets of guidelines have been adopted (30, 58) or are under consideration for adoption in other countries. The debates within the scientific community ultimately stimulated concern on the part of the public, and this in turn has led to passage or current consideration for passage of legislative acts to regulate recombinant DNA activities at the local, state, and federal level in the United States (29; J. M. McCullough, this volume).

In light of the past debates and our current knowledge about the benefits and potential biohazards of recombinant DNA activities, it is fitting to ask, "Should there be legislation to enable regulation of recombinant DNA activities in both the public and private sectors? If not, why not, and, if so, why and at what levels of government and in what manner?" We can also ask, "What are the consequences of any of the possible actions with regard to enactment of legislative authority to regulate recombinant DNA activities?" In arriving at my answers to these questions, I will consider the safety of recombinant DNA activities, the impact of regulation or lack thereof on society's perception and support of science and scientists, and the benefits, if any, accruing to society and science due to legislatively mandated regulation of these activities.

SAFETY OF RECOMBINANT DNA RESEARCH

Since 1973, there has been a gradual change in what I call the "median" opinion of biologists concerning the risks associated with recombinant DNA activities. At the outset of the recombinant DNA debate, one talked about "potential biohazards" (6, 7, 18, 46), and some considered the risks to be real rather than just potential (13, 47, 55). More recently, the phraseology "conjectural biohazards" (14) has been used, and "biohazard committees" are referred to as "biosafety committees" in the currently proposed revisions of the NIH Guidelines for Recombinant DNA Research, which reflect these changed assessments of risk in reducing the required levels of physical and biological containment for certain experiments.

We can ask, "What is the basis of this change in scientific opinion, and is the information upon which these reassessments depend valid?" Until the present, probably 95% of all recombinant DNA research has made use of *Escherichia coli* K-12 host-vector systems (48). It was the use of these systems that led to the initial concern that introduction of foreign DNA might pose a threat to human health (6, 46). Six different factors have entered into the discussions which have led to the general opinion that use of the *E. coli* K-12 host-vector systems is much safer than originally believed.

The first was a consideration of information which was available prior to 1973 but was not well appreciated by those initially involved in the recombinant DNA debate. This information was on such topics as infectious diseases, public health, epidemiology, mammalian physiology, endocrinology, gastroenterology, immunology, etc., and was highly relevant in evaluating risks (24, 28, 43). When those engaged in the recombinant DNA debate ultimately became aware of the necessary knowledge in these disciplines, it became apparent that many of the oft cited scenarios for harm and/or catastrophe were highly improbable (43).

The second factor was the availability of an increasing amount of data about the properties of *E. coli* K-12 that was not known to the early recombinant DNA debaters. When this information, on the inability to endow K-12 with pathogenicity by use of conventional genetic crosses with known pathogens (26, 38, 50) and on the relatively poor ability of K-12 to colonize the intestinal tracts of humans and other animal species (1, 19, 21, 27, 28, 49), became generally known, it was realized that the risks in using *E. coli* K-12 hosts were less than initially perceived.

The third factor was a result of studies in different laboratories to assess the biological containment afforded by various *E. coli* K-12 host-vector systems in recombinant DNA research. These experiments not only verified and substantiated the levels of biological containment of these host-vector systems (3, 8, 21, 22, 27) but also provided previously unavailable information on the likelihood for transmission of recombinant DNA (1, 2, 19, 21, 25, 44) and the ability of various animal host defense mechanisms to thwart the spread of organisms and/or vectors containing recombinant DNA (19, 33, 37).

The fourth factor was a consideration of principles of evolution which led to the belief that the addition of a new trait to an organism would not often provide a selective advantage or lead to a situation in which the new genetic information would be totally compatible (i.e., coadaptive) with that of the host organism (4, 24). The observation that organisms and/or vectors containing recombinant DNA are less fit than the same organism or vector lacking the foreign DNA, at least under laboratory conditions (9, 15), seems to support these beliefs.

The fifth factor has been extensively debated and concerns the question whether gene flow does or does not occur between lower and higher organisms in nature (10, 13, 18, 24, 47). The observed expression of genes from lower eucaryotes in *E. coli* (51) and the demonstration that foreign DNA taken up by *E. coli* can result in formation of recombinant DNA molecules in vivo (12) indicate that gene flow can occur. However, I do not believe this factor is particularly relevant in deciding on the issue of safety. In addition, gene flow in nature requires that both donor and recipient share the same ecological niche, and there is no such restriction in the formation of in vitro recombinant molecules.

The sixth factor has been the absence of any noticeable biohazard associated with the construction of any recombinant during the past several years (14). While reassuring, it should be noted that biohazards were not looked for and, indeed, all experiments were conducted under levels of physical and biological containment that we now agree were overrestrictive and which would preclude manifestation and/or detection of any biohazard should they have arisen. Thus, it is scientifically invalid to use the experience of 5 years of research as a justification to argue that biohazards do not exist. It will thus be necessary to design and conduct experiments expressly for the purpose of determining whether the introduction of foreign DNA does or does not result in manifestation of biohazards, and such experiments have not been done (43).

It was on the basis of the first four factors that I concluded in April of 1977 "that the introduction of foreign DNA sequences into EK1 and EK2 host-vectors offers no danger whatsoever to any human being with the exception ... that an extremely careless worker might under unique situations cause harm to him—or herself" (see 20). I still believe that this conclusion is valid since it is substantiated by more data than were available a year ago. I also believe, although with few data to validate my belief, that use of the *E. coli* K-12 host-vectors for recombinant DNA research poses no threat to nonhuman organisms in the biosphere. I should emphasize, however, as I did last November (20), that these conclusions were stated in reference to the conduct of those experiments then permitted by the NIH Guidelines for Recombinant DNA Research (41) and only pertained to the use of the *E. coli* K-12 host-vector systems. It should be evident that the recombinant DNA technology will ultimately be used with a diversity of microbial and eucaryotic hosts in conjunction with an equal diversity of viral and plasmid vectors. We can also anticipate that the technology will ultimately be used to clone genes from highly virulent pathogens and/or genes that specify potent toxic substances since such experimentation will ultimately contribute great benefits to infectious disease control and improvement in public health. While we can surmise that the use of these other systems and the conduct of experiments not now permitted might pose little more risk than currently permitted experiments with the *E. coli* K-12 host-vector systems, there is a conspicuous lack of experimental data to justify such beliefs (20).

In conclusion, the scientific community now has sufficient information to substantiate the belief that recombinant DNA activities with *E. coli* K-12 host-vectors are without hazard, at least to humans and other warm-blooded animals. This information has also led the scientific community to believe, although certainly it does not know for a fact, that other recombinant DNA activities may be equally safe. In the absence of data to substantiate the safety of any and all uses and applications of recombinant DNA technology, it is only prudent that users of this technology be conservative and cautious until more information is known. If the only perceived risk of such activities were to the human species, I would argue that the issue of safety or the lack thereof is not particularly relevant to the issue of whether legislation is or is not needed to mandate adherence to a prescribed set of rules for the conduct of these activities. This belief is based on experience concerning laboratory-acquired infections, which have shown a low probability for transmission to family members, much less to members of the general public (42), and on the human attribute of being careful when the consequence of carelessness is harm to oneself. The use of recombinant DNA technology is not, however, restricted to constructing chimeric organisms that, if harmful, would be harmful to humans; indeed, some who believe that biohazards might be manifest consider that harm to organisms other than humans in the biosphere is a real possibility. In this instance, motivation of users of the technology would not always be present because of self-interest, and I thus believe that legislation mandating compliance with appropriate standards of performance of recombinant DNA activities would be prudent.

IMPACT OF LEGISLATED REGULATION OF RECOMBINANT DNA ACTIVITIES ON SOCIETY'S PERCEPTION AND SUPPORT OF SCIENCE AND SCIENTISTS

In the early stages of the recombinant DNA debate in 1973 through 1975, it was the scientists actively engaged in using the newly discovered recombinant DNA technology who initially expressed concern about the potential biohazards of certain experimental uses of the technology (6, 7, 46). Society's reaction to this concern among the scientific community was to infer that if the scientific community was concerned about the potential hazards of the research then these hazards must indeed exist. This inference (14) was certainly popularized in the news media. In the subsequent years, it was the scientific community that also began to amass the relevant knowledge and to collect appropriate data to enable the formulation of cogent arguments about the absence of risks and the safety of recombinant DNA research permitted by the NIH Guidelines (43). These activities took place concurrently with numerous experiments using recombinant DNA technology that contributed knowledge about biological processes which was not obtainable prior to the discovery of the technology (40). Thus, the scientific community not only contributed information to substantiate an assessment of decreased risks but also provided the information to justify the claims of the numerous benefits of recombinant DNA research. However, most claims for the increased safety of recombinant DNA research were made shortly after the introduction of various legislative bills in the U.S. Congress (McCullough, this volume) and in state legislatures which, in general, were written to emphasize the risks and minimize the benefits while proposing highly restrictive regulations that would impede use of the technology. This sequence of events confused society, and indeed some wondered whether scientists were not heralding the safety of recombinant DNA research because of self-interest. In many respects, we were, since I for one certainly did not want the use of recombinant DNA technology to be regulated essentially out of existence. I was thus concerned that we were about to "embark on excessive regulation of an important area of biomedical research based almost solely on fear, ignorance, and misinformation" (see 20). I and other scientists acting as individuals, as well as several scientific societies (see 31), have therefore endeavored to see that if legislation was enacted it was written in a manner to acknowledge the substantial benefits from the research, to indicate that the biohazards although conceivable have yet to be demonstrated, and to establish regulations that would foster rather than hinder the conduct of the experimentation under appropriate uniform regulations. It is my opinion that the bill introduced on 28 February 1978 by Congressmen Harley O. Staggers and Paul G. Rogers (H.R. 11192), as amended slightly by the House Committee on Interstate and Foreign Commerce and with the yet to be approved amendments proposed by the House Committee on Science and Technology (McCullough, this volume), is a reasonable and just bill in keeping with the above-stated aims. Although the drafting of a reasonable and just bill has been achieved, it is my opinion that society's support of and respect for science and scientists has diminished, in that society increasingly perceives science as self-serving and scientists as espousing a holier than thou attitude.

Because of the above considerations, I very much believe that acceptance and support of a bill such as H.R. 11192 by the scientific community would improve society's perception and trust of science and scientists. This would certainly be beneficial.

BENEFITS OR LACK THEREOF ACCRUING TO SOCIETY AND SCIENCE DUE TO LEGISLATION REGULATING RECOMBINANT DNA ACTIVITIES

Whether the enactment of legislation governing recombinant DNA activities is a benefit or detriment to society and science very much depends on the content of the legislative act. The issues of paramount importance, in my opinion, concern the provision for rapid revision of regulations in response to new informaton, the degree of delegation of responsibility for decision making to local biosafety committees, and the guarantee of uniform national regulations and enforcement provisions.

As is well recognized, new information from use of recombinant DNA technology is appearing with ever-increasing rapidity. It is therefore evident that new information about new host-vector systems, about alternate means to achieve physical and/or biological containment levels, about unexpected biohazards, etc. needs to be considered promptly to result in appropriate rule changes. Thus, any legislation enacted needs to delegate authority for making rule changes necessary to improve safety and/or to achieve the benefits of new developments in technology. The Staggers-Rogers bill (H.R. 11192), as well as the new version of Senator Edward M. Kennedy's bill (S. 1217) introduced 1 March 1978 but then tabled, delegates such authority to the Secretary of Health, Education, and Welfare and allows for rapid change of rules dependent upon appropriate new information.

It is also important that any legislation enacted delegate considerable responsibility for surveillance and decision making to locally constituted and broadly representative institutional biosafety committees. This would permit prompt consideration and approval for initiation of new projects under levels of containment that are appropriate to the experiment and in conformance with rules established by the responsible federal agency. Both the Staggers-Rogers and Kennedy bills would permit the Secretary of Health, Education, and Welfare to decentralize initial decision making, and the NIH is currently considering such an approach in the proposed revisions of its Guidelines for Recombinant DNA Research.

Probably the most important feature of any legislation governing recombinant DNA activities is the means to assure uniform standard rules that govern all recombinant DNA activities in the public and private sectors wherever conducted throughout the United States. Thus, the provision for preemption of all local regulations unless necessary to protect public health and the environment as contained in H.R. 11192, but not in the new S. 1217, is, in my opinion, essential. The arguments in support of a strong preemptive clause are numerous and have been stated by many individuals (53, 54). They are based on the microbiological fact that microorganisms know no geographical boundaries, and consequently a hodge-podge of local regulations makes absolutely no sense from a scientific point of view. There are other considerations, however, of equal importance.

The scientific community, in initially calling for caution in the conduct of certain recombinant DNA experiments, has indicated a desire that all scientists, irrespective of geographical and political boundaries, adhere to a similar set of guidelines for the conduct of recombinant DNA activities (56). Certainly, international agreements for adherence to uniform standards for conduct of recombinant DNA activities require that a country like the United States adopt a uniform set of standards itself.

The lack of federal legislation with a strong preemptive clause causes uncertainty, and such uncertainty is not conducive for private firms to invest in developing the technology. Thus, the option for local governments to adopt regulations that are more stringent than those in other jurisdictions can preclude or inhibit use of the technology within that local jurisdiction. It is also evident that the potential for different local requirements can lead to the flow of both scientists and dollars to areas where regulations are less stringent. Obviously, neither consequence of non-uniformity of regulations governing recombinant DNA technology is in the best interest of either science or society.

The enactment of federal legislation without a clause preempting adoption of local regulation has often been defended by citing the experience of Cambridge, Mass., in reviewing the benefits and potential biohazards of recombinant DNA research and in deciding on local requirements for performance of these activities in addition to the requirements of the federal government (35). While I would acknowledge that Cambridge didn't err too badly, I would not agree that their review of how to regulate recombinant research was just or that their decisions were in the best interest of science or society, much less the citizens of Cambridge. As a consequence of their decisions, some scientists left Cambridge permanently, others left temporarily to do experiments elsewhere, and still others decided not to commence recombinant DNA research. Cambridge's wrong decisions were based on several factors. First, scientists from the Boston area who could have had a meaningful and useful input into the deliberations were not consulted. Second, many of the most knowledgeable experts about certain disciplines relevant to the discussion cannot be found in the Boston area and obviously had no input. Indeed, I would argue that no locality or state within the United States of America has the diversity of people with sufficient relevant expertise to deal with all aspects of the benefits and potential biohazards of recombinant DNA activities. It is even debatable whether this expertise exists within the confines of the United States since some of the individuals most knowledgeable about certain apsects of recombinant DNA activities are Europeans, and only the federal government has been able to bring these individuals to the United States to contribute their knowledge to arrive at establishing federal guidelines for conducting recombinant DNA activities (7, 23, 43). It should thus be apparent that no local government can afford to consult all the relevant experts to reach a valid and just decision on the types of regulations necessary, nor can the individual scientists, whose main responsibility is to conduct research and to train new generations of scientists, justify traveling all over the world to give advice to all government jurisdictions that wish to consider adopting their own regulations. As a consequence, most local and state governments, with the exception of a few, that have either enacted or are proposing enactment of local regulations have done a much more abysmal job of it than has Cambridge, Mass.

In conclusion, for all of the above stated reasons, it is apparent, especially with the eagerness of local and state governments to adopt a myriad of differing regulations governing recombinant DNA activities, that federal legislation with a strong preemptive clause must be adopted to allow uniform and reasonable regulations for the performance of recombinant DNA activities throughout the country.

CONCLUSION

I thus favor passage of federal legislation such as the bill introduced by Congressmen Staggers and Rogers provided that such federal legislation contains the preemptive clause as now written. I believe that passage of such a bill will be beneficial in (i) allowing the federal government to be positive in supporting research while being cautious in providing for the public health and welfare and protection of the environment should biohazards be encountered, (ii) permitting a framework for being able to change rules and regulations upon learning new information stemming from research, (iii) facilitating the attainment of a uniform international agreement on the use of recombinant DNA technology, (iv) improving society's trust, perception, and support of science and scientists, and (v) creating an environment that is conducive for capitalization of the knowledge learned from basic research using recombinant DNA technology so that this knowledge can be used to solve problems of society for the betterment of humanity.

REFERENCES

1. **Anderson, E. S.** 1975. Viability of, and transfer of a plasmid from, *E. coli* K-12 in the human intestine. Nature (London) **255**:502–504.
2. **Anderson, E. S.** 1978. Plasmid transfer in *Escherichia coli.* J. Infect. Dis. **137**:686–687.
3. **Armstrong, K. A., V. Hershfield, and D. Helinski.** 1977. Gene cloning and containment properties of plasmid ColE1 and its derivatives. Science **196**:172–174.
4. **Ayala, F. J.** 1977. The stability of biological species, p. 90–97. *In* Research with recombinant DNA. National Academy of Sciences, Washington, D.C.
5. **Beers, R. F., Jr., and E. G. Bassett (ed.).** 1977. Recombinant molecules: impact on science and society. Tenth Miles International Symposium. Raven Press, New York.
6. **Berg, P., D. Baltimore, H. W. Boyer, S. N. Cohen, R. W. Davis, D. S. Hogness, D. Nathans, R. Roblin, J. D. Watson, S. Weissman, and N. D. Zinder.** 1974. Potential biohazards of recombinant DNA molecules. Science **185**:303.
7. **Berg, P., D. Baltimore, S. Brenner, R. O. Roblin, and M. Singer.** 1975. Summary statement of the Asilomar conference on recombinant DNA molecules. Proc. Natl. Acad. Sci. U.S.A. **72**:1981–1984.
8. **Blattner, F. R., B. C. Williams, A. E. Blechl, K. Denniston-Thompson, H. E. Faber, L. A. Furlong, D. J. Granwald, D. O. Kiefer, D. D. Moore, J. W. Shumm, E. L. Sheldon, and O. Smithies.** 1977. Charon phages: safer derivatives of bacteriophage lambda for DNA cloning. Science **196**:163–169.
9. **Cameron, J. R., and R. W. Davis.** 1977. The effects of *Escherichia coli* and yeast DNA insertions on the growth of lambda bacteriophage. Science **196**:212–215.
10. **Campbell, A.** 1978. Tests for gene flow between eucaryotes and procaryotes. J. Infect. Dis. **137**:681–685.
11. **Chang, A. C. Y., and S. N. Cohen.** 1974. Genome construction between bacterial species *in vitro:* replication and expression of *Staphylococcus* plasmid genes in *Escherichia coli.* Proc. Natl. Acad. Sci. U.S.A. **71**:1030–1034.
12. **Chang, S., and S. N. Cohen.** 1977. *In vivo* site-specific recombination promoted by the *Eco* RI restriction endonuclease. Proc. Natl. Acad. Sci. U.S.A. **74**:4811–4815.
13. **Chargaff, E.** 1976. On the dangers of genetic meddling. Science **192**:938–940.
14. **Cohen, S. N.** 1977. Recombinant DNA—fact or fiction? Science **195**:654–657.
15. **Cohen, S. N., F. Cabello, M. Casadaban, A. C. Y. Chang, and K. Timmis.** 1977. DNA cloning and plasmid biology, p. 35–58. *In* W. J. Whelan and J. Schultz (ed.), Molecular cloning of recombinant DNA. Academic Press Inc., New York.
16. **Cohen, S. N., and A. C. Y. Chang.** 1974. A method for selective cloning of eukaryotic DNA fragments in *Escherichia coli* by repeated transformation. Mol. Gen. Genet. **134**:133–141.
17. **Cohen, S. N., A. C. Y. Chang, H. W. Boyer, and R. B. Helling.** 1973. Construction of biologically functional bacterial plasmids *in vitro.* Proc. Natl. Acad. Sci. U.S.A. **70**:3240–3244.
18. **Curtiss, R., III.** 1976. Genetic manipulation of microorganisms: potential benefits and biohazards. Annu. Rev. Microbiol. **30**:507–533.
19. **Curtiss, R., III.** 1978. Biological containment and cloning vector transmissibility. J. Infect. Dis. **137**: 668–675.
20. **Curtiss, R., III.** 1978. Testimony before the Subcommittee on Science, Technology and Space of the Committee on Commerce, Science and Transportation, United States Senate. Regulation of Recombinant DNA Research, p. 50–69. U.S. Government Printing Office, Washington, D.C.
21. **Curtiss, R., III, J. E. Clark, R. Goldschmidt, J. C. Hsu, S. C. Hull, M. Inoue, L. J. Maturin, Sr., R. Moody, and D. A. Pereira.** 1977. Biohazard assessment of recombinant DNA molecule research, p.

375–387. *In* S. Mitsuhashi, L. Rosival, and V. Krčméry (ed), Plasmids: medical and theoretical aspects. Avicenum, Prague.

22. **Curtiss, R., III, D. A. Pereira, J. C. Hsu, S. C. Hull, J. E. Clark, L. J. Maturin, Sr., R. Goldschmidt, R. Moody, M. Inoue, and L. Alexander.** 1977. Biological containment: the subordination of *Escherichia coli* K-12, p. 45–56. *In* R. F. Beers, Jr., and E. G. Bassett (ed.), Recombinant molecules: impact on science and society. Tenth Miles International Symposium. Raven Press, New York.
23. **Curtiss, R., III, W. Szybalski, D. R. Helinski, and S. Falkow.** 1976. Workshop on design and testing of safer prokaryotic vehicles and bacterial hosts for research on recombinant DNA molecules. ASM News **42:**134–138.
24. **Davis, B. D.** 1977. Epidemiological and evolutionary aspects of research on recombinant DNA, p. 124–135. *In* Research with recombinant DNA. National Academy of Sciences, Washington, D.C.
25. **Dougan, G., J. H. Crosa, and S. Falkow.** 1978. Mobilization of the *Escherichia coli* plasmid ColE1 (colicin E1) and ColE1 vectors used in recombinant DNA experiments. J. Infect. Dis. **137:**676–680.
26. **Formal, S. B., and R. B. Hornick.** 1978. Invasive *Escherichia coli.* J. Infect. Dis. **137:**641–644.
27. **Freter, R.** 1978. Possible effect of foreign DNA on pathogenic potential and intestinal proliferation of *Escherichia coli.* J. Infect. Dis. **137:**624–629.
28. **Gorbach, S. L.** 1978. Recombinant DNA: an infectious disease perspective. J. Infect. Dis. **137:**615–623.
29. **Grobstein, C.** 1977. The recombinant DNA debate. Sci. Am. **237:**22–33.
30. Guidelines for the Handling of Recombinant DNA Molecules and Animal Viruses and Cells. 1977. Minister of Supply and Services, Ottawa.
31. **Halvorson, H. O.** 1977. ASM continues scrutiny of DNA legislation. ASM News **43:**392–393.
32. **Hershfield, V., H. W. Boyer, C. Yanofsky, M. A. Lovett, and D. R. Helinski.** 1974. Plasmid ColE1 as a molecular vehicle for cloning and amplification of DNA. Proc. Natl. Acad. Sci. U.S.A. **71:**3455–3459.
33. **Hoskins, L. C.** 1978. Host and microbial DNA in the gut lumen. J. Infect. Dis. **137:**694–698.
34. **Jackson, D. A., R. H. Symons, and P. Berg.** 1972. Biochemical method for inserting new genetic information into DNA of Simian Virus 40: circular SV40 DNA molecules containing lambda phage genes and the galactose operon of *Escherichia coli.* Proc. Natl. Acad. Sci. U.S.A. **69:**2904–2909.
35. **Krimsky, S.** 1977. Recombinant DNA research: public must regulate. Chem. Eng. News, 30 May 1977, p. 36–41.
36. **Lobban, P. E., and A. D. Kaiser.** 1973. Enzymatic end-to-end joining of DNA molecules. J. Mol. Biol. *78:* 453–471.
37. **Maturin, L., Sr., and R. Curtiss III.** 1977. Degradation of DNA by nucleases in intestinal tract of rats. Science **196:**216–218.
38. **Minshew, B. H., J. Jorgensen, M. Swanstrum, G. A. Grootes-Renvecamp, and S. Falkow.** 1978. Some characteristics of *Escherichia coli* strains isolated from extraintestinal infections of humans. J. Infect. Dis. **137:**648–654.
39. **Morrow, J. F., S. N. Cohen, A. C. Y. Chang, H. W. Boyer, H. M. Goodman, and R. B. Helling.** 1974. Replication and transcription of eukaryotic DNA in *Escherichia coli.* Proc. Natl. Acad. Sci. U.S.A. **71:** 1743–1747.
40. **National Academy of Sciences.** 1977. Research with recombinant DNA: an Academy forum. National Academy of Sciences, Washington, D.C.
41. **National Institutes of Health.** 1976. Recombinant DNA research: guidelines. Fed. Regist. **41:**37902–37943.
42. **Pike, R. M.** 1976. Laboratory associated infections: summary and analysis of 3921 cases. Health Lab. Sci. **13:**105–114.
43. Proceedings from a workshop on risk assessment of recombinant DNA experimentation with *Escherichia coli* K-12. 1978. J. Infect. Dis. **137:** 613–714.
44. **Richmond, M. H.** 1977. *Escherichia coli* K-12 and its use for genetic engineering purposes, p. 429-443. *In* R. F. Beers, Jr., and E. G. Bassett (ed.), Recombinant molecules: impact on science and society. Tenth Miles International Symposium. Raven Press, New York.
45. **Scott, W. A., and R. Werner (ed).** 1977. Molecular cloning of recombinant DNA. Ninth Miami Winter Symposium. Academic Press Inc., New York.
46. **Singer, M., and D. Soll.** 1973 Guidelines for DNA hybrid molecules. Science **181:**1114.
47. **Sinsheimer, R. L.** 1975. Troubled dawn for genetic engineering. New Sci., p. 148–151.
48. **Sinsheimer, R. L.** 1977. Recombinant DNA. Annu. Rev. Biochem. **46:**415–438.
49. **Smith, H. W.** 1975. Survival of orally administered *E. coli* K-12 in alimentary tract of man. Nature (London) **255:**500–502.
50. **Smith H. W.** 1978. Is it safe to use *Escherichia coli* K-12 in recombinant DNA experiments? J. Infect. Dis. **137:** 655–660
51. **Struhl, K., J. R. Cameron, and R. W. Davis.** 1976. Functional genetic expression of eucaryotic DNA in *Escherichia coli.* Proc. Natl. Acad. Sci. U.S.A. **73:**1471–1475.
52. **Thomas, M., J. R. Cameron, and R. W. Davis.** 1974. Viable molecullar hybrids of bacteriophage lambda and eucaryotic DNA. Proc. Natl. Acad. Sci. U.S.A. **73:**4579–4583.
53. **U.S. House of Representatives.** 1977. Science policy implications of DNA recombinant molecule research. Hearings before the Subcommittee on Science, Research and Technology of the Committee on Science and Technology. U.S. Government Printing Office, Washington, D.C.
54. **U.S. Senate.** 1978. Regulation of recombinant DNA research. Hearings before the Subcommittee on Science, Technology, and Space of the Committee on Commerce, Science, and Transportation. U.S.

Government Printing Office, Washington, D.C.
55. **Wald, G.** 1976. The case against genetic engineering. Sciences (N.Y. Acad. Sci.) **16:**6–11.
56. **Weissman, C., and H. P. Green.** 1977. National and international efforts to develop guidelines: should there be voluntary or enforced rules of conduct, p. 190–191. *In* Research with recombinant DNA. National Academy of Sciences, Washington, D.C.
57. **Wensink, P. D., D. J. Finnegan, J. E. Donelson, and D. S. Hogness.** 1974. A system for mapping DNA sequences in the chromosomes of Drosophila melanogaster. Cell **3:** 315–325.
58. **Working Party on the Practice of Genetic Manipulation.** 1976. Report. Her Majesty's Stationery Office, London.

Voluntary Compliance and Surveillance: an Alternative to Legislation

GEORGE S. GORDON

Office of Environmental Affairs, U.S. Department of Commerce, Washington, D.C. 20230

This presentation deals with regulation, legislation, and the possibility of voluntary private sector compliance with whatever recombinant DNA research guidelines exist now or will exist in the future. Current activity on Capitol Hill with regard to recombinant DNA legislation is covered in the article by James M. McCulloch (this volume), and I will, therefore, explore various approaches to recombinant DNA activity regulation and then summarize what has happened over the past few months in the exploration by the Office of Environmental Affairs, within the Department of Commerce, working with industrial representatives and other agencies of the government, of the possibility of a workable voluntary program of recombinant DNA regulation in the private sector.

At this stage in the progress of research, legislation, and the general state of the research art in recombinant DNA, it may be a truism to point out that there is considerable uncertainty concerning the future and that no clear directions are apparent in either the setting of legislation or the long-term status of recombinant DNA research itself.

Whether or not legislation is the ultimate answer to our problems or a voluntary system applied to recombinant DNA activity in the private sector can be developed, there are three main questions which must be answered. First, how can the benefits of recombinant DNA research be realized while the human race is protected from some potential biological catastrophe? Second, how can we protect those doing the research, the public at large, and the environment from hazards while still respecting scientists' freedom to conduct research in a responsible manner? Finally, to what degree can this protection be accomplished by self-regulation and to what degree must we rely on public authority?

There are three basic trends (I hesitate to say basic assumptions or general agreements), which I think can be used as a basis for discussion. First, as time goes on, there seems to be developing a feeling among scientists that there is a considerably lower level of risk in doing this kind of research than was thought back in 1974 when the first warnings were raised about possible catastrophic effects. There are some scientists who do not believe this, but there seem to be more who do. Second, in the past 5 years most industrial organizations have demonstrated that they can be trusted to carry out this kind of research in good faith and with a sense of public responsibility. Aside from fundamental concern with their own employees and the public at large, there is much to be lost and little to be gained by any deliberate attempt to flout agreed-on regulations. Third, there seems to be general agreement that the National Institutes of Health (NIH) guidelines (with potential revisions) are a good basis for operating in the recombinant DNA field. I must add the caveat that although the NIH guidelines, as far as physical containment and experimental procedures are concerned, are acceptable to industrialists in general, there are parts which might have to be modified to meet specific requirements of industrial concern.

With an eye toward the serious questions being raised by proposed legislation in the recombinant DNA field, the Senate Subcommittee on Science, Technology and Space held a series of hearings last November in an attempt to clarify some of the issues. As Senator Stevenson pointed out at the hearings, bills and amendments had been introduced in both houses and yet Congress was at that time farther from a legislative position than it had been many months before. Particularly in view of the drastically changed views of many scientists regarding the potential hazards resulting from this research, Senator Stevenson urged that there be renewed inquiry into the complicated issues involved. As he said, "the issue is delicate, the subject matter is intricate and the stakes are high." He is on record as favoring some kind of legislation while at the same time seeking to determine the degree to which self-regulation can be expected to work. Again, to quote the Senator, his chief aim in holding the hearings was to "help decrease the polarization of opinion and the controversy which has surrounded the issue of recombinant DNA research." Several points were raised during the hearings, among which were the following. (i) The protection of proprietary information is an absolute necessity, whether the research is done in universities or by private industry. Premature disclosure can jeopardize patent rights, particularly in foreign countries. Decisions, however, as to what is private or public information may sometimes be difficult. (ii) There is a problem of nonuniformity of standards and regulations, with the danger of proliferation of a multitude of local and state regulations. (iii) Legislation, if it is deemed necessary, should be minimal, reasonable, and flexible. (iv) A major concern to a great many people who are involved with recombinant DNA activity is that of providing adequate education in the fundamentals of what it is and what the real dangers, if any, are in recombinant DNA activity.

The difficulty in bridging the government-industry interface is well known, not only in this country but also abroad. For example, in England the Genetics Manipulation Advisory Committee, which is advisory to the Health and Safety Commission, has attempted to reach agreement on disclosure of industrial proposals without jeopardizing proprietary interests. Agreement has not yet been achieved. In France, there is a declaration of agreement under which all laboratories, including governmental and industrial, would submit recombinant DNA research plans for review of safety procedures. The pharmaceutical industry has agreed to voluntary compliance in this area. In this country, we have tended to use the word regulation widely without really facing the question of what *is* "regulation." We have frequently heard the terms "licensing," "registration," "standardization," "sanctions," "inspection," and "monitoring" used, but without really facing up to the issue of what they are and how broadly any of them should be or can be applied.

Government staffs are limited, and there is also a limit to the willingness and ability of industry to put up with heavy additional paper work and surveillance. On the other hand, recombinant DNA activity, in the face of the highly charged public feeling that now exists, obviously cannot be left completely free from any controls. Starting with the Asilomar Conference in 1974, the possibility of dire results from recombinant DNA activity has been repeatedly put before the American people. Although the likelihood of catastrophic consequences seems to be diminishing as the months go by, it is still a very lively issue that must be faced.

In approaching the problem of regulation of recombinant DNA research, either in the area of federally funded research in places such as universities or in the private sector, there are essentially three ways of trying to resolve the dilemma. First, there is the approach of developing new federal legislation which, if carefully drawn, could take care of the problems without going so far as to inhibit and stifle the very research it is regulating.

A second approach would be to explore the possibility of using existing statutes, laws that are already on the books and might be applied to adequate regulation of ongoing research without further change. A subcommittee of the Interagency Committee on Recombinant DNA Research, which is chaired by D. Fredrickson, The Director of NIH, was charged with examining existing legislative authority, to see whether any single statute or combination of statutes would suffice to permit the regulation of this type of research. The assumption was made that the following regulatory requirements would have to be met: (i) review of research by an institutional biohazards committee before it is undertaken; (ii) compliance with physical and biological containment standards and prohibitions according to the NIH guidelines; (iii) registration of this research with a national registry for the time the research is undertaken; and (iv) enforcement of these requirements through monitoring, inspection, and sanctions.

The Subcommittee concluded that the present law could permit meeting some of the above requirements, but no single legal authority or combination of authorities currently exists that would clearly reach all research and other uses of recombinant DNA techniques. It looked at four existing laws: the Occupational Safety and Health Act of 1970, the Toxic Substances Control Act, the Hazardous Materials Transportation Act, and the Public Health Service Act. The last of these four, the use of Section 361 of the Public Health Service Act, seemed to show the most promise and has been the subject of much discussion within the administration. The Subcommittee carefully reviewed this section, which is directed to organisms that are communicable and cause human disease. To use Section 361 for regulation of recombinant DNA research, the assumption would have to be made that products of this research may cause human diseases and that these may be communicable. Also, this section does not apply to plants or animals or the general environment. The Subcommittee concluded that Section 361 therefore lacks the required authority.

The Occupational Safety and Health Act has only limited access to many of the laboratories which might be involved in recombinant DNA research and does not cover self-employed persons.

Under the Toxic Substances Control Act the Environmental Protection Agency has been directed to control chemicals that may present undue risk of injury to health or the environment. Although most recombinant DNA can be called chemicals, the Act explicitly exempts chemical substances used in small quantities for scientific experimentation.

The Department of Transportation and the Center for Disease Control have been given considerable authority, under the Hazardous Materials Transportation Act, to regulate the shipment of hazardous materials, but there are many gaps as far as its application to regulation of recombinant DNA activities is concerned.

Now we come to the third major possible approach to the regulation process, that of voluntary compliance with some agreed-on set of rules for the protection of public health which would apply particularly to research and other recombinant DNA activity carried on in the private sector. This is an interesting idea, so long as, again, it meets the requirements of prereview, compliance with some sort of guidelines, and the inclusion of some sort of enforcement through monitoring, inspection, and sanctions. Early in November 1977, the Department of Commerce, at the request of the Office of Science and Technology Policy in the Executive Office of the President, started to explore with the private sector and with other agencies of the federal government the feasibility of establishing a surveillance program under which companies or others doing nonfederally supported recombinant DNA research could participate by agreeing to follow the NIH guidelines.

Since then, the Office of Environmental Affairs in the Department of Commerce has discussed the approach with industrial representatives, with NIH, and with the Office of Science and Technology Policy. I should like to mention some of the issues which emerged from our discussions, which have a vital bearing on the opinions of both industrial people and government representatives as to whether or not a system of voluntary compliance should be undertaken.

First, regarding the NIH guidelines, while most of the industrialists and the government personnel believe that compliance with the safety and containment provisions of the NIH guidelines is both possible and highly desirable, there is some feeling that other parts of the guidelines should be somewhat modified to protect legitimate proprietary interests. Just what sort of modification of the guidelines with regard to disclosure of information might be necessary in order not to jeopardize vital proprietary information has not been resolved, but it is a subject which must be considered for either a voluntary or a legislative approach.

A second major issue that emerged was that of having to cope with differing and sometimes conflicting local and state regulations with regard to recombinant DNA research. Companies engaged in recombinant DNA activity, even if they are in compliance with NIH guidelines as far as safety and containment are concerned, are still subject to local and state ordinances. Concern with this aspect of the problem has led several industrialists either currently doing this kind of research or interested in doing it at some future time to express doubts with regard to the workability of a voluntary program and has produced in some, though not all, a leaning toward federal legislation, with the possibility of preemption of these local regulations. The question of preemption has, of course, emerged as a major issue in the discussions that have already taken place in Congress. In addition, the Federal Interagency Committee on Recombinant DNA Research has publicly recommended preemption of state and local laws, which can only be done by new federal legislation.

Another issue is that there is general agreement that some sort of sanctions are necessary to ensure compliance with NIH guidelines. So far, however, there is no close agreement as to what form these sanctions should take. Aside from the presence of an innate sense of responsibility, any temptation to stray from the guidelines would be countered by the threat of public exposure and consequent loss of proprietary information and reputation in the eyes of the rest of the industry and the public. There is always, however, the possibility of violation of the guidelines by the irresponsible few. It may be that in such cases legal sanctions are the only answer.

Another unresolved point in looking at the question of voluntary versus legislative regulation of recombinant DNA activity is that of by whom and how the monitoring shall be done. The use of local or institutional biohazards committees for general surveillance and as a bridge between an institution and the public has apparently been quite successful in some cases and questionable in others. What the role of such local committees might be under a general voluntary system is by no means clear and would have to be the subject of more discussion.

By far the most difficult issue to deal with, however, is that of ensuring that any system of recombinant DNA surveillance, whether under a legal system or a voluntary system, should provide maximum assurance that proprietary information not be divulged. Such information could include the source of an inserted DNA molecule, the expected or actual results of experiments, the intent of research protocols, or descriptions of experiments to be performed. In the case of industrial firms doing this kind of work, publications of information of this sort could possibly divulge the nature and use of possible products, the nature of markets, and even the loss of patent rights. It is questionable whether safeguarding of proprietary information can be guaranteed under existing statutes, no matter who in the federal government may be the recipient of this type of information.

Many industrialists have adopted a "wait and see" attitude, watching closely what is happening in the halls of Congress.

Are High-Yielding Microbial Strains Pure Haploids?[1]

MARIJA ALAČEVIĆ

Faculty of Technology, University of Zagreb, Zagreb, Yugoslavia

Genetic recombination has not been used widely for increasing the yields of microbial products. In industrial laboratories, the general approach to successful strain improvement has been to treat the microorganism of interest with suitable mutagens and to select high-yielding mutants from the surviving population. Some of these high-yielding strains do not appear to be haploids.

Strains of *Saccharomyces* that are used in industrial fermentations have been found to be diploids and polyploids. Truly polyploid strains have been shown to be stable and to have desirable fermentation properties under various production conditions (1). In fungi (*Aspergillus, Penicillium, Cephalosporium*), diploidy can be induced by chemicals (such as camphor) or by means of the parasexual cycle. On the other hand, aneuploids, which are quite common in ascomycetes, are not stable. Aneuploids in *Aspergillus* were found to grow poorly and to sector out haploid subclones or nondisjunctional diploids (5). Duplications have also been reported to be the reason for mitotic nonconformity. Such aneuploids or partial diploids were unstable, and their instability is sometimes masked if they carry a system of balanced lethals (3; A. Upshall et al., this volume).

The difference between high tetracycline-producing strains of *Streptomyces rimosus* and their low-yielding parents was investigated. Crossing two low-producing auxotrophs often resulted in prototrophs with improved productive capacity. However, when spores of these prototrophs were plated on nonselective sporulating media, a considerable genotypic and phenotypic heterogeneity was observed. This could be due to the fact that these strains are not true haploids and that they harbor tandem duplications of various parts of the genome, as suggested by Anderson and Roth (2). An additional piece of evidence which suggests that these high-producing prototrophs are merodiploids is the difficulty with which auxotrophs are isolated from these prototrophs. Most of the mutants obtained have the same auxotrophic requirements (1, 4). This reappearance of the same auxotrophic requirements upon successive mutagenesis may be due to either segregation of the parental marker from the merodiploid or the action of the mutagen on that part of the remaining genome that is not duplicated (6).

An interesting picture emerged when recombination between two stable mutants was examined. In size, the colonies of most of the recombinants were similar to their parents but did not give higher antibiotic yields. However, a few very large and irregularly shaped prototrophs were detected which interfered in the selective genetic analysis. When restreaked on the sporulation medium, a segregation pattern analogous to that of heteroclones was noted (J. Pigac, Ph.D. thesis, University of Zagreb, Zagreb,Yugoslavia, 1975). The colonies did not breed true to their phenotype (cf. 7). This further suggests that they are indeed merodiploids. When the low-producing parents were mated, the yields of the antibiotic by the resulting merodiploids increased as much as 20-fold. One may infer from these observations that these merodiploids possess tandem duplication of that part of the genome which effects antibiotic synthesis. Many high-yielding industrial strains are very unstable, probably

[1] The text includes modifications made by the Editors to fit the scope of the symposium.

because of their partial diploidy. This strain degeneration may occur as a result of the loss of duplicated parts of the genome during repeated transfer or inadequate storage.

REFERENCES

1. **Alikhanian, S.** 1970. Applied aspects of microbial genetics. Curr. Top. Microbiol. Immunol. **53:**91–148.
2. **Anderson, R. P., and J. R. Roth.** 1977. Tandem genetic duplication in phage and bacteria. Annu. Rev. Microbiol. **31:**473–505.
3. **Ball, C.** 1973. The genetics of *Penicillium chrysogenum.* Prog. Ind. Microbiol. **12:**47–72.
4. **Hošťálek, Z., M. Blumauerová, and Z. Vaněk.** 1974. Genetic problems of the biosynthesis of tetracycline antibiotics. Adv. Biochem. Eng. **3:**13–67.
5. **Käfer, E.** 1977. Meiotic and mitotic recombination in *Aspergillus* and its chromosomal aberrations. Adv. Genet. **19:**33–131.
6. **Malik, V. S.** 1978. Genetics of applied microbiology. Adv. Genet., vol. 20.
7. **Sermonti, G.** 1969. Genetics of antibiotic-producing microorganisms. Wiley-Interscience, London.
8. **Windisch, S. E., and I. Neumann.** 1972. Yeast breeding and taxonomy of *Saccharomyces,* p. 877–880. *In* G. Terui (ed.), Fermentation technology today. Society of Fermentation Technology, Osaka.

Mapping and Plasmid Control in *Streptomyces griseus*, Producer of Cephamycin

YAIR PARAG

Department of Genetics, The Hebrew University, Jerusalem, Israel

Streptomyces griseus is an industrially important organism because it produces several useful antibiotics (5, 18, 19). The streptomycin-producing *S. griseus* was among the first actinomycetes to be investigated genetically (2–4). The early studies failed to reveal recombination but established the existence of heterokaryosis, which was believed to be a stage prior to recombination in the sexual reproduction cycle (2). Since the first discovery of recombination in actinomycetes in *S. coelicolor* (17), this nonindustrial species has supplied most of our knowledge on the genetic systems in this group (6, 7, 16). It has a circular chromosome and at least two plasmids carrying genetic information (1, 14). Haploid recombinants can be selected after matings. The haploid recombinants are the outcome of temporary partial diploids, merozygotes. The merozygotes give rise to partially prototrophic colonies, heteroclones, which are heterozygous for a part of the chromosome, are unstable, and give rise to the haploid recombinants. Thus, genetic analysis can be done by direct selection of haploid recombinants and by selection for heteroclones followed by isolation of single spores from single heteroclones (heteroclone analysis). Three fertility types are recognized: (i) IF, which carries a free plasmid (SCPl); (ii) NF, in which the plasmid DNA is integrated to the chromosome at the 9 o'clock position; and (iii) UF, which does not contain this plasmid. Matings between UF and NF are especially fertile and give 100% recombinants. Recombination has been detected recently in several industrial actinomycetes and in some of them genetic maps, similar to the one of *S. coelicolor*, have been plotted (for review, see K. F. Chater, this volume).

The strain initially used in this study was the cephamycin A and B producer *S. griseus* NRRL 3851 (13). The other strains, the growth conditions, and the crossing techniques used have been described elsewhere (12).

CROSSES

All the crosses involved one strain resistant to streptomycin and the other resistant to acriflavine. The results of ca. 20 crosses, each on at least three selective media, showed a recombination frequency of about 10^{-6} putative recombinants, which appeared as prototrophs (or streptomycin-resistant) on the selective media and were transferred to identical ("secondary") selective media. The colonies surviving the transfer were tested for their phenotypes. Some instability was observed, and its nature and extent were examined (12). Single-spore isolates from single-spore recombinants showed phenotypes different from their progenitors. When such putative recombinants were characterized on the different media, spores from each medium on which growth had occurred were transferred to new master plates. The phenotypes of such secondarily selected recombinants frequently differed from the original putative recombinants and showed further segregation of markers. These results showed clearly that the original recombinants were unstable as a result of heterozygosity. Putative recombinants, sensitive to the antibiotic but with marks of heterozygosity, were used for selection of the masked alleles from possible heterozygotes (12). In most cases, only one new recombinant type was obtained from the suspected heterozygote; when more than one class was obtained, such a putative recombinant was a source of heteroclone analysis (12). One such heteroclone analysis, described elsewhere (12), gave the first indication of gene order but, more importantly, showed that the vast majority of the spores collected from the putative recombinant (= heteroclone) were true haploid recombinants. This gave us a new tool for a true haploid analysis. In principle, it involved the collection of spores from ca. 100 putative-recombinant colonies and the plating of them on the same selective medium on which the progenitor colonies were selected. Usually, 15 to nearly 100% of these spores germinated to true haploid recombinants showing the expected prototrophy. One 8-point cross, 14013/12, *str-6 rib-1 ade-9 gly-3* × 301/N6/5 *acr-7 arn-1 met-7 ura-8,* was the basis for the arrangements of main indicating markers, and a series of further 5-point crosses, each time introducing one new marker, gave us the 14-point genetic map (Fig. 1). Hopwood's "4 on 4" procedure (7) could not be used. First, in the above 8-point cross *acr* and *ade* gave 100:0 segregation, and *gly* and *rib* closely approached this ratio. Second, all the other crosses involved 301/N6/5, which suddenly showed instability for the streptomycin resistance marker. In the 5-point crosses any order which required four exchanges for the predominant recombinant

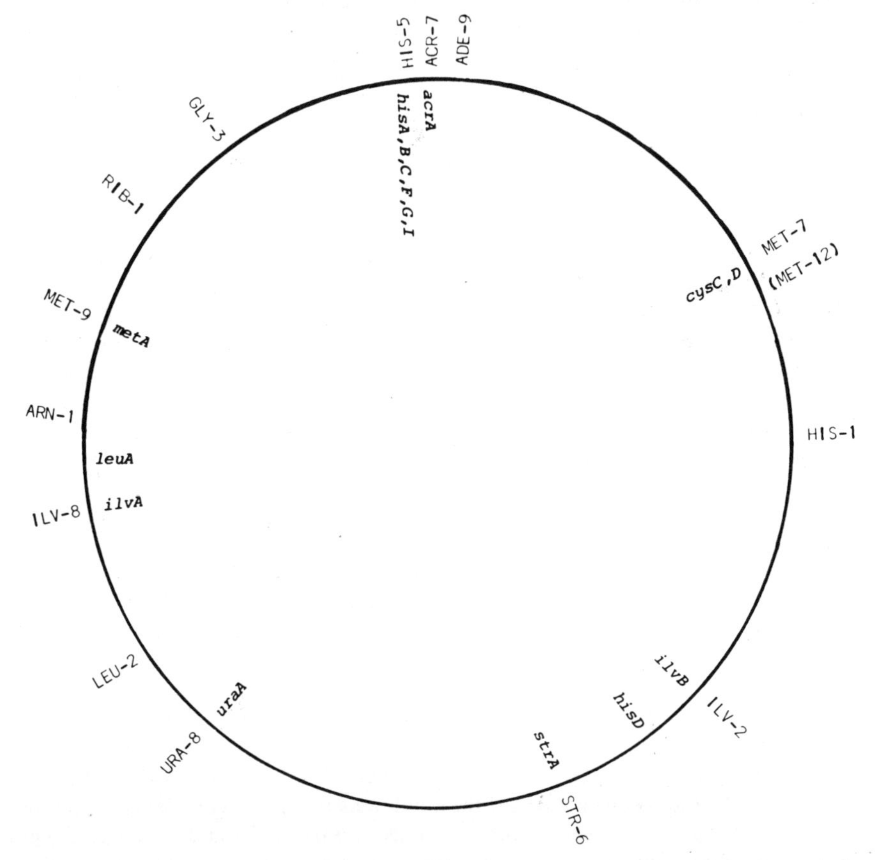

FIG. 1. *Genetic map of* S. griseus *NRRL 3851. Outer circle:* S. griseus *markers. Inner circle:* S. coelicolor *relevant markers (8).*

class was overruled; the one or two remaining possible orders were plotted on the 8-point map from the major cross, and the order requiring the minimal number of exchanges for the major recombinant classes was favored. The position of the unknown marker was also confirmed by the position in the gradient of allele frequencies (15). When two loci were found to be in the same interval, the order within the interval was determined by comparing their recombination frequencies with the indicator loci. There was an apparent asymmetry in many crosses. Alleles from 301 or its progeny (like 301/N6/5) showed an excess of segregation (up to 100: 0 in *acr* and *ade* in the 8-point cross). It is possible that this asymmetry is related to the donor-recipient relationship.

FERTILITY MUTANTS

Mutations for apparent higher recombination frequency were selected by the method described by Hopwood et al. (8). Single colonies of auxotrophic, streptomycin-resistant strains, previously treated with nitrosoguanidine, were grown on a lawn of a complementary auxotrophic, streptomycin-sensitive strain. The normal fertility strains did not give enough recombination to yield prototrophs that could grow on the selective medium. High-recombination mutants, or any other type of mutation that may increase the frequency of prototrophs on contact between these two strains, will yield a cluster of minute prototrophic colonies representing the original colony on the master plate. Such mutants were detected on the minimal medium, but not on the streptomycin selective medium. Such mutants, regardless of the mechanism underlying the appearance of prototrophy, were called fertility mutants (FR). An additional FR mutant, 1466, was found on a plate on which newly selected nutritional mutants were scored for auxotrophy: in the region of contact between this strain and the neighboring strains, there appeared a zone of prototrophic growth.

The FR mutants were of two types, one represented by strain 14015 and the other, by strain 1466. Both mutants, when mated with FR^+ (strain 301 *arn-1 acr-7,* or its derivatives) and plated heavily on the selective medium, showed normal recombination frequency, and the recombinants showed the same features as in ordinary crosses, with one exception: in heteroclone analysis, all the alleles of strain 1466 were represented in three heteroclones, whereas only three of the five alleles of strain 301 were represented, which may indicate a recipient-donor relationship (11). When strain 14015 was mated and plated sparsely (10^{-3} dilution), it gave a high frequency of prototrophs which, on further tests, broke down into two parental types.

Strain 1466 (11) shows autolysis (Fig. 2A), and the autolytic character (*lss*) segregates in crosses independently of the other markers. The lytic factor is particulate or attached to particles (Fig. 2B) and is heat stable and resistant to Pronase. Lysis of strain 1466 was effected by the centrifuged pellet of strain 301 which itself did not show any lysis (lss^+). When spores of strains 1466 (*lss*) and 301 (lss^+) were mixed in a 1:1 ratio, there was no lysis of the resulting mycelium and there was normal sporulation, with equal proportions of the two types of spores. When they were mixed in a 100:1 proportion, extensive lysis occurred, and the ratio of the surviving spores increased 100-fold in favor of strain 1466. It appears that the lss^+ strain prevents, or confers resistance to, the lysis in mixed cultures. This remedial effect was cancelled when the *lss* partner was predominant. Moreover, the lss^+ partner was overgrown by the *lss* partner under the conditions in which lysis occurs. When the pellet was suspended in saline, the lytic effect disappeared, and no plaques were found when the pellet or the original broth with the lysed mycelium was plated on the lawn of strain 1466. Electron microscopy of the pellet of strain 1466 revealed long threads which sometimes broke down to short subunits (11). It is not clear whether these virus-like particles (VLP) cause the lysis or are released from the lysed cells (11). In the cross 1466 *val str lss* FR × 301 *arn acr*, when the density was lowered to 10^2 to 10^3 spores per plate, a few healthy colonies appeared and yielded ca. 1% apparent recombinants (*arn*$^+$ *ilv*$^+$ *str-6 acr-7*). However, different types of single-spore analysis of these recombinants showed that the majority of spores from a single recombinant colony were of the two parental types. If spores were collected from the area on which a pellet of 301, FR$^+$, *lss*$^+$ induced rapid lysis of 1466, FR, *lss*, the frequency of recombinants increased. The emerging picture of two phenomena in crosses involving 1466 is of an ordinary recombination mechanism in normal frequency and of a lysis-related high frequency of recombination or of possible heterokaryosis in which recombination of unknown type occurs. The second type is probably responsible for the FR phenotype of this strain and may be related to the high recombination in protoplast fusion (10).

GENETIC CONTROL OF ANTIBIOTIC SYNTHESIS

Among ca. 100 nutritional mutants, no correlation was found between the nutritional requirement and the antibiotic production level, even among the *met/cys* mutants. Mutant *met-12* and the level of antibiotic production segregate independently in crosses (13). In strain 301, the antibiotic production was increased sixfold by one-step mutagenic treatment. Electrophoretic and high-pressure liquid chromatography data indicated that the production of cephamycin B was increased and that of cephamycin A was lost in this strain. The progeny of the cross of strains 301 and 1462 (normal production) segregated as expected from a single gene segregation, although the significance of the results was reduced because of the high frequency of progeny that produced no antibiotic. In a cross of strain 301 with 1425 (also a high producer), only a few of the progeny reached the level of antibiotic production of the

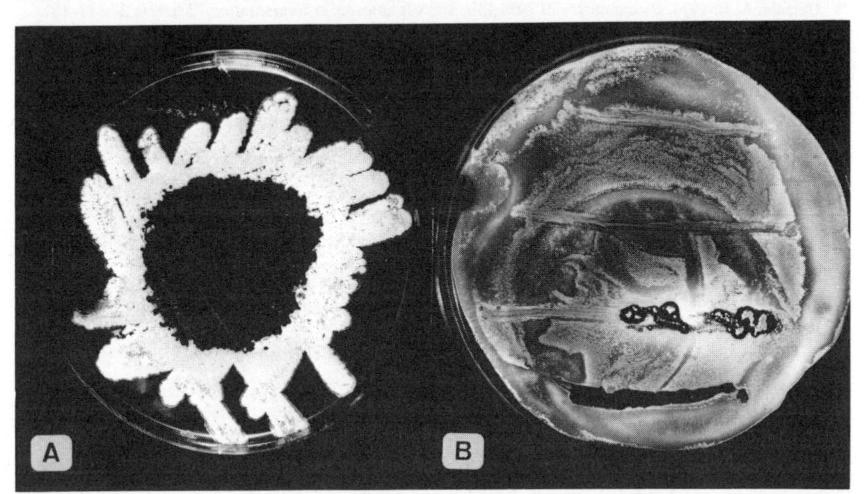

FIG. 2. *Autolysis of* S. griseus *strain 1466. (A) A plate 3 days after inoculation showing lysis in center of lawn. (B) Effect of fraction of lysate of strain 1466 on strain 1466. Four-day shake cultures were centrifuged at low speed; the supernatant was centrifuged at high speed. (Upper line) Supernatant of low-speed centrifugation; (2nd line) supernatant of high-speed centrifugation; (3rd line) pellet of low-speed centrifugation; (4th line) pellet of high-speed centrifugation.*

two parents, and the majority showed a significant decrease of this property. Among 100 auxotrophic mutants, 35% lost antibiotic activity, and the activity was lost in up to 17% of untreated single colonies of the wild-type strain. The antibiotic-producing ability of the wild-type strain and of strain 1462 was lost to differing degrees upon ethidium bromide treatment. Hopwood and Wright (9) suggested that sensitivity (or resistance) of *Streptomyces* strains to the SCPl-carrying strain, IF, of *S. coelicolor* indicates lack (or existence) of a similar plasmid in the tested strain. The wild-type strain NRRL 3851 and the progenitors of the FR mutants were resistant to *S. coelicolor* strain 12 (IF, carrying plasmid SCPl). Blocks with mycelia of this strain (IF) caused strong inhibition of aerial mycelium formation and sporulation of 14015 type mutants, similar to the inhibition shown against *S. coelicolor* 1190 (UF, $SCPl^-$). On the other hand, a single-spore isolate of the wild-type strain that failed to show any antibiotic activity in fermentation flasks showed sensitivity to IF similar to that of 14015. A correlation between criteria that indicate plasmid loss and the loss of antibiotic production may exist, and the control of antibiotic production by plasmids, as indicated by experimental results, is an attractive hypothesis.

ACKNOWLEDGMENT

This research was partially supported by a grant from the National Council for Research and Development of Israel.

REFERENCES

1. **Bibb, M. J., R. F. Reeman, and D. A. Hopwood.** 1977. Physical and genetical characterisation of a second sex factor, SCP2, for *Streptomyces coelicolor* A3(2). Mol. Gen. Genet. **154:**155–166.
2. **Bradley, S. G., D. L. Anderson, and L. A. Jones.** 1959. Genetic interactions within heterokaryons of Streptomycetes. Ann. N.Y. Acad. Sci. **81:**811–823.
3. **Bradley, S. G., and J. Lederberg.** 1956. Heterokaryosis in Streptomyces. J. Bacteriol. **72:**219–225.
4. **Braendle, D. H., and W. Szybalski.** 1959. Heterokaryotic incompatibility, metabolic cooperation, and genic recombination in *Streptomyces.* Ann. N.Y. Acad. Sci. **81:**824–851.
5. **Demain, A. L.** 1974. Biochemistry of penicillin and cephalosporin fermentations. Lloydia **37:**147–167.
6. **Hopwood, D. A.** 1967. Genetic analysis and genome structure in *Streptomyces coelicolor.* Bacteriol. Rev. **31:**373–403.
7. **Hopwood, D. A.** 1973. Genetics of the Actinomycetales, p. 131–153. *In* G. Sykes and F. A. Skinner (ed.), The Actinomycetales: characteristics and practical importance. Academic Press, London.
8. **Hopwood, D. A., K. F. Chater, J. E. Dowding, and A. Vivian.** 1973. Advances in *Streptomyces coelicolor* genetics. Bacteriol. Rev. **37:**371–405.
9. **Hopwood, D. A., and H. M. Wright.** 1973. Transfer of a plasmid between *Streptomyces* strains. J. Gen. Microbiol. **77:**187–195.
10. **Hopwood, D. A., H. M. Wright, M. J. Bibb, and S. N. Cohen.** 1977. Genetic recombination through protoplast fusion in *Streptomyces.* Nature (London) **268:**171–174.
11. **Parag, Y.** 1977. Genetics of *Streptomyces griseus:* fertility mutants, antibiotic production, and relation to possible plasmid. Actinomycetes Related Organisms **12(2):**14–20.
12. **Parag, Y.** 1978. Genetic recombination in *Streptomyces griseus.* J. Bacteriol. **133:**1027–1031.
13. **Parag, Y.** 1978. Genetic recombination and possibly plasmid controlled antibiotic production in *Streptomyces griseus* NRRL 3851, a cephamycin producer, p. 47–50. *In* E. Freersken, I. Tarnok, and J. H. Thomin (ed), Genetics of the Actinomycetales. Proceedings of the International Colloquium, Forschungsinstitut Borstel, 1976. Fischer Verlag, Stuttgart.
14. **Schrempf, H., H. Bujard, D. A. Hopwood, and W. Goebel.** 1975. Isolation for covalently closed circular deoxyribonucleic acid from *Streptomyces coelicolor* A3(2). J. Bacteriol. **121:**416–421.
15. **Schupp, T., R. Hütter, and D. A. Hopwood.** 1975. Genetic recombination in *Nocardia mediterranei.* J. Bacteriol. **121:**128–136.
16. **Sermonti, G.** 1969. Genetics of antibiotic-producing microorganisms. Wiley-Interscience, London.
17. **Sermonti, G., and I. Spada-Sermonti.** 1955. Genetic recombination in *Streptomyces.* Nature (London) **176:**121.
18. **Stapley, E. O., M. Jackson, S. Hernandez, S. B. Zimmerman, S. A. Currie, S. Mochales, J. M. Mata, H. Woodruff, and D. Hendlin.** 1972. Cephamycins, a new family of β-lactam antibiotics. I. Production by actinomycetes including *Streptomyces lactamdurans* sp. n. Antimicrob. Agents Chemother. **2:**122–131.
19. **Umezawa, H. (ed.).** 1967. Index of antibiotics from actinomycetes. University of Tokyo Press, Tokyo.

Acquisition of New Metabolic Capabilities: What We Know and Some Questions That Remain

GEORGE D. HEGEMAN

Department of Biology, Indiana University, Bloomington, Indiana 47401

Because much of what is known about microbial genetics has been learned by studying mutants that have lost normal metabolic functions, and perhaps also because one tends to think of the products of 2 billion years or so of evolution as unimprovable, until recently little considerations has been given by biologists to the study of forward evolutionary change in microbial metabolism. That such evolution has occurred is clear from consideration of the chronology of emergence of the major biological groups. Many students of the fossil record believe that the first cellular organisms appeared 1.5 to 2 billion years ago, and there are reports of fossils identified as procaryotic in rocks that were formed 3.5 billion years ago (25). However, the land plants and most animals appeared only 0.5 to 0.25 billion years ago. Since higher plants have evolved whole new classes of compounds (e.g., the terpenes), and yet these are readily mineralized by bacteria and fungi in the soil today, it seems that acquisitive metabolic evolution in the microbial world has occurred recently.

I would like to preface this discussion with a more precise definition of "new metabolic capabilities" than is provided by the title. We shall be concerned in most cases with the gaining of a new function—typically, the ability to use a previously unusable source of carbon and energy—and in a few cases with the regaining of a previous function after loss of the function by genetic deletion. We shall *not* be concerned with recombinational processes or net contributions to the genome of an organism through genetic exchange, but rather with mutation and genetic reorganization that occurs in the background of a single strain. This seems appropriate since it is arguable what role is played by genetic exchange in organisms that normally reproduce asexually. Indeed, there exist a number of lines of evidence that recombination has played little or no part in the evolution even of the *Enterobacteriaceae* where the capacity for genetic exchange is known to occur quite generally (24). Groups of phenotypically very similar bacteria often show great structural differences for isofunctional proteins (1, 6), which suggests parallel evolution unaccompanied by genetic exchange.

Interest in the gain of metabolic function by microbes is relatively recent, and has been heightened because of the number and quantities of novel products (e.g., plastics and pesticides) that human industry has added to the environment in the past 40 years, coupled with the realization that much of this debris is at best unsightly and at worst toxic. There are a number of recent reviews and symposia on the evolution of metabolic capability (5, 18, 20, 26). Both the variety of experimental approaches and the number of specific cases studied have grown impressively in recent years. The questions that most workers seek to answer include:

1. From where do "new" enzymes come?
2. By what mechanism(s) does the new enzyme(s) become available?
3. How do organisms adapt to the selection for new function?
4. How do the new activities become functionally integrated with other aspects of metabolism?

In the scheme put forward by Horowitz to describe the evolution of biosynthetic abilities, the need for a stepwise accretion of enzymes in the retrograde sense was recognized (16). This scheme postulates the existence of intermediates free in the environment and their sequential exhaustion as the driving force to select each stepwise addition to the chain of catalysis. An analogous scheme for catabolic abilities is much more difficult to imagine. Such pathways are useless to cells unless present intact so that it is hard to imagine them "growing" in the sense of Horowitz (16). Furthermore, intermediates in such pathways are often unstable, toxic in higher concentrations, or unable to penetrate cells at a useful rate (15, 18). Given these theoretical difficulties, it is fortunate that cases of acquired catabolic ability comprise a vast majority of examples studied experimentally.

Answers to the first of the above questions are not often found. In cases where an answer is possible, it most often appears that a preexistent enzyme that has substrate ambiguity is recruited to perform a novel role (15, 18), becoming available by loss of normal regulation (7, 15, 18, 20) and thereby also answering the second question. Generally, the enzyme is preadapted to its "new" role because of some ability to bind the novel substrate and by possession of the appropriate catalytic potential. In one of the most ingenious experimental approaches to this question, Campbell and co-workers isolated strains that had reacquired the ability to grow at the expense of lactose in *lac* deletion mutants of *Escherichia coli* K-12 (4). "Revertants" (actually physiological suppressor mutants) were found to have a weak lactase activity, and all mapped at the same novel locus. Subsequent study of the "new" lactase protein revealed its general affinity for galactose-like monosaccharides (3). It has also been shown that the "evolved β-galactosidase" is an exception to the rule that recruitment in a new function involves loss of normal regulation; its synthesis is induced by lactose (10).

It is to the third question that some of the most interesting experimental answers have been obtained. In the best-studied cases, this has been done by growing the microbes in such a way that a single enzyme or a single known pathway must change in its properties or effective catalytic rate as an adaptive response to an imposed condition(s) of stress (Table 1). In most of these recent examples, a single mutational

TABLE 1. *Examples of directed evolution using known enzymes*

Reference	Stress	Enzyme change(s)	Mutational basis
Clarke and co-workers (reviewed in 7)	New but related compounds as substrates	Specificity K_m V_{max} Increased levels	Structural genes Regulator; prometer
Hegeman and Root (14)	Inhibitory analog plus normal substrate	Increased levels	Promoter or attenuator
Lin and co-workers (reviewed in 20)	Fast growth selection	K_m	Structural genes (13)
	Fast growth selection	Constitutive permease	Regulatory
Francis and Hansche (8)	Medium pH Chemostat	Optimal pH K_m	Structural gene Structural gene
Hansche (11)	Chemostat	Degree of association with cell wall	?
Hansche, Beres, and Lange (12)	Chemostat	Increased levels	Gene duplication (dosage)

event leads directly to a recognizable, better-adapted phenotype. Although it may be argued that most of these cases in fact represent selection for improvement of some existing function under stressful conditions, it is usually true that the conditions are so stringent that a truly "new" functional phenotype is obtained. If this is accomplished, the answer to the third question is that one or, at most, a few mutations in structural or regulatory genes can result in a key enzyme with a higher affinity for a new substrate, a higher V_{max}, a changed pH-rate profile, or an altered rate of expression. The only case among these examples were more than one mutational step is required to achieve a marked adaptive change is that of the ribitol dehydrogenase which is recruited to permit growth on xylitol (20). By sequence analysis of the dehydrogenase, it was determined that two or more amino acid substitutions had taken place (13). Even though in such cases intermediate strains may occur that have accumulated one or two mutations but are not yet capable of markedly faster growth, they are at least not immediately counterselected, even if the mutant enzyme is nonfunctional (2).

It has been suggested by several workers that duplicated genes that may or may not be translated and that are free to undergo mutational alteration would provide the basis for rapid enzyme evolution, especially under primitive conditions (19, 23). However, such cases have not been observed. The suggestion that gene duplications provide the raw material for the development of new microbial pathways in the form of redundant and partially preadapted proteins has been made (16). However, evidence directly supporting this idea is scant. Although duplications occur frequently in bacteria (9), they seem unstable, and, aside from the ribosomal cistrons, bacterial DNA is devoid of redundancy. These objections do not pertain to eucaryotes, in which evidence for the role of duplications in evolution is available (12, 21).

The failure to find structural similarity among enzymes catalyzing sequential reactions in the same pathway also suggests an origin for enzymes of the same pathway other than gene duplication; relatedness between enzymes of similar function in parallel pathways has, however, been observed (22). Perhaps this should not be surprising since successive steps in a pathway often involve catalytic mechanisms that are very different in successive steps, although exceptions may occur (17). In fact, most well-studied cases of evolution of a new function are consistent with the idea that a patchwork assembly process is the most likely pattern for enzyme accretion (7, 15, 18, 20).

It has frequently been suggested that tandem gene duplications play a key role in the origin of new enzyme proteins during growth of a pathway (11, 12, 16, 21, 23). Such a mechanism of growth would preserve coordinacy of expression and facilitate inheritance of the entire suite of genes that encode the pathway during genetic transfer. However, Hansche and co-workers, using β-glycerophosphate in a chemostat, found that duplications of the acid phosphatase gene in yeast which arise in response to limitation of growth by phosphate are nontandem (12).

One of the most striking and general findings made in experimental attempts to answer the third question is that a given strain responds to a particular stress in a stereotypical fashion. Repetition of selection experiments tends to evoke repeatedly the same kinds of mutations in the same genes and in the same order (4, 10, 11, 18, 23). This suggests that the evolutionary choices available are relatively few and rather steeply graded in terms of the adaptive possibilities that they can afford.

There are fewer answers to the fourth question than to any of the other three. It is clear from a few cases (e.g., 7) that an old function is often lost as the new one is gained. It is also true that mutation to new function can produce a preadaptive result; unforseen ability to use several new compounds may accompany selection for the ability to use a particular one (7). It has been shown in yeast that mutational change under selection can affect the distribution of the enzyme within the cell. An increase in the amount of acid phosphatase that associates with the cell wall enhances the ability of the cell to exploit low concentrations of exogenous organic phosphate esters (11).

The applications of the knowledge so far gained about the way new metabolic capabilities evolve have not been great. As a practical matter, efforts devoted to isolating high-yielding strains from nature or screening strains often produce greater returns than efforts devoted to strain improvement. This is particularly true when the pathways involved are not known well. However, it is clear that higher yields of enzymes are generally available by the application of some of the techniques mentioned here and that the student of structure and function in enzyme catalysis may prepare a whole "family" of enzymes differing in various specified properties by using general techniques, provided the appropriate selections are possible (13). Similarly, it now seems possible, by use of the techniques of directed evolution, to select strains that have enzymes with desirable properties.

One of the most pressing questions we now face is whether, or to what extent, microbes can be "trained" to mineralize undesirable chemical residues. The answers depend partly on the extent to which the residue in question resembles a normally used substrate, the ingenuity of the experimenter and his care in observing the rules of the genetic game, and the ability of the microbial product of genetic manipulation to survive in soil and surface waters. Many "experiments" of this kind are presumably going on in the soil at this moment, and we can only hope that many of them are successful. Since microbes are great opportunists, it seems likely that some of them will succeed.

REFERENCES

1. **Ambler, R. P., T. E. Meyer, and M. D. Kamen.** 1976. Primary structure determination of two cytochromes C_2: close similarity to functionally unrelated mitochondrial cytochrome C. Proc. Natl. Acad. Sci. U.S.A. **73:**472–75.
2. **Andrews, K. J., and G. D. Hegeman.** 1976. Selective disadvantage of nonfunctional protein synthesis in *Escherichia coli.* J. Mol. Evol. **8:**317–328.
3. **Arraj, J. A., and J. H. Campbell.** 1975. Isolation and characterization of the newly evolved (*ebg*) β-galactosidase of *Escherichia coli* K12. J. Bacteriol. **124:**849–856.
4. **Campbell, J. H., J. A. Lengyel, and J. Langridge.** 1973. Evolution of a second gene for β-galactosidase in *Escherichia coli.* Proc. Natl. Acad. Sci. U.S.A. **70:**1841–1845.
5. **Carlile, M. J., and J. J. Skehel (ed.).** 1974. Evolution in the microbial world: 24th Symposium of the Society for General Microbiology. Cambridge University Press, Cambridge.
6. **Champion, A. B., and J. C. Rabinowitz.** 1977. Ferredoxin and formyltetrahydrofolate synthetase: comparative studies with *Clostridium acidiurici, Clostridium cylindrosporum* and newly isolated uric acid-fermenting strains. J. Bacteriol. **132:**1003–1020.
7. **Clarke, P. H.** 1974. The evolution of enzymes for the utilization of novel substrates. Symp. Soc. Gen. Microbiol. **24:**183–217.
8. **Francis, J. C., and P. E. Hansche.** 1972. Directed evolution of metabolic pathways in microbial populations. Modification of the acid phosphatase pH optimum in *Saccharomyces cerevisiae.* Genetics **70:**59–73.
9. **Glansdorf, N., and G. Sand.** 1969. Duplication of a gene belonging to an arginine operon of *Escherichia coli* K12. Genetics **60:**257–268.
10. **Hall, B. G., and D. L. Hartl.** 1974. Regulation of newly evolved enzymes. I. Selection of a novel lactase by lactose in *Escherichia coli,* Genetics **76:**391–400.
11. **Hansche, P. E.** 1975. Gene duplications as a mechanism of genetic adaptation in *Saccharomyces cerevisiae.* Genetics **79:**661–674.
12. **Hansche, P. E., V. Beres, and P. Lange.** 1978. Gene duplication in *Saccharomyces cerevisiae.* Genetics

88:673–687.
13. **Hartley, B. S.** 1974. Enzyme families. Symp. Soc. Gen. Microbiol. **24:**151–182.
14. **Hegeman, G. D., and R. T. Root.** 1976. The effect of a non-metabolizable analogue on mandelate catabolism. Arch. Microbiol. **110:**19–25.
15. **Hegeman, G. D., and S. L. Rosenberg.** 1970. Evolution of bacterial enzyme systems. Annu. Rev. Microbiol. **24:**429–462.
16. **Horowitz, N. H.** 1965. The evolution of biochemical synthesis, p. 15–24. *In* V. Bryson and H. J. Vogel (ed.), Evolving genes and proteins. Academic Press Inc., New York.
17. **Jeffcoat, R., and S. Dagley.** 1973. Bacterial hydrolases and aldolases in evolution. Nature (London) New Biol. **241:**186–187.
18. **Jensen, R. A.** 1976. Enzyme recruitment in evolution of new function. Annu. Rev. Microbiol. **30:**409–425.
19. **Koch, A. L.** 1972. Enzyme evolution. I. The importance of untranslatable intermediates. Genetics **72:** 297–316.
20. **Lin, E. C. C., A. J. Hacking, and J. Aguilar.** 1976. Experimental models of acquisitive evolution. BioScience **26:**548–554.
21. **Ohno, S.** 1970. Evolution by gene duplication. Springer Verlag, Heidelberg.
22. **Patel, R. N., R. B. Meagher, and L. N. Ornston.** 1973. Relationships among enzymes of the β-ketoadipate pathway. II. Properties of crystalline β-carboxy *cis,cis*-muconate lactonizing enzyme from *Pseudomonas putida.* Biochemistry **12:**3531–3537.
23. **Rigby, P. W. J., B. D. Burleigh, and B. S. Hartley.** 1974. Gene duplication in experimental enzyme evolution. Nature (London) **251:**200–204.
24. **Sanderson, K. E.** 1976. Genetic relatedness in *Enterobacteriaceae.* Annu. Rev. Microbiol. **30:**327–349.
25. **Schopf, J. W.** 1974. Paleobiology of the precambrian. The age of the blue-green algae. Evol. Biol. **7:**1–43.
26. **Wilson, A. C., S. S. Carlson, and T. J. White.** 1977. Biochemical evolution. Annu. Rev. Biochem. **46:** 573–639.

Experimental Evolution of Amidases with New Substrate Specificities

PATRICIA H. CLARKE

University College, London, England

Since the evolution of new metabolic activities requires new enzymes, it is important to know where they come from and what mutational events give rise to them. In considering the evolution of enzymes in general, it is reasonable to suppose that over long periods of time there have been many gene duplication events followed by mutational divergence of the duplicated genes. Successive gene duplications with subsequent independent mutation could have led to the families of related enzymes that can be recognized today. Many comparative studies have been made on amino acid sequences and three-dimensional structures of enzymes. One example is the group of nucleotide-requiring dehydrogenases. It was suggested that an ancestral dehydrogenase gene gave rise to the family of dehydrogenases with similar catalytic reactions but different substrate specificities (13). Comparisons between the serine proteases also indicated common ancestry for enzymes with different substrate specificities (11). During evolution, it appears that the shape of the protein molecule has remained more constant than the total amino acid sequence. The differences in substrate specificities can be related to the configuration of the peptide chain around the active site, and the amino acid residues in this region are of critical importance. Many other amino acid differences are found in other regions of the protein. In comparing enzymes of similar function that have had long periods of independent evolution, the differences in amino acid sequences cannot be interpreted solely in terms of differences in specificity.

It can be asked whether mutations leading to changes in enzyme specificity can be studied experimentally in microbial systems. Microorganisms adapt rapidly to environmental challenges, and among the adaptive responses is the appearance of new enzyme activities. This offers an opportunity to explore the molecular changes accompanying the new enzyme activities. In some cases, the enzymes are the same as those of the parent strain, but they are produced in larger amounts, thus enabling a relatively poor substrate to be utilized. A mutation may affect the regulatory control, so that enzymes that are normally inducible become constitutive. This may enable the mutant strain to utilize a compound that is a substrate but not an inducer of an enzyme. In other cases, it has been possible to select mutants in which the enzyme activity is related to alterations in the enzyme itself, and these mutants can be used to study the evolution of enzyme specificity.

Microbial mutants with novel metabolic activities can be selected in several ways. Mutants with new catabolic activities can be isolated either because they grow on substrates not utilized by the parent or because they grow faster. Chemostat culture can be used for the selection of faster-growing mutants. This method was used to select xylitol-utilizing mutants of *Klebsiella aerogenes* (11, 12) in which the growth-limiting enzyme is ribitol dehydrogenase which has low activity towards xylitol. One class of mutants produces increased amounts of ribitol dehydrogenase, probably by gene duplication, whereas another produces altered enzymes. It was thought at one time that several mutations were needed to obtain mutant enzymes with improved xylitol dehydrogenase activity, but it now appears that a single mutation may be sufficient (12). Chemostat selection has also led to the isolation of mutants of *Saccharomyces cerevisiae* producing an acid phosphatase with altered properties (8).

Another approach in looking for strains with new metabolic activities is to isolate mutants with novel growth phenotypes and then to determine what enzymes and regulatory systems are involved. One way of doing this is to remove a normal metabolic enzyme and then to see if it can be replaced, which has the advantage that the subsequent metabolic pathways are likely to remain intact. This procedure was used with a strain of *Escherichia coli* that was unable to grow on lactose because the *lac* genes had been deleted. After several steps, a mutant was obtained that produced a new lactose hydrolase. This was described as evolved β-galactosidase, and the enzyme was shown to be determined by gene *ebgA* which maps on the chromosome at 59 min, quite distant from the classical *lac* genes (5). The complexity of the original isolation led to the suggestion that multiple mutations were needed to get an active *ebgA* enzyme, but subsequent work has shown that a single mutation will suffice.

In the parent *E. coli lac* deletion strain the second lactose enzyme is regulated by the gene $ebgR^{+}$. It can be induced by lactose, but the $ebgA^{0}$ enzyme has a very low affinity for lactose and the activity in the induced state is not sufficient to allow growth on lactose. Mutations in both *ebgA* and *ebgR* are needed to produce lactose-utilizing mutants. The mutation in the regulatory gene can be to $ebgR^{-}$, giving a constitutive phenotype, or to $ebgR^{+u}$, which allows lactose to induce higher levels of the *ebgA* enzymes (10). Several different $ebgA^{+}$ enzymes have been identified and shown to differ from each other and from the $ebgA^{0}$ enzyme with respect to both K_m and V_{max} for lactose (9). When three different *ebgA* variants were compared in the same $ebgR^{-}$ background, the growth rate was proportional to the in vivo lactose hydrolase activity. A strain carrying $ebgR^{+u}$ with $ebgA^{+a}$ gave comparable results. It was clear from these studies that growth on lactose occurred only when a threshold lactose hydrolase activity had been reached. The in vivo lactose hydrolase activity was calculated for strains carrying various combinations of *ebgR* and *ebgA* alleles from the rates of synthesis and the activities of the purified enzymes. The estimated lactose hydrolase activities in all the $ebgA^{+}$ strains carrying the $ebgR^{+}$ regulatory gene fell below the calculated threshold value of 5.8 units per mg of protein and accorded with the requirement for both regulator and structural gene mutations. In this system there was a clear demonstration of the barrier that had to be crossed in terms of the in vivo enzyme activity needed to achieve the new growth phenotype (10).

My colleagues and I have been working with an inducible aliphatic amidase of *Pseudomonas aeruginosa.* The wild-type strain grows on acetamide and propionamide, which are good substrates and good inducers, and these are the only aliphatic amides that support growth. Mutants have been obtained that are able to grow at the expense of a number of amides that cannot be utilized by the wild-type strain. Some of these have regulatory gene mutations only, but others have mutations in the structural gene for the amidase, *amiE,* and produce enzymes with altered substrate specificities.

One of the main reasons for selecting this system was that amides are readily available and that the homologous series of aliphatic amides made it possible to examine the properties of amides with substituent groups. Also, amides are potential sources of both carbon and nitrogen for growth and can be incorporated in different combinations into the selective media. This made it possible to isolate various classes of mutants (Table 1).

One of the key observations in these studies was that some amides were not

TABLE 1. *Amide media used for mutant selection*

Medium[a]			Class of mutants[b]	Mutant strain[c]
	Carbon	Nitrogen		
S/F	Succinate	Formamide	Constitutive, butyramide sensitive	C11
			Constitutive, butyramide resistant	C1
			Formamide inducible	F6
B	Butyramide		Constitutive, butyramide resistant	CB4
			Constitutive, B amidase	B6
S/L	Succinate	Lactamide	Catabolite repression resistant	
			Constitutive	L10
			Inducible	L11
V	Valeramide		Constitutive, V amidase	V9
S/Ph	Succinate	Phenylacet-amide	Constitutive, Ph amidase	PhB3
				PhV1, PhV2
				PhF1, PhA1
AI	Acetanilide	$(NH_4)_2SO_4$	Constitutive, AI amidase	A13

[a] Minimal salt media contained amides either as the major carbon source or as the nitrogen source only.

[b] Examples are given of some of the classes of mutants isolated from these amide selective media. Unless otherwise indicated, the mutants produced A amidase (wild type).

[c] Mutant strains listed include those shown in Fig. 1, which indicates the number of mutational steps involved in the selection of each mutant. Strain LAm1, which is not included here, was a mutant with a defective amidase enzyme.

inducers but competed with the inducing amides and prevented enzyme synthesis. This was termed amide-analog repression (2). Cyanoacetamide, thioacetamide, and butyramide repress amidase induction by acetamide and *N*-acetylacetamide. This property of butyramide is significant in restricting the range of aliphatic amides available as growth substrates. The wild-type A amidase hydrolyzes butyramide very poorly (2% of the acetamide rate), but the real barrier to the utilization of butyramide by the wild type is the absence of induction. Some constitutive strains can grow on butyramide, and these produce enough of the wild-type enzyme to hydrolyze butyramide at a rate that will support growth. The threshold for amidase activity is reached by producing high levels of a rather poor enzyme.

Other constitutive strains are capable of producing high levels of A amidase but are unable to grow on butyramide because they have retained the sensitivity to amide-analog repression and amidase synthesis is repressed in the presence of butyramide. Strain C11 is constitutive, but unable to grow on butyramide, and gives rise to two classes of butyramide-utilizing mutants. Strain CB4 belongs to the group of constitutive mutants that are resistant to butyramide repression, and strain B6 belongs to the group producing amidases with altered substrate specificities (6). The existence of these two classes of butyramide-utilizing mutants led us to suggest that there was a critical value for amidase activity on a particular amide and that the bacteria would grow only if this critical activity could be reached (7).

Figure 1 is a family tree of some of our amidase mutants. The mutant amidases can be distinguished from one another, and from the wild-type enzyme, by their physicochemical properties as well as by their substrate specificities. Strain B6 was the parent of valeramide-utilizing mutants (V group) and of phenylacetamide-utilizing mutants (PhB group). Phenylacetamide-utilizing mutants PhV1 and PhV2 were derived from strain V9; PhA1 and PhF1 were derived from other parental strains. It can be seen that the evolution of a phenylacetamidase from an acetamidase

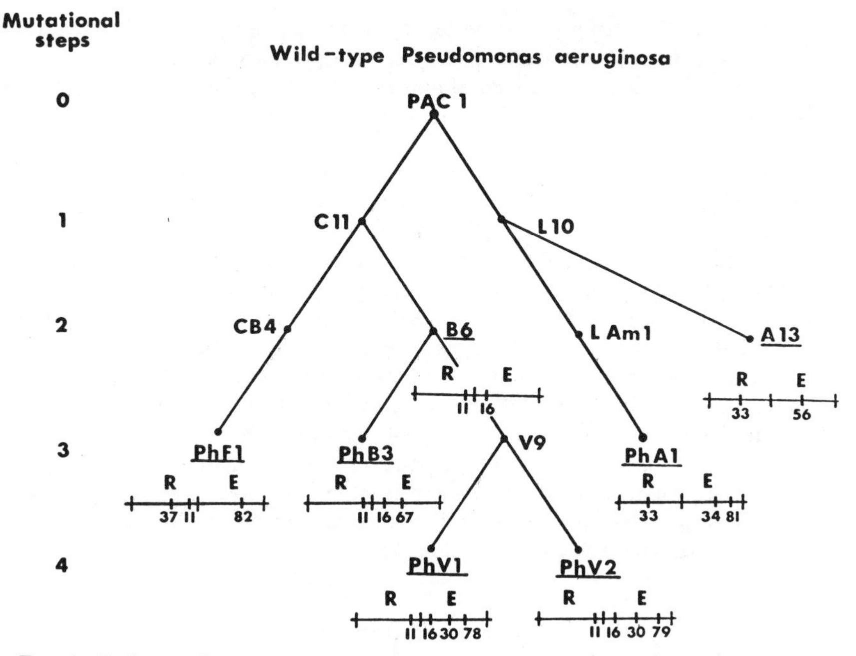

FIG. 1. *Evolution of mutant amidases. Strains producing amidases with altered substrate specificities are underlined. The mutations in the structural and regulatory genes,* amiR *and* amiE, *are shown but not to scale, and the map order of some mutations is provisional. The amide growth phenotypes of the mutants are given in Table 1.*

in *P. aeruginosa* could be achieved by several different routes. We were not able to proceed directly from the wild-type strain, but mutant CB4 produces the wild-type A amidase and a single mutational step gave rise to the PhF1 enzyme (6).

The five Ph mutants shown in Fig. 1 have different phenotypes and are assumed to have acquired between one and three mutations in the structural gene *amiE*. The PhV2, PhA1, and PhF1 amidases are very thermolabile, and the mutants show only trace growth on acetamide. PhB3 and PhV1 are unable to grow on acetamide, and their amidases are more thermostable. The absence of growth on acetamide made it possible to use PhB3, V2, and V5 for transductional crosses with amidase-negative mutants. The transductants were selected on acetamide plates, and some produced A amidase whereas others produced B amidase, which could be readily identified by its higher rate of butyramide hydrolysis. This indicated that the *amiE16* mutation of strain B6 had been retained in the subsequent selection of the PhB3 and PhV1 mutants. The *amiR* and *amiE* genes are closely linked, and the *amiE16* mutation of strain B6 appeared to be located towards the end of the structural gene nearest to the regulator gene (1).

Amidase is a hexameric protein with identical subunits of about 35,000 daltons (4). The mutant amidases are the same as the wild type in this respect although some are less stable and dissociate more readily into their subunits. Amidases can be purified relatively easily. The cell-free extracts are treated to remove nucleic acids, and a brief heat treatment removes a considerable amount of bacterial protein. The remaining soluble protein is fractionated with ammonium sulfate, and the fraction containing amidase activity is subjected to chromatography on DEAE-Sephadex. With the very thermolabile amidases the heat treatment is omitted.

The differences in physicochemical properties indicated that the mutant amidases differed from one another in the amino acid substitutions that had taken place. Attempts were made to identify these substitutions by hydrolyzing the enzymes with various proteolytic enzymes and analyzing the peptides obtained. The A and B amidases were oxidized with performic acid and digested with chymotrypsin, and the peptides were separated by gel filtration on Sephadex G25. The preparations were almost identical, as had been expected, but, in one of the fractions examined by high-voltage electrophoresis at pH 6.5, a peptide appeared in the B amidase hydrolysate that was not present in the starting amidase hydrolysate. This peptide was also present in hydrolysates of PhB3 and PhV1 amidases. It presented a problem since no comparable peptide appeared in the A amidase hydrolysate. The significance of this peptide became clearer when digests were made with thermolysin. Peptides were separated by gel filtration into several pools, and the neutral peptides from pool 2 showed a difference between the A amidase and the mutant PhB3 and PhV1 amidases. A peptide difference was also present in thermolysin digests of B6 and V9 amidases. As shown below, the peptide from the mutants contained a phenylalanine residue in place of a serine in the A amidase:

Wild-type A amidase

Met, Arg, His, Gly_2, Asx_3, Ile, Ser_3, Val, Thr

B6, V9, PhB3, and PhV1 amidases

Met, Arg, His, Gly_2, Asx_3, Ile, Phe, Ser_2, Val, Thr

The substitution of a phenylalanine residue for a serine residue provides an additional cleavage site for chymotrypsin, and this accounts for the extra peptide in the mutant enzymes. Its occurrence in the chymotryptic digests from B6, PhB3, V9, and PhV1 amidases indicates that this is the substitution determined by the *amiE16* mutation. The peptide was sequenced and could be aligned with the N-terminal sequence (A. P. Paterson, Ph.D. thesis, University of London, London, England):

A amidase Met-Arg-His-Gly-Asp-Ile-Ser-Ser-Ser-
B amidase Met-Arg-His-Gly-Asp-Ile-Phe-

The differences in the properties of the A and B amidases are not very great. Both are thermostable enzymes and indistinguishable by immunodiffusion cross-reactions. The B amidase is slightly less active than the A amidase on acetamide but more active on propionamide. The K_m value for butyramide is 10-fold lower and the V_{max} is 10-fold greater. A small change in the conformation of the enzyme protein could be enough to make it easier for butyramide to reach the active site. In the mutation to change A amidase to B amidase, this has been achieved by the substitution of a phenylalanine residue for a serine residue. We had previously shown that the substitution of an isoleucine residue for a threonine residue was sufficient to allow the mutant A13 to utilize acetanilide as a growth substrate (3). In this experimental system it has been shown that single amino acid substitutions can improve the activity of an enzyme on a very poor substrate and that successive single-step mutations can introduce progressively greater changes in enzyme specificity.

REFERENCES

1. **Betz, J. L., J. E. Brown, P. H. Clarke, and M. Day.** 1974. Genetic analysis of amidase mutants of *Pseudomonas aeruginosa.* Genet. Res. **23:**335–359.
2. **Brammar, W. J., P. H. Clarke, and A. J. Skinner.** 1967. Biochemical and genetic studies with regulator mutants of the *Pseudomonas aeruginosa* 8602 amidase system. J. Gen. Microbiol. **47:**87–102.
3. **Brown, P. R., and P. H. Clarke.** 1972. Amino acid substitution in an amidase produced by an acetanilide-utilizing mutant of *Pseudomonas aeruginosa.* J. Gen. Microbiol. **70:**287–298.
4. **Brown, P. R., M. J. Smyth, P. H. Clarke, and M. A. Rosemeyer.** 1973. The subunit structure of the aliphatic amidase of *Pseudomonas aeruginosa.* Eur. J. Biochem. **34:**177–187.
5. **Campbell, H. H., J. A. Lengyel, and J. Langridge.** 1973. Evolution of a second gene for β-galactosidase in *Escherichia coli.* Proc. Natl. Acad. Sci. U.S.A. **70:**1841–1845.
6. **Clarke, P. H.** 1974. The evolution of enzymes for the utilization of novel substrates. Symp. Soc. Gen. Microbiol. **24:**183–217.
7. **Clarke, P. H., and M. D. Lilly.** 1969. The regulation of enzyme synthesis during growth. Symp. Soc. Gen. Microbiol. **19:**113–157.
8. **Francis, J. C., and P. E. Hansche.** 1972. Directed evolution of metabolic pathways in microbial populations. I. Modification of the acid phosphatase pH optimum in *S. cerevisiae.* Genetics **70:**59–73.
9. **Hall, B. G.** 1976. Experimental evolution of a new enzymatic function. Kinetic analysis of the ancestral (ebg^0) and evolved (ebg^+) enzymes. J. Mol. Biol. **107:**71–84.
10. **Hall, B. G., and N. D. Clarke.** 1977. Regulation of newly evolved enzymes. III. Evolution of the *ebg* repressor during selection for enhanced lactase activity. Genetics **85:**193–201.
11. **Hartley, B. S.** 1974. Enzyme families. Symp. Soc. Gen. Microbiol. **24:**151–182.
12. **Hartley, B. S., I. Altosaar, J. M. Dothie, and M. Neuberger.** 1976. Experimental evolution of a zylitol dehydrogenase, p. 191–200. *In* R. Markham and R. W. Horne (ed.), Structure-function relationships of proteins. North-Holland Publishing Co., Amsterdam.
13. **Rossman, M. G., D. Moras, and K. W. Olsen.** 1974. Chemical and biological evolution of a nucleotide-binding protein. Nature (London) **250:**194–199.

Importance of Regulatory Mutations in Channeling the Evolution of Metabolic Pathways in Bacteria

E. C. C. LIN

Department of Microbiology and Molecular Genetics, Harvard Medical School, Boston, Massachusetts 02115

Cells have a much larger catalytic potential than is ordinarily expressed because enzymes hardly ever possess absolute specificity. This full biochemical potential is partially masked by well-coordinated genetic control mechanisms so that most of the proteins are fabricated only when they can ordinarily be useful to the organism. The price of such economy is the curtailment of biochemical versatility.

An example is presented here to illustrate how a latent metabolic potential can be progressively expressed by a series of mutations that permit the critical enzyme to be synthesized at high constitutive levels under normally noninducing conditions (for further discussion of the roles of regulatory mutations in biochemical evolution, see 8, 10–12). The predominance of regulatory mutations over mutations that modify enzyme structure in this particular case has to do with the fact that an enzyme which functions exclusively as a reductase in the wild-type strain is used by the mutants only as a dehydrogenase, and thus an anaerobic enzyme is converted to an aerobic one.

Wild-type strains of *Escherichia coli,* such as K-12, B, ML, and W, are unable to grow on L-1,2-propanediol as a sole source of carbon and energy. Several series of mutants capable of growing aerobically on propanediol at progressively more rapid rates have been partially characterized. All of these mutants exploit the L-fucose system by alterations of gene expression, and, with one exception, all of the mutations map at the fucose locus (*fuc*). During the course of the experimental evolution of new strains adapting to propanediol utilization, genes of the fucose system undergo drastic regulatory changes, eventually leading to inactivation, or perhaps even deletion, of the residual elements that are superfluous for the novel pathway.

In wild-type cells, L-fucose is metabolized by the sequential action of fucose permease (5), fucose isomerase (4), fuculose kinase (7), and fuculose 1-phosphate aldolase (3). The products of the last reaction are dihydroxyacetone phosphate and L-lactaldehyde. Under aerobic conditions, L-lactaldehyde is oxidized to L-lactate by a nicotinamide adenine dinucleotide-dependent lactaldehyde dehydrogenase (13), and the product is in turn oxidized to pyruvate by a flavine-linked lactate dehydrogenase (9). Under anaerobic conditions, lactaldehyde is reduced by a reduced nicotinamide adenine dinucleotide-dependent oxidoreductase to propanediol, which is excreted into the medium (1, 14; Fig. 1). For each mole of fucose fermented, 1 mol of propanediol is eliminated. The disposal of two hydrogens in this manner allows the further metabolism of dihydroxyacetone phosphate without the need for an exogenous hydrogen acceptor. The propanediol lost into the medium is irretrievable even when molecular oxygen becomes available (1).

In wild-type cells, fucose permease, fucose isomerase, fuculose kinase, and fuculose 1-phosphate aldolase are inducible, irrespective of whether growth was aerobic or anaerobic. In contrast, the enzyme following each branch point can be fully induced only under the proper respiratory conditions. Lactaldehyde dehydrogenase is highly

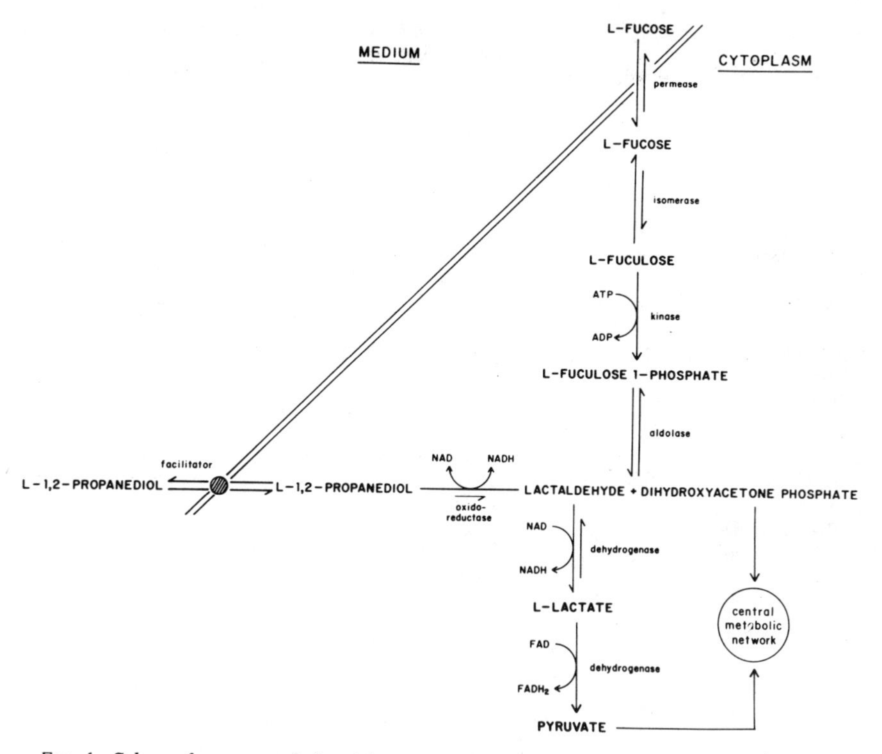

FIG. 1. *Scheme for propanediol and fucose metabolism in* E. coli. *The enzyme catalyzing the interconversion of propanediol and lactaldehyde is referred to as propanediol oxidoreductase since the actual role of this protein depends upon the strain in which it is found. FAD and $FADH_2$, the oxidized and reduced forms of flavine adenine dinucleotide; NAD and NADH, the oxidized and reduced forms of nicotinamide adenine dinucleotide.*

inducible only aerobically (presumably by lactaldehyde itself), and the product, L-lactate, in turn induces its dehydrogenase (1). The opposite holds for propanediol oxidoreductase, which is highly inducible only under anaerobic conditions (1; Table 1).

A basic feature of the mutants that can grow well on propanediol is their ability to produce the oxidoreductase at a high basal or constitutive level, despite the presence of molecular oxygen. More than one mutational step is required to attain this condition. In Table 2, it can be seen that in an early-stage mutant, strain 413, the aerobic basal level of the enzyme is elevated by an order of magnitude. This genetic modification, however, does not change the inducibility of the enzyme: in both parental strain 1 and mutant strain 413, the presence of fucose can cause another 5-fold increase of the enzyme. In strain 418, derived from strain 413 by further selection on the novel substrate, the basal level of the enzyme is elevated another 3-fold, but the presence of fucose can now only increase the level of the enzyme by a factor of two. Nonetheless, it might be noted that, when grown on casein hydrolysate with fucose, the specific activity of the enzyme in strain 418 is about 20 times the level observed in wild-type strain 1.

An unexpected consequence of continued selection for growth on propanediol is

TABLE 1. *Activities of fucose pathway enzymes in crude extracts of wild-type* E. coli *as a function of growth conditions*[a]

Carbon source	O_2	Specific activity[b]					
		Fucose permease	Fucose isomerase	Fuculose kinase	Fuculose 1-phosphate aldolase	Propanediol oxidoreductase	Lactaldehyde dehydrogenase
CAA[c]	+	0	36	27	0	20	93
CAA + fucose	+	55	500	440	240	80	100
Fucose	+	58	520	420	130	42	280
Glucose	+					6	15
CAA + pyruvate	−					89	0
Fucose	−	64	750	310	270	240	32
Glucose	−					0	0

[a] Adapted from reference 5.
[b] Fucose permease activity is expressed in nanomoles per minute per milligram of dry weight at 37°C; all other activities are expressed in nanomoles per minute per milligram of protein at 25°C.
[c] Casein hydrolysate.

TABLE 2. *Activities of fucose pathway enzymes in extracts of various mutants grown aerobically on different media*[a]

Strain	Carbon source	Specific activity[b]					
		Fucose permease	Fucose isomerase	Fuculose kinase	Fuculose 1-phosphate aldolase	Propanediol oxidoreductase	Lactaldehyde dehydrogenase
1	CAA[c]	0	36	27	0	20	93
	CAA + fucose	55	500	440	240	80	100
413	CAA	0	20	40	10	210	160
	CAA + fucose	50	460	420	190	990	160
	Propanediol	0	10	0	0	260	330
418	CAA	0	10	30	0	750	130
	CAA + fucose	64	380	330	180	1,400	180
	Propanediol	0	20	0	0	790	300
421	CAA	0	15	20	390	1,300	120
	CAA + fucose	0	15	20	380	1,200	130
	Propanediol	0	10	0	180	1,300	300
3	CAA	0	43	80	490	410	130
	CAA + fucose	0.08	30	46	490	390	110
	Propanediol		43	57	490	420	260
430	CAA				0	750	380
	CAA + fucose	0	20	48	0	680	420
	Propanediol				0	720	410

[a] Adapted from reference 6, and A. J. Hacking et al., manuscript in preparation.
[b] Fucose permease activity is expressed in nanomoles per minute per milligram of dry weight at 37°C; all other activities are expressed in nanomoles per minute per milligram of protein at 25°C.
[c] Casein hydrolysate.

the inevitable emergence of a mutant that no longer grows on fucose. Strains 421 and 3 are two such examples derived from separate wild-type clones. Enzymatic examination of these two mutants revealed several common features: (i) the noninducibility of fucose permease, fucose isomerase, and fuculose kinase, (ii) the constitutivity of fuculose 1-phosphate aldolase, now unconditionally a gratuitous enzyme, and (iii) a high constitutive level of propanediol oxidoreductase. All of these changes are likely to be the result of a particular kind of mutation that further increases the synthesis of the oxidoreductase. (Strain 421 was derived in one step from strain 413 [6]. Unfortunately, the immediate precursor of strain 3 was not isolated.) The attribution of the low levels of the permease, the isomerase, and the kinase to a mutation in a regulatory protein is supported by the observation that three independent fucose-positive revertants of strain 3 produce these three proteins constitutively (Table 3).

The mutants discussed above were selected by serial transfer of cells into media containing initially 12 to 20 mM L-1,2-propanediol. Mutants like strains 3 and 421 exhibit a doubling time of about 2 h under these conditions. The method of selection was then changed by culturing cells in media in which the propanediol concentration was not allowed to exceed 0.5 mM. Strain 430 was obtained from strain 3 in this way.

Four phenotypic changes were detected in strain 430: (i) the fuculose 1-phosphate aldolase gene is inactivated; (ii) the constitutive propanediol oxidoreductase activity level is increased; (iii) lactaldehyde dehydrogenase is synthesized constitutively and at an elevated level; and (iv) at low concentrations of propanediol, the rate of specific transport across the cell membrane is enhanced (A. J. Hacking et al., manuscript in preparation). The last three changes seem to act in concert in the trapping of propanediol by hastening its rate of entry into the cell and its conversion to an ionized metabolite, L-lactate. The significance of the loss of fuculose 1-phosphate aldolase is not clear; either it is associated with the increase in the constitutive level of propanediol oxidoreductase, or it is the result of a separate mutation that abolishes a deleterious enzymatic activity during the growth on propanediol because of the cul-de-sac accumulation of fuculose 1-phosphate. In any event, strain 430 no longer

TABLE 3. *Activities of fucose pathway enzymes in crude extracts of revertants of strain 3 grown aerobically*[a]

Strain no.	Carbon source	Specific activity[b]			
		Fucose permease	Fucose isomerase	Fuculose kinase	Fuculose 1-phosphate aldolase
54	CAA[c]	0.3	110	110	470
	CAA + fucose	0.3	100	120	500
55	CAA	8.0	230	160	500
	CAA + fucose	7.0	240	160	440
56	CAA	15.0	600	550	370
	CAA + fucose	15.0	630	400	420

[a] Adapted from reference 5.
[b] Fucose permease activity is expressed in nanomoles per minute per milligram of dry weight; all other activities are expressed in nanomoles per minute per milligram of protein.
[c] Casein acid hydrolysate.

reverts to a strain that can grow on fucose, and consequently in time those genes of the fucose system not contributing to the utilization of propanediol can be expected to become extinct by deletion.

From the enzymatic profiles in wild-type, mutant, and pseudorevertant strains studied under various growth conditions, it would seem that the fucose system comprises a regulon with at least four operons coding, respectively: (i) fucose permease (see partial constitutivity of this protein versus full constitutivity of the isomerase and kinase in strain 56 [Table 3]), (ii) fucose isomerase and fuculose kinase, (iii) fuculose 1-phosphate aldolase, and (iv) propanediol oxidoreductase. This would then explain the extraordinarily complex changes in the patterns of gene expression as the aerobic constitutive level of propanediol oxidoreductase is raised. (The structural gene for lactaldehyde dehydrogenase is probably under the control of an independent regulator protein, since the enzyme can be induced by propanediol in strains 413 and 418 without concomitant induction of the first three operons described above.)

Our working model is that the expression of the propanediol oxidoreductase operon itself requires the participation of a complex regulatory protein that responds to two kinds of signals: a specific inducer and an indicator of the respiratory state. Since this protein also participates in the expression of the first three operons of the fucose system, and since the enzymes of the trunk pathway are synthesized under both aerobic and anaerobic conditions, the base sequences of the promoter-operator regions of these operons must necessarily differ from that of the propanediol oxidoreductase operon. As the regulatory protein undergoes structural modification to allow increased aerobic expression of the propanediol oxidoreductase operon in the absence of the inducer, the behavior of the protein towards the other operons changes in incidental fashions.

The evolution of the propanediol pathway provides a striking contrast to several previously studied systems in which the structural gene of an enzyme mobilized for a new function by virtue of its side specificity belongs to a system under simpler repressor control, such as that found in the lactose system of *E. coli.* In these cases, the constitutive synthesis of the critical enzyme can result from a random mutation that destroys repressor function. The synthesis of the other gene products that share the same control, whether or not they are necessary for the new pathway, also becomes constitutive. The only price imposed on the mutant may be the wasteful synthesis of several proteins, when neither the original nor the novel substrate is present in the environment. With sufficient time, however, gene duplication might occur, which would then permit the reestablishment of repression for one set of genes according to the original pattern, as the other set of genes undergoes reorganization to fit the new requirement. Thus, the mutant may gain a new metabolic function without sacrificing the original pathway.

Repressor-controlled catabolic systems might be genetically more adaptable: the probability of their mobilization for a new role is high because any number of random mutations in the structural gene of the repressor can achieve the goal. Hence, such systems should be evolutionarily quite opportunistic. Conversely, a highly elaborate regulatory mechanism, such as the one postulated for the fucose pathway, restricts its evolution for a new role, and, if by necessity it is the only available system in the genome, the acquisition of a new function might entail the complete sacrifice of the old, as demonstrated in this study.

Ending with a note pertinent to a symposium on the genetics of industrial microorganisms, it might be pointed out that the availability of 1,2-propanediol mutants is also a matter of practical importance. The compound, commonly known as propylene glycol, is produced commercially in gigantic amounts by alkaline hydrolysis of the corresponding chlorohydrin. The large volumes of brinewater generated by this process can contain up to 20 mM propanediol and 2 M sodium chloride. To prevent environmental pollution, mixed cultures of propanediol-degrading mutants of various halophilic bacteria, tolerant of alkaline surroundings, have been successfully employed by a decontaminating plant capable of treating more than 1,000 liters of wastewater per min with a propanediol-removing efficiency of 90% (2). Advancing our knowledge of the role of regulatory mutations in channeling the evolution of metabolic pathways should be helpful in designing the selection of microbes for more complex tasks.

ACKNOWLEDGMENTS

This investigation was supported by grant PCM74-07659 from the National Science Foundation and by Public Health Service grant 5 RO1 GM11983 from the National Institute of General Medical Sciences.

REFERENCES

1. **Cocks, G. T., J. Aguilar, and E. C. C. Lin.** 1974. Evolution of L-1,2-propanediol catabolism in *Escherichia coli* by recruitment of enzymes for L-fucose and L-lactate metabolism. J. Bacteriol. **118:**83–88.
2. **Dow Chemical Co.** 1971. Treatment of wastewater from the production of polyhydric organics. U.S. Environmental Protection Agency, Washington, D.C. (Available from Superintendent of Documents, U.S. Government Printing Office, Washington, D.C. 20402.)
3. **Ghalambor, M. A., and E. C. Heath.** 1962. The enzymatic cleavage of L-fuculose 1-phosphate. J. Biol. Chem. **237:**2427–2433.
4. **Green, M., and S. S. Cohen.** 1956. The enzymatic conversion of L-fucose to L-fuculose. J. Biol. Chem. **219:**557–568.
5. **Hacking, A. J., and E. C. C. Lin.** 1976. Disruption of the fucose pathway as a consequence of genetic adaptation to propanediol as a carbon source in *Escherichia coli.* J. Bacteriol. **126:**1166–1172.
6. **Hacking, A. J., and E. C. C. Lin.** 1977. Regulatory changes in the fucose system associated with the evolution of a catabolic pathway for propanediol in *Escherichia coli.* J. Bacteriol. **130:**832–838.
7. **Heath, E. C., and M. A. Ghalambor.** 1962. The purification and properties of L-fuculose kinase. J. Biol. Chem. **237:**2423–2426.
8. **Hegeman, G. D., and S. L. Rosenberg.** 1970. The evolution of bacterial enzyme systems. Annu. Rev. Microbiol. **24:**429–462.
9. **Kline, E. S., and H. R. Mahler.** 1965. The lactic dehydrogenases of *E. coli.* Ann. N.Y. Acad. Sci. **119:** 905–919.
10. **Lerner, S. A., T. T. Wu, and E. C. C. Lin.** 1964. Evolution of catabolic pathway in bacteria. Science **146:** 1313–1315.
11. **Lin, E. C. C.**1970. Evolution of catabolic pathways in bacteria, p. 89–102. *In* Miami Winter Symposia, vol 1. North-Holland Publishing Co., Amsterdam.
12. **Lin, E. C. C., A. J. Hacking, and J. Aguilar.** 1976. Experimental models of acquisitive evolution. BioScience **26:**548–555.
13. **Sridhara, S., and T. T. Wu.** 1969. Purification and properties of lactaldehyde dehydrogenase from *Escherichia coli.* J. Biol. Chem. **244:**5233–5238.
14. **Sridhara, S., T. T. Wu, T. M. Chused, and E. C. C. Lin.** 1969. Ferrous-activated nicotinamide adenine dinucleotide-linked dehydrogenase from a mutant of *Escherichia coli* capable of growth on 1,2-propanediol. J. Bacteriol. **98:**87–95.

Author Index

Subject Index